Integrated Pest Management: Current and Future Strategies

Council for Agricultural Science and Technology, Ames, Iowa, USA
Printed in the United States of America

Cover design by Lynn Ekblad, Different Angles, Ames, Iowa
Graphics and layout by Richard Beachler, Instructional Technology Center, Iowa State University, Ames

ISBN 1-887383-23-9
ISSN 0194-4088
06 05 04 03 4 3 2 1

Library of Congress Cataloging-in-Publication Data

Integrated Pest Management: Current and Future Strategies.
p. cm. -- (Task force report, ISSN 0194-4088 ; no. 140)
Includes bibliographical references and index.
ISBN 1-887383-23-9 (alk. paper)
1. Pests--Integrated control. I. Council for Agricultural Science and Technology. II. Series: Task force report (Council for Agricultural Science and Technology) ; no. 140.
SB950.I4573 2003
632'.9--dc21 2003006389

Task Force Report

No. 140 June 2003

Council for Agricultural Science and Technology

Ames, Iowa, USA

Task Force Members

Kenneth R. Barker (Chair), Department of Plant Pathology, North Carolina State University, Raleigh

Esther Day, American Farmland Trust, DeKalb, Illinois

Timothy J. Gibb, Department of Entomology, Purdue University, West Lafayette, Indiana

Maud A. Hinchee, ArborGen, Summerville, South Carolina

Nancy C. Hinkle, Department of Entomology, University of Georgia, Athens

Barry J. Jacobsen, Department of Plant Sciences and Plant Pathology, Montana State University, Bozeman

James Knight, Department of Animal and Range Science, Montana State University, Bozeman

Kenneth A. Langeland, Department of Agronomy, University of Florida, Institute of Food and Agricultural Sciences, Gainesville

Evan Nebeker, Department of Entomology and Plant Pathology, Mississippi State University, Mississippi State

David A. Rosenberger, Plant Pathology Department, Cornell University–Hudson Valley Laboratory, Highland, New York

Donald P. Schmitt, Department of Plant and Environmental Sciences, University of Hawaii, Honolulu

John C. Snyder, Department of Horticulture, University of Kentucky, Lexington

Ann Sorensen, American Farmland Trust, DeKalb, Illinois

Walter R. Stevenson, Department of Plant Pathology, University of Wisconsin, Madison

Stephen M. Stringham, Department of Entomology, North Carolina State University, Raleigh

Scott M. Swinton, Department of Agricultural Economics, Michigan State University, East Lansing

David W. Watson, Department of Entomology, North Carolina State University, Raleigh

Philip Westra, Department of Bioagricultural Science and Pest Management, Colorado State University, Ft. Collins

Mark E. Whalon, Center for Integrated Plant Systems, Michigan State University, East Lansing

Tom Whitson, Department of Plant Sciences, University of Wyoming, Laramie

Contributors

Harold D. Coble, North Carolina State University, Raleigh, and U.S. Department of Agriculture

Larry Pedigo, Iowa State University, Ames

Invited Reviewers

Gerald Anderson, U.S. Department of Agriculture, Agricultural Research Service, Sidney, Montana

Larry F. Grand, Department of Plant Pathology, North Carolina State University, Raleigh

Dennis Keeney, Emeritus Professor, Iowa State University, Ames

Kimberly Mann, U.S. Department of Agriculture, Agricultural Research Service, Sidney, Montana

Clyde Sorenson, Department of Entomology, North Carolina State University, Raleigh

Neal Spencer, U.S. Department of Agriculture, Agricultural Research Service, Sidney, Montana

John C. Stewart, U.S. Department of Agriculture, Animal and Plant Health Inspection Service, Raleigh, North Carolina

Ron Stinner, Department of Entomology, North Carolina State University, Raleigh

Turner B. Sutton, Department of Plant Pathology, North Carolina State University, Raleigh

Acknowledgments

CAST recognizes and appreciates the financial support of the U. S. Department of Agriculture–Agricultural Research Service (Agreement Number 59-0790-3-107) to partially assist in the development and completion of this report.

The authors thank Dr. Richard E. Stuckey, former executive vice president of CAST, and Dr. Kayleen A. Niyo, former managing scientific editor, for their considerable assistance during the development of this report.

Contents

Figures

Tables

Textboxes

Foreword

Following a recommendation by the CAST National Concerns Committee, the CAST Board of Directors authorized preparation of a new report on integrated pest management as an update of its 1982 report on that topic.

Dr. Kenneth R. Barker, Department of Plant Pathology, North Carolina State University, Raleigh, served as chair for the report. A highly qualified group of scientists served as task force members. The group included individuals with expertise in pest management, entomology, plant pathology, animal and range science, agronomy, environmental science, horticulture, agricultural economics, and bioagricultural science.

The task force prepared an initial draft of the report and revised all subsequent drafts. Contributors provided additional input and invited reviewers read selected sections. The CAST Executive Committee and Editorial and Publications Committee reviewed the final draft, and the authors reviewed the proofs. The CAST staff provided editorial and structural suggestions and published the report. The task force authors are responsible for the report's scientific content.

On behalf of CAST, we thank the chair, authors, contributors, and invited reviewers who gave of their time and expertise to prepare this report as a contribution by the scientific community to public understanding of the issue. We also thank the employers of the scientists, who made the time of these individuals available at no cost to CAST. CAST thanks all members who made additional contributions to assist in the preparation of this document. The members of CAST deserve special recognition because the unrestricted contributions they have made in support of CAST also have financed the preparation and publication of this report.

This report is being distributed widely; recipients include Members of Congress, the White House, the U.S. Department of Agriculture, the Congressional Research Service, the Food and Drug Administration, the Environmental Protection Agency, and the Agency for International Development. Additional recipients include media personnel and institutional members of CAST. Individual Members of CAST may receive a complimentary copy upon request for a $3.00 postage and handling fee. The report may be reproduced in its entirety without permission. If copied in any manner, credit to the authors and to CAST would be appreciated.

L. J. Koong
President

Teresa A. Gruber
Executive Vice President

Linda M. Chimenti
Managing Scientific Editor

Interpretive Summary

In 1982, the Council for Agricultural Science and Technology (CAST) published *Integrated Pest Management*, a report that focused on the decreased use of pesticides. Now, new technologies and rigorous government policies are bringing much promise, as well as many challenges, to integrated pest management (IPM). The current report emphasizes the development and implementation of biointensive IPM strategies and tactics as well as the continuing challenges. Issues addressed include the rapidly evolving pests in natural and agroecosystems, the availability of innovative pest management tools, and the need to integrate those tools via a systems approach.

Deployment in Cropping Systems

The availability of both effective pest management strategies and the systems for their deployment is fundamental for successful IPM programs. Key strategies include

- prevention or avoidance of pests and/or pathogens,
- decrease of pest populations,
- decrease of crop or animal susceptibility,
- eradication of pests,
- a combination of strategies, or
- pursuing no action.

Primary tactics (tools) used in IPM encompass

- host resistance,
- use of natural enemies or biological controls,
- cultural practices, and
- pesticides.

Reliable identification, diagnosis, and monitoring of pest populations are essential for successful IPM implementation.

Diverse strategies and tactics that may be used to enhance the activities of soil- and foliage-inhabiting "beneficial" organisms include

- crop rotations,
- antagonistic plants or other organisms,
- trap crops,
- refugia, and
- cover crops.

With the pressure for decreased pesticide use and rapid growth of biointensive and organic food industries, divergent pest management practices are being adapted for certain production systems. Biointensive IPM production involves one or more management tools, including chemical inputs, when needed, to restrict pest populations below economic thresholds. In contrast, few or no synthetic pesticides or fertilizers or genetically modified organisms (GMOs) are permitted in organic production. Still, IPM must be central for sustainable organic farming.

Natural Ecosystems

Invasive pests are becoming an enormous problem in U.S. terrestrial and aquatic habitats. Natural areas and waterways are being invaded by both nonnative and native pests and weed species. Increasing international jet travel has made introductions of pests nearly limitless. Although prevention is the most cost-effective control method, other tactics—including herbicides, mechanical methods, and habitat manipulation—must be used for previously introduced pests. Biological controls also have promise.

Food Animals, Wildlife, and Companion Animals

Integrated pest management is recognized as the preferred best management practice to minimize pest problems on dairy and beef cattle and to protect farm workers, consumers, and the environment. In swine production, effective management of arthropod pests is important in minimizing the risk of diseases and preventing related poor growth and poor feed conversion. For poultry IPM, houseflies are key pests of dry-waste systems and other types of production that allow manure accumulation.

Because of increasing human population and intensified land-use practices, wildlife damage control is an increasingly important part of U.S. wildlife management. Control programs focus on problem-species ecology, control-method application, and control evaluation.

With approximately one-half of U.S. households owning pets, pests that affect companion animals are an important problem. In addition to insecticides, acaricides, and other available control tactics, future IPM options may include innate animal physiology to suppress the ectoparasites and to limit their deleterious effects on pet and pet-owner health.

Urban Areas

Attaining acceptable levels of pest control without exposing people or the environment to excessive risks from pesticides is the major goal of urban IPM; qualitative factors such as aesthetics and peace of mind usually substitute for the quantitative economics involved in decision making for agricultural IPM. The components of urban IPM are inspection, monitoring, situation-specific decision making, application of control techniques, and record keeping.

Economics

Economic analyses have been used to evaluate expected profitability, social welfare impacts, effect of IPM research on financial returns, and policies affecting pest management adoption. Most of these assessments have shown net benefits from IPM use, especially from pest-resistant or pest-tolerant crop varieties. Pioneering benefit-cost analyses have tackled the problems of multiple-season effects as well as the valuation of health and environmental effects for individual IPM methods and programs. The complexity of IPM methods and the difficulty of placing values on environmental and health attributes may account for the scarcity of comprehensive economic assessments.

Education and Delivery Systems

Pest management personnel need training in a variety of disciplines. Many universities have an IPM component within traditional departments such as entomology or plant pathology.

Striking changes in IPM delivery systems have occurred since the 1960s. Initially, IPM information was disseminated primarily through print media and verbal communications. Now, diverse IPM delivery systems are used, including many comprehensive Internet web sites. With so much information available, IPM resource personnel are needed to select the appropriate information for specific situations.

Assessments

Because of increasing public concern about pesticides, the approaches used to evaluate IPM are becoming increasingly important. The environmental and social parameters essential in assessing the consequences of pesticide use include

- health impacts on farm workers, consumers, and the general public;
- lethal and chronic effects on other nontarget biota;
- direct or indirect effects on natural and agroecosystems;
- calculation of air, soil, and water pollution; and
- costs versus benefits to producers and to society for decreasing pesticide use.

Enlisting producers' cooperation in compiling data on the parameters through surveys, sampling, and other means continues to be a challenge.

Future Challenges and Directions

This comprehensive report concludes with a consideration of seven key issues for the future of IPM.

- **Gene technology constraints**. Future crops may be modified for increased compatibility with IPM systems. Key unresolved issues include the extent to which genetically engineered crops will be used in production systems, the rate at which they will be adopted, their compatibility with IPM systems, their acceptance or lack thereof by the public, and their ultimate beneficiaries.
- **Genetic diversity and pest adaptability**. The ecological elasticity of many animal and crop pests allows them to adapt to almost limitless habitats. Genetic diversity within most crop pests often limits the utility of plant varieties developed with resistance to one or more pests.
- **Ecologically based IPM**. Interest is growing in shifting the focus of IPM from pesticide management to a biointensive systems approach based on biological knowledge of pests and their interactions with crops. Significant funding investments will be required to build the ecological knowledge base needed for the multitude of

cropping systems, pets, environments, and pest complexes.

- **Systems approaches**. A major goal for maximizing the benefits of IPM and related cropping systems is to increase understanding of the interactions of microflora/fauna in natural and agricultural ecosystems. Effective collaborative research and extension programs as well as coordination with funding and support agencies are necessary; increased research on numerous field and vegetable crops and animals is warranted.
- **Evolving pool of trainers and speed of technology transfer**. Distance education programs may help expand students' access to college-level IPM courses. The increasing rate of development and the introduction of new technologies challenges the agricultural community's capacity to provide the necessary training and information to incorporate new tools into existing programs.
- **Government policy and regulations**. Implementation of IPM for urban pests likely will be the policy for most public properties at the federal, state, and local levels. Two future regulatory issues will have impacts on IPM: the international phase-out of methyl bromide, and the phasing in of the Food Quality Protection Act. The new National IPM Initiative, which focuses on economic and environmental risk reduction, soon will be introduced. The need will continue for additional funding to support research on an IPM systems approach.
- **Assessments of IPM**. Surveys suggest that the greatest shortcoming in most current IPM programs is the limited use of a systems approach. The ongoing national and international debates on the future of genetically modified organisms will affect policies related to production, marketing, and use of these new products.

The prospects for increased adoption of IPM and related cropping systems are excellent, despite challenges that include public perceptions of new technologies, limited financial resources, and an inadequate infrastructure. As the earth's carrying capacity for humankind is stretched, and as thousands of invasive pests are encountered worldwide, research, agriculture, industry, government, and communities must work together. Integrated pest management offers an effective option for production of the increasing amounts of food and fiber supplies needed to sustain the nation and the world.

Executive Summary

Historical Perspectives and Evolution of Integrated Pest Management Concepts

The idea of using multiple tactics and strategies for disease control was introduced by the German botanist Julius Gotthele Kühn in the 1880s. Although the term *integrated pest management* (IPM) was used first by the entomologists Smith and van den Bosch (1967), the concept was grasped in the late 1950s and 1960s by entomologists beginning to identify the problems of insect resistance to pesticides and of detrimental ecological effects caused by widespread use of insecticides. Stern and colleagues (1959) used the phrase *integrated control* to describe pest control involving a combination and integration of biological and chemical tactics. They deployed chemical tactics in a manner to minimize disruption of biological control. The concepts *economic injury level* and *economic thresholds*, both crucial to IPM decision making, also were introduced by these researchers. While development of the concepts of multiple pest-control tactics and their integration set the stage for IPM, Rachel Carson's 1962 *Silent Spring* ignited widespread debate about pesticides and undoubtedly helped catalyze the development of IPM along with the environmental movement.

The need for IPM increased in the 1970s and 1980s as pesticides posing hazards to human health and the environment were removed from the market. Simultaneously, decision-support software programs resulted in significant savings for growers by decreasing their reliance on pesticides. Formal involvement of plant pathologists, nematologists, and weed scientists with entomologists in IPM programs resulted mainly from numerous U.S. government-supported projects. Regrettably, funding for IPM research and implementation remained essentially static in the 1970s and the 1980s. The 1994 U.S. Department of Agriculture's (USDA) IPM Initiative heralded a new era for IPM. The initiative, which was adopted after the Clinton Administration's 1993 commitment to the decreased use of high-risk pesticides and implementation of IPM in 75% of all crop production by the year 2000, renewed interest and funding from various IPM agencies. Corresponding increases in IPM adoption for certain crops have occurred over the last decade, but a multidimensional integration of methods for controlling all types of pests remains a challenge.

Numerous federal and technological developments continue to affect IPM. At the federal level, the 1996 Food Quality Protection Act (FQPA) limits the use of certain pesticides. The problems of pest resistance and of new races of pests overcoming genetic host-plant resistance also pose challenges. New technologies for host resistance or tolerance that are related to genetically modified organisms (GMOs), and the higher-quality pest management offered through precision agriculture, constitute opportunities as well as challenges.

The recent books *Ecologically Based Pest Management: New Solutions for a New Century* (Overton 1996; see Chapter 1) and *Pest Management at the Crossroads* (Benbrook et al. 1996; see Chapter 14) map out enormous tasks for the pest control disciplines. Movement beyond individual pests and crops by building and participating in broadly based, competent IPM teams is necessary if pest control in diverse agroecosystems is to be understood. Additionally, administrators, grant managers, and others must find ways to encourage scientists to participate in interdisciplinary teams in which long-term, complex research and technology transfer are end products.

Integrated Pest Management Toolbox

The development and deployment of effective pest management strategies and tactics into suitable production systems are fundamental to successful IPM programs. Not surprisingly, tools effectively used for managing pests of crops or animals often are useful for managing pests in rural or urban settings. Key IPM strategies include prevention or avoidance of pests and/or pathogens, decreased pest populations, decreased host susceptibility, and pest eradication. These strategies may be used independently but are more often combined in a holistic approach. Primary tactics used in IPM encompass breeding for host-plant resistance and using natural enemies or biological controls (Figure S.1), cultural practices, and pesticides. Reliable identification, diagnosis, and monitoring of pest populations are essential.

New and Emerging Technologies

New technologies based on computers, cell phones, the World Wide Web, and "smart machines" are affecting IPM and agriculture generally. These new technologies are being applied to agriculture and to IPM by private companies and through public and private research. Revolutions in information systems and genetic technologies (*bioinformatics*) are affecting IPM as a component of integrated crop production.

Figure S.1. A new biological control, the celery looper virus, is being tested for use against costly crop pests such as the cotton bollworm. Photo by Scott Bauer, Agricultural Research Service, U.S. Department of Agriculture, Beltsville, Maryland.

Three new technologies affecting IPM adoption are precision farming tools, transgenic plants, and integrated production systems. Precision farming involves the use of computers, global positioning systems (GPS), receivers, computer-generated field maps, decision-support models, and specialized variable-rate equipment to modify input levels and to monitor final crop yield. Specialized crop harvesters equipped with GPS and yield monitors allow farmers to generate yield maps and to monitor variations in yield, based on inputs. By tracking this information through geographical information systems, computerized decision-support models simulating crop-pest interactions can be developed to automate the threshold-based decision rules of classic IPM.

With recent publicity regarding animal cloning and transgenic crops (e.g., Starlink corn), issues related to transgenic organisms have been publicized widely. Although transgenic plant technology is in its infancy, precision genetics holds great promise for bringing new, effective tools to IPM. For example, herbicide-tolerant and insect-resistant crops developed through genetic engineering have been commercialized and used throughout the United States (Figure S.2).

An integrated, whole-farm systems approach to agricultural research and management, including IPM, remains an important goal. The complex system of sustainable agriculture, which demands consideration of integrated factors, processes, and institutions, reflects this need. Although computer-based decision-support systems have been developed for various aspects of agriculture and for IPM, the complexity of farm-crop and IPM systems limits interdisciplinary collaborations so essential to integration. Nevertheless, integrated information and operational systems offer improved prediction and management capabilities that enhance understanding of these systems and their economic and environmental effects.

Crop Production Systems

Cropping system concepts for IPM have expanded to encompass very diverse strategies and tactics. Crop rotations, antagonistic plants or other organisms, trap crops, refugia, and cover crops may be used to enhance the activities of beneficial organisms. Indeed, one goal of a rotation or change in cropping system is to eliminate harmful pests from the soil so that it can support a healthy crop. To bring this about, IPM systems must address the fact that significant levels of predation and/or parasitism promote diversity and sustainability. Thus, cropping systems must relate to IPM goals as well as to soil and crop health.

Although certain crops have low-intensity inputs and values, steps for achieving IPM adoption are similar for all crops. The decision to manage pests on crops by specific means should be made only after the problem and magnitude have been identified and a cost/benefit analysis of various management tactics has been conducted. Cultural practices and host resistance have been the primary IPM tools on corn and soybean, but significant changes are occurring on these crops, including the use of genetically engineered herbicide tolerance, transgenic insect resistance, and precision-agriculture tools. Thus, a new philosophy and approach to IPM is forthcoming. Traditionally, high pesticide-usage has been central to vegetable, fruit, and cotton production. Today, IPM offers promise for decreasing chemical use on these crops. Perennial fruit crops pose special challenges for IPM, exacerbated by consumer demands for blemish-free products. But growers have become adept at monitoring pest populations, delaying pesticide applications until pest numbers reach economic thresholds, and timing applications to maximize pesticide efficacy. Certification of planting stock as pest-free, especially virus-free, remains an IPM cornerstone for fruit and other vegetatively propagated crops such as potato.

Figure S.2. A corn eastern gamagrass hybrid. Eastern gamagrass is a native grass with a gene pool that has a lot to offer corn, including resistance to cold and insects. Photo by Scott Bauer, Agricultural Research Service, U.S. Department of Agriculture, Beltsville, Maryland.

Integrated pest management is beginning to diverge significantly among production systems. Biointensive IPM production involves, in concert with cultural and other management tools, one or more chemical inputs, when needed, to restrict pest populations below economic thresholds. In contrast, IPM practices for organic production involve a very restricted list of synthetic pesticides and fertilizers, and no GMOs. The animal waste products often used in organic production, however, can pose considerable environmental risks. Nevertheless, organic products have an annual growth rate of more than 20% in the United States and Europe.

Integrated pest management concepts for forests often involve divergent approaches as well. Ecosystem management for sustaining forest resources has much to offer but may not include consideration of pest problems. Yet many forests are encountering severe damage from pollutants, endemic pests, and invasive species. The guiding principles used to prevent loss of forest resources have much in common with other ecology-based management systems.

The twofold increase in U.S. agricultural production on fewer acres over the last five decades is viewed by some people as evidence that U.S. agriculture is sustainable. But heavy reliance on chemical pesticides and fertilizers and poor soil management have had significant negative effects. Simultaneously, rapidly increasing world populations and decreasing amounts of arable land translate into demands for even greater productivity per land unit. A key development related to enhancing interest in agricultural sustainability has been widespread recognition that environmentally sound practices must be profitable if they are to be adopted and certain synthetic chemicals may be necessary to ensure sustainability of intensive farming systems. Scientifically based IPM programs are an essential component of sustainable agricultural systems for the future.

Rangeland and Pasture

In 1997, more than 589 million acres (a.) of pasture and rangeland existed in the United States. Regrettably, invasive weeds and other pests have become increasingly troublesome on these important lands. For example, more than 17 million a. of federally owned rangeland was infested with invasive weeds in 1997, an amount that has increased at 15% annually. The most effective IPM strategies encompass multiple tactics (e.g., physical, mechanical, chemical, and biological controls) as well as public awareness and education. Prevention is the least costly means of invasive-weed management. Once established, control of unwanted species, such as leafy spurge, may cost more than $500/a. per year (yr). Much research remains to be done, including the development of a competitive systems approach, a combination of biological and herbicide management, and plant rehabilitation.

Natural Areas

Natural areas, or areas set aside in national and state parks and elsewhere for management of natural resources, are being invaded by nonnative plant and other pest species. Introductions of invasive pests have been both accidental and intentional, and with greatly improved transportation, the potential for long-distance anthropogenic, or human-influenced, introductions has become virtually limitless. Until recently, regulation of weeds in natural areas has been limited; often, the problem has been addressed by cities, counties, states, and other institutions or individuals. The 1999 Presidential Executive Order on invasive species and the establishment of state and regional Exotic Pest Plant Councils provide a framework for progress in IPM related to invasive plant species in natural areas. Many plant species considered invasive by land managers often are available in the nursery and landscape business. Thus, prohibiting the sale of invasive species might contribute to more effective IPM for natural areas. Although prevention is the most cost-effective tactic for invasive plant control, other tactics such as cultural methods and herbicidal and biological controls must be used for established pests.

Aquatic Vegetation

Examples of nonnative, invasive aquatic weeds include alligatorweed, water hyacinth, hydrilla, Eurasian watermilfoil, torpedo grass, water lettuce, and giant water fern. Management of aquatic vegetation is essential to navigation, irrigation, flood control, and natural and recreational resource protection. Both native and nonnative plant species often are problems in aquatic habitats. Aquatic plant managers consider it necessary to manage these pests at the lowest population level (often referred to as *maintenance control*) because of difficulties associated with (1) defining aquatic weed populations at or below economic injury levels, (2) the great reproductive potentials and the rapid growth rates of invasive aquatic plant species found in U.S. waters, and (3) the return of uncontrolled populations to manageable levels. Numerous principles central to successful aquatic IPM have been documented, and a range of control tactics is available.

Food Animals

Livestock and poultry industries are characterized by production and marketing systems ranging from large, vertically integrated companies that dominate poultry and swine production to small, independent growers selling products directly to consumers. Each segment represents a unique set of IPM challenges, in terms of both the pests and the management approaches compatible with the production environment. But cultural practices, biological controls, monitoring, and the judicial use of pesticides all play important roles in IPM programs for livestock and poultry production. Because livestock and poultry production are considered minor pesticide use commodities, the availability of labeled pesticides and the introduction of new chemistries are limited. The relative paucity of pesticide options and the associated complexity of managing pesticide resistance have combined to emphasize the need for a balanced IPM approach for animal agriculture. Indeed, the dairy and beef cattle industries recognize IPM as the preferred set of best management practices to minimize pest problems and to protect farm workers, consumers, and the environment.

Effective management affecting livestock and poultry pests decreases disease risk and production loss resulting from poor growth, development, and/or feed conversion. Strict biosecurity, as practiced by confinement livestock and poultry producers, limits the movement of people and equipment and prevents the spread of many common ectoparasites between and within farms. Isolation and prophylactic treatment of all newly purchased stock ensures introduction of pest-free animals into the herd or flock. Injectable oral and pour-on parasiticides have largely replaced sprays for the management of mites, lice, and bot flies in cattle and swine. The precision of these application methods has improved applicator safety, improved control, and decreased the environmental risks associated with sprays or dips.

Houseflies are the primary pests of dry-waste systems often associated with certain types of poultry production where manure is allowed to accumulate beneath cages or slats. Similarly, houseflies and stable flies are associated with cattle feed lots and dairies where they breed in manure, silage, and wastage around feed bunkers and silos. The two principal IPM objectives for managing these and other waste-breeding flies in poultry and livestock facilities are (1) decreasing moisture in the fly-breeding substrate and (2) effecting biologic control. An implicit goal of all such fly IPM programs is decreased reliance on pesticides.

Mammalian Wildlife-Damage Control

Because of expanding human populations and intensified land-use practices, wildlife-damage control is becoming increasingly important. Wildlife IPM includes preventive as well as curative practices. Cultural practices coupled with exclusion, repellents, or toxicants and other lethal control measures address most types of mammalian wildlife damage. As professionals become increasingly aware of nonbiological considerations, public education, legal, and social considerations are influencing wildlife-damage prevention programs. Such programs usually focus on four issues: problem definition, problem-species ecology, control-method application, and control evaluation. Key wildlife pests include ungulates, rodents, and mammalian predators. Accurate damage identification of wildlife and assessment of losses to specific predators are equally important to the implementation of appropriate control practices.

Companion-Animal Ectoparasites, Associated Pathogens, and Diseases

Pet ownership in the United States is increasing, as is the significance of pets in American life. Along with dogs and cats come their ectoparasites and disease agents. Management of these ectoparasites relies primarily on insecticides and acaricides supplemented with mechanical, cultural, and biological controls. More than one-half of U.S. households contain pets, totaling more than 60 million dogs and nearly 70 million cats. The three most economically important ectoparasites are fleas, ticks, and mosquitoes. (Endoparasites are treated with drugs and parasiticides; however, sanitation plays a role in exposure of the pet.) Fleas transmit tapeworms and cause anemia and flea-allergy dermatitis. Ticks transmit several pathogens, causing illnesses such as Lyme disease, Rocky Mountain spotted fever, and ehrlichiosis. The primary health effect of mosquitoes is transmission of dog heartworm. Nationwide, costs of flea treatment combined with costs for testing and treating ancillary flea problems such as flea-allergy dermatitis and tapeworms amount to $3.8 billion per yr. Costs of treating tick-transmitted disease agents amount to approximately $108 million per yr. Nationally, more than $600 million is spent annually on heartworm testing, prophylaxis, and treatment. Thus, pet ectoparasites are exceptionally significant in terms of their effects on companion animals and the economy.

Diseases associated with animal ectoparasites fall into two categories: those caused by the arthropod itself and those induced by arthropod-vectored pathogens. In addition to continued use of insecticides, acaricides, and other available control tactics, options for controlling animal ectoparasites in the future likely will include manipulating innate animal physiology to suppress ectoparasites, minimizing deleterious effects on pet and pet-owner health.

Urban Areas

Although its roots lie in contemporary agriculture, urban IPM features a human rather than an economic factor in the pest management equation. For example, qualitative aspects such as aesthetics, health, and peace of mind substitute in many ways for the quantitative aspects involved in agricultural IPM decision making. Regardless of whether a pest is a pathogen, an insect, a weed, or a rodent, if it damages homes, structures, clothing, food, or landscape plantings, or harms, annoys, or otherwise interferes with people and their activities, it is an urban pest. General concerns in the urban IPM community include public attitudes, perceptions, and prejudices regarding pests and pesticides and their effects on human activity and the environment. A massive push to decrease human exposure to toxicants of all kinds is underway. Rights are being legislated: rights to be informed of pesticide use, residues, and toxicity levels; rights to pollution-free public environments in which to work, visit, or study; and rights to be free from harmful parasites, diseases, and nuisance pests. Implementation of IPM in schools and other public facilities is one of the fast-est-growing segments of pest management. Attaining acceptable levels of pest management without exposing people and the environment to excessive risks from pesticides is the major goal of urban IPM.

Economics

Economics has played a central role in IPM technology assessment and policy analysis. Economic analysis has been applied to evaluate expected profitability (ex ante and ex post adoption), social welfare effects, research returns, and policies affecting pest management generally. Economic analyses have played an important role in the development of threshold-based IPM and associated software to aid pest management decision making.

Benefit-cost analysis (BCA) can answer the question of whether IPM is worthwhile from the viewpoint of producer, consumer, and/or society at large. Such an analysis can be financial, including only cash costs and benefits, or economic, including the opportunity cost of alternatives not pursued and the external effects of a pest management practice on other parts of society. The BCA becomes considerably more complex when involving IPM effects over more than one season or estimating the value of the seemingly priceless (e.g., clean water, biodiversity, or more-stable crop yields).

An economic threshold offers a guideline for deciding whether pest control (usually by pesticide) is needed and, if so, which method is preferred. This threshold simply identifies the pest level at which treatment would become profitable. More sophisticated thresholds have been developed that take into account several factors: (1) yield damage from future generations of pests allowed to reproduce, (2) environmental and health costs, and (3) spatial patterns of pest dispersion.

Although most biological-pest control methods remain to be analyzed economically, considerable economic work has been done on crops genetically modified for pest resistance or herbicide tolerance. National survey results for 1997 indicated that planting herbicide-resistant corn or soybean or planting *Bacillus thuringiensis* (*Bt*) corn did not affect profit level for U.S. farmers. Cotton growers did increase profitability, however, by using both herbicide-resistant and *Bt* varieties.

Several approaches to decreasing pesticide reliance have emerged from economic research. Appropriately designed crop insurance can insulate farmers against the risks not only of variable crop yield, quality, and price, but also of adopting threshold-based IPM practices. Additionally, public cost-sharing programs can decrease the costs of adopting IPM practices. Surveys indicating that consumers are willing to pay for foods with decreased pesticide residues have led firms to develop ecolabels certifying production process quality. In certain regions of the United States, ecolabeling is making IPM a prerequisite for growers wishing to obtain production contracts for fruits and vegetables.

The daunting problems associated with defining and measuring IPM have impeded comprehensive assessments of IPM research, and no major IPM extension assessment has been completed in the United States for two decades. The failure to measure benefits comprehensively will hamper attempts to garner future public support for IPM programs.

Education and Delivery Systems

Conceptualizations of IPM education, adoption, and related evaluation have evolved concurrently. Most IPM educational programs have been interdisciplinary, with only a few institutions offering degrees or minors in IPM. Although most U.S. universities offer an IPM component in traditional departments such as entomology or plant pathology, the University of Florida recently established the only Ph.D. degree program in plant medicine—a long-discussed need in the United States.

Very diverse IPM delivery systems are being used, including many comprehensive web sites and traditional outlets involving printed materials and personal interactions. With nearly limitless information available, IPM resource personnel are needed to identify which information applies in given situations (Figure S.3).

The 1994 USDA IPM Initiative had a goal of 75% implementation of IPM (on the basis of acres of cropland) by the year 2000. Although at least certain aspects of IPM were practiced on 75% or more of land planted to several crops, in-depth or biointensive IPM was adopted on less than 20% of cropland in 2000.

With increasing public concerns about pesticides and related issues, approaches used in IPM evaluations are becoming increasingly important. The environmental and social parameters essential in assessments of the consequences of pesticide use include health of farm workers, consumers, and the general public; lethal and chronic effects on other nontarget biota; direct or indirect effects on natural and agroecosystems; parameters to calculate air, soil, and water pollution; and costs versus benefits to producers and society for decreased pesticide use. Enlisting producer cooperation in compiling these data by survey, sampling, and other means continues to be a challenge.

Future Challenges and Directions

In this era of increasing public pressure for safe, high-quality food, new technologies promise much for improved IPM programs. Although traditional genetic resistance to pests continues to be an invaluable tool for protecting crops against many very damaging pests, genetic engineering is providing an important new approach to the development of host resistance. When issues such as food and environmental safety and grower/society acceptance are resolved, DNA technology should have great potential to develop value-added traits such as herbicide tolerance and pathogen and insect resistance to crops. This technology also offers exceptional accuracy in identification and characterization of pest species and communities. Precision agriculture also should lead to decreased pesticide use and enhanced profitability. Systems approaches can be used to integrate information, production, and IPM systems and to bring pest management to a new level.

Figure S.3. **A program manager for the Agricultural Research Service's areawide pest management effort discusses the placement of codling moth pheromone dispensers in a pear orchard. Photo by Scott Bauer, Agricultural Research Service, U.S. Department of Agriculture, Beltsville, Maryland.**

Close attention still must be directed at the management of pesticide resistance. To date, more than 500 species of arthropods, 100 species of plant pathogens, and 100 biotypes of weeds have developed such resistance. For many pests, their rapid windborne dissemination compounds the importance of this worldwide problem.

As new species and strains of pests continue to appear in the United States, interest in invasive pests increases. Annual costs and damages due to nonnative pest species have been estimated at $123 billion, or higher. Invasive plants and other pests spread rapidly into natural areas, aquatic habitats, forests, pastures and rangelands, and other agroecosystems. Jet-based global travel and global trade undoubtedly are important factors contributing to the increase in nonnative pests.

Federal government policies, bans, regulations, resource allocations, and special programs affect IPM programs. At the state level, various Departments of Education and Health have created policy statements on implementation of IPM for schools and other public properties. The phasing out of methyl bromide by the year 2005 and the phasing in of the FQPA pose significant challenges to IPM adoption. This act redefines how pesticides are regulated and will restrict the availability of certain pesticides. Fruits and vegetables, because of their preponderance in children's diets, will be hard hit by the FQPA. The new USDA Regional IPM Centers are addressing the FQPA requirements as well as advancing improved IPM strategies and systems.

Government policies on specific food groups or products such as organic foods and GMOs continue to be refined. Crop insurance for IPM and labeling of IPM also could have important effects on the future of IPM. The most recent important federal legislation to affect IPM is the 2002 farm bill, including the new IPM Initiative. That new initiative focuses on risk reduction—economic risk and environmental risk—with decisions on the specific use of related funds being made at the state level.

For significantly enhanced future development and full adoption of IPM nationally, a well-coordinated and supported federal IPM entity that focuses on the goals of the IPM initiative is crucial. Application of findings from in-depth ecological research on agroecosystems coordinated with extensive economic assessments that document the benefits of IPM will contribute immeasurably to this goal.

1 History of Integrated Pest Management Concepts

The concept of using multiple strategies and tactics for plant disease control was introduced in the mid-1880s by the German botany professor Julius Gotthele Kühn. The term *integrated pest management* (IPM), clearly related to Kühn's work, was first used by entomologists Smith and van den Bosch in 1967 and was based on the integrated control concept. Researchers in the Departments of Entomology at the Universities of California–Riverside and –Berkeley had defined *integrated control* as applied pest control that combines and integrates biological and chemical controls, the latter being used as necessary and in the manner least disruptive to the former (Stern et al. 1959). In fact, widespread concerns about the effects of pesticides on the environment and health played a key role in the development of IPM. Carson's 1962 book *Silent Spring* contributed greatly to the worldwide debate on the potentially negative aspects of pesticides and to the environmental movement (Waddell 2000).

Although definition of the term *IPM* first appeared primarily in entomological lexicons, integration of several control strategies has been the basic tenet of plant disease control since the inception of the applied science of plant pathology. Although slow to adopt IPM concepts at first, pathologists, nematologists, and weed scientists now are important partners in the development and implementation of those concepts.

It can be argued that in the late 1950s and early 1960s the formalized IPM concept first was grasped by entomologists after they began identifying the problems of pest resistance to insecticides and of adverse ecological effects associated with their widespread use. From the 1940s to the mid-1960s, the lack of a pesticide "silver bullet" such as that afforded to entomologists and weed scientists by synthetic insecticides and herbicides had forced plant pathologists to emphasize nonpesticide control strategies. These strategies included genetic host resistance to pathogens, cultural controls such as rotation and tillage, and pathogen-free seed or planting stock. More recently, fungal and bacterial pathogen resistance to fungicides and bactericides, weed resistance to several classes of herbicides, loss of registration of several nematicides, paucity of new synthetic chemical products, and increased awareness of ecological consequences of pesticide use have led to an ever increasing emphasis on implementation of IPM practices by all crop protection disciplines.

The formal involvement of other pest control professionals with entomologists in IPM programs was the result mainly of the Huffaker Project, cofunded in 1972 by the National Science Foundation, the U.S. Environmental Protection Agency (EPA), the U.S. Department of Agriculture (USDA), and the USDA's Cooperative Extension Service (CES). The last provided funding for Extension IPM programs in every state in 1975. Although these high-visibility projects promoted adoption of IPM terminology in all pest control disciplines, plant disease control personnel never embraced the concept of pathogen (pest) eradication/control based solely on pesticide use, a common practice of entomologists and weed scientists. Plant pathologists endorsed integrated disease management by means of the application of fundamental knowledge of loss potential and pathogen biology, ecology, and disease epidemiology. In comparison, the highly effective synthetic organochlorine and organophosphate insectides and herbicides available during the post-WWII period made chemicals the first choice for control of insect and weed pests in the 1950s and the 1960s.

Since the Huffaker Project and the CES initiative, plant pathologists, entomologists, nematologists, weed scientists, agronomists, horticulturists, crop consultants, farmers, and others have contributed significantly to the implementation of IPM programs for many crops and cropping systems. Cooperative Extension Service funding of pilot projects using scouts and encouraging development of pest management cooperatives has been key to the development of a cadre of independent crop consultants that promote and implement IPM in every state. Another key to building interdisciplinary teams for IPM has been the acceptance by other crop protection disciplines of the entomological definition of the term *pest* to describe not only arthropods and rodents but also plant pathogens, weeds, and nematodes.

Although many working definitions of IPM are available, that adopted by the National Coalition of IPM will be used in this chapter. The coalition's definition of IPM is "a sustainable approach to managing pests by combining biological, cultural, physical, and chemical tools in a way that minimizes economic, health, and environmental risks" (Jacobsen, B. J. 2002. Personal communication). This definition reflects considerable evolution of the IPM concept, from its beginnings in 1959 (Stern et al. 1959) to a more recent call for ecologically based pest management (EBPM) (Overton 1996), or biointensive IPM (Benbrook et al. 1996). In 1998, the USDA's IPM Program Coordinator articulated the prevention, avoidance, monitoring, and suppression (PAMS) concept: *prevention*, or the practice of keeping a pest population from infesting a crop or field; *avoidance*, or the practice of using cultural techniques to avoid losses to pests that could be present in a field; *monitoring*, or the practice of observing pest populations, weather, and nutrients; and *suppression*, or the practice of using cultural, physical, biological, or pesticidal means to suppress pest populations (Coble 1998). The PAMS concept provides broad descriptors for assessing IPM implementation across various crops and cropping systems.

Evolution of the Integrated Pest Management Concept

The 1959 paper "The Integration of Chemical and Biological Control of the Spotted Alfalfa Aphid" (Stern et al. 1959) formalized the concept not only of integrated control but also of equilibrium population and introduced the concepts of economic injury level (EIL) and economic threshold (ET). The *EIL* is the pest population level capable of producing damage the cost of which exceeds that of control. The *ET* is the pest population level (just below the EIL) at which control options are used to prevent the pest from reaching the EIL. These concepts provided for a decision structure focusing on (1) restricting pest populations below levels causing economic damage and (2) using chemical or other methods of intervention to prevent economic damage. Although these concepts worked well for pests that could be counted and whose damage could be described easily, other pest disciplines were confronted with pests whose populations were difficult to count. Moreover, certain pest populations did not cause predictable damage, and damage exceeded ETs before symptoms became visible. Thus, most pest control practitioners did not identify readily, based on pest population monitoring, with the EIL or the ET concepts.

Smith and van den Bosch (1967) not only introduced the term *integrated pest management* but made a strong case for integration of all control strategies and for application of ecological principles to pest control in agricultural production systems. These researchers also introduced the concept of the *key pest*, "a serious, perennially occurring species that dominates control practices because in the absence of deliberate control by man, the pest population usually remains above the economic injury levels" (p. 295). Examples of key insect pests include coddling moth on pome fruits in the western United States, boll weevil on cotton (Figure 1.1), alfalfa weevil on alfalfa, flea beetle and looper on crucifers, root maggot on sugar beet, and Hessian fly on wheat. Examples of diseases that might be considered key pests are apple scab in the Midwest and eastern United States, damping-off diseases, potato/tomato early and late blights, seedborne smuts, viral diseases transmitted through vegetative propagation, early and late leaf spot of peanuts, sigatoga of bananas, root-knot and cyst nematodes, Dutch elm disease, chestnut blight, and many other diseases both native and exotic. Unlike the insect pests mentioned by Smith and van den Bosch (1967), some of the diseases just noted do not require chemical controls. In fact, most key pathogens are managed by integration of host resistance, sanitation, certification, fungicides, or nematicides; by rotation or of selection of planting time and tillage; and, most recently, by use of selected biological control products (Figure 1.2). The term *key pest* is irrelevant to most

Figure 1.1. A cotton boll weevil. Photo from the Agricultural Research Service, U.S. Department of Agriculture, Beltsville, Maryland.

agricultural weed pest problems because a complex of weeds occurs in most agricultural fields. The term is appropriate, however, for most nematode problems.

Many key insect pests and invasive weed species are exotic and require control because their natural enemies did not migrate with them. But for most plant pathogens, the absence of biological control agents is less important than the uniformity of host susceptibility. In other pest management disciplines, a second divergence from entomology is that they rarely deal with secondary pests.

Secondary pests are those organisms whose populations normally are maintained below ET by biological agents but which may be disrupted by pesticide application aimed at key pests. Again, the lexicon for IPM was designed for entomology, not for other pest

Figure 1.2. The leaf beetle *Diorhabda elongata* is the first approved biological control agent for saltcedar in the United States. Photo by Bob Richard, APHIS. Agricultural Research Service, U.S. Department of Agriculture, Beltsville, Maryland.

management disciplines although weeds released from competition because of herbicidal control of other weeds may have parallels to secondary pests described by entomologists. Other concepts introduced by Smith and van den Bosch (1967), including *occasional pests*, *potential pests*, and *migrant pests*, have parallels in the lexicons of other disciplines. Pedigo (1972) introduced the term *preventive pest management* to cover practices that prevent or avoid pest outbreaks or maintain pest populations below the ET. Examples of such practices include the use of resistant varieties, crop rotations, selected planting or harvesting dates, and many other cultural controls. In the proceedings of a 1970 national conference on pest management concepts, Rabb and Guthrie (1970, p. 2) stated that "pest management deals primarily with populations, communities, and ecosystems; thus, the basic discipline involved is ecology." This perspective illustrates entomology's move toward adopting the conceptual foundation of other pest management disciplines.

The period from the late 1960s to the present has established IPM as a desirable pest control strategy from the perspectives of both scientific pest control disciplines and national policy. In 1969, the National Academy of Sciences (NAS) published a study entitled "Principles of Plant and Animal Pest Control," which emphasized systems analysis and computer simulation modeling and encouraged increased disciplinary involvement in IPM. The term *integrated control* was used widely in that publication, and the modern IPM concept was envisioned because quantitative ecology is integral to describing the complex of interactions among the multitude of species interacting in crop-pest systems in agriculture. In 1972, a report issued by President Nixon's Council on Environmental Quality (CEQ) focused on IPM; the $12.5 million Huffaker Project was the direct result. The CEQ report broadened the scope of IPM, as is evident in the final paragraphs of the report's summary:

> Integrated pest management is an approach which maximizes natural controls of pest populations. An analysis of potential pest problems must be made. Based upon knowledge of each pest in its environment, and its natural enemies, farming practices are modified (such as changes in planting or harvesting schedules) to affect the potential pests adversely and to aid natural enemies of the pests. If available, seed which has been bred to resist the pests should be planted.
>
> Once these preventive measures are taken, the fields are monitored to determine the levels of pests, their natural enemies, and important environmental factors. Only when the threshold level at which significant crop damage for the pest is likely to be exceeded should suppressive measures be taken. If these measures are required, then the most suitable technique or combination of techniques, such as biological controls, use of pest-specific diseases, and even selective use of pesticides, must be chosen to control a pest while causing minimum disruption of its natural enemies.
>
> This approach differs markedly from the traditional application of pesticides on a fixed schedule. (CEQ 1972)

In 1977, while instructing the CEQ to recommend actions to aid development and implementation of

IPM techniques, President Carter used this description of IPM in his 1979 environmental message:

> IPM uses a systems approach to reduce pest damage to tolerable levels through a variety of techniques, including natural predators and parasites, genetically resistant hosts, environmental modifications, and when necessary and appropriate, chemical pesticides. Integrated Pest Management strategies generally rely first upon biological defenses against pests before chemically altering the environment. (Bottrell 1979)

During the Carter Administration, the 17-university Consortium for Integrated Pest Management (CIPM) was formed and funded to develop and to implement IPM for alfalfa, corn, cotton, and soybean. The Huffaker and the CIPM projects provided the funding to develop new knowledge and a cadre of IPM–trained scientists from all pest control disciplines. Beginning in 1978, the funding of IPM Extension projects in every state resulted in broad recognition of IPM and its concepts at state and national levels (Figure 1.3). Perhaps the most important outcome of federal and state funding for IPM was the establishment of private consultancies and scouting services for producers. The basis of these new businesses was the development of science-based IPM technologies adapted to local situations by means of Extension education programs. The axiom that IPM is site specific and both knowledge and information-intensive bears repeating. Many of the first IPM consultants gained experience from Extension employment in pilot IPM demonstration programs.

In 1979, the Experiment Station Committee on Policy (ESCOP) commissioned the Intersociety Consortium for Plant Protection to produce a report on IPM research. The report, authored by Apple and colleagues (1979), reflected the interdisciplinary composition of its writers and provided a new, tactic-neutral definition for IPM. The report stated that IPM is "the optimization of pest control in an economically and ecologically sound manner. This is accomplished by the use of multiple tactics in a compatible manner to maintain pest damage below the economic injury level while providing protection against hazards to humans, animals, plants, and the environment" (p. 1). The report offered the interesting insights that "the agroecosystem is not characterized by 'biological balance,' but by 'biological imbalance.' And, thus, humankind's role is not to maintain biological balance of the agroecosystem (in ecological terms), but to maintain a dynamic state of imbalance that optimizes crop production" (p. 18). The report emphasized the need and potential for both descriptive and predictive models in IPM systems.

Although authors such as Cate and Hinkle (1994) have cited Apple and colleagues' report as the beginning of the change in focus from natural control to efficient pesticide use in IPM, the report also reflects both the limited applicability of "classical biological control" to species other than arthropod pests and the broad base of pest management tactics adopted by plant pathologists, weed scientists, and nematologists. Also in 1979, the Office of Technology Assessment published an IPM report endorsing the ESCOP report and, again, supporting systems analysis and simulation models to establish ET values.

Figure 1.3. An Extension agent, an agronomist, and a farmer work together as a team to figure out the best methods for growing wheat; timely and appropriate weed control measures are critical to conserving scant soil moisture in a dryland cropping system. Photo by Scott Bauer, Agricultural Research Service, U.S. Department of Agriculture, Beltsville, Maryland.

Formula funding from CES for IPM education (Smith-Lever 3[d] funds) was available to Extension crop protection specialists in each state from 1978 onward whereas widespread research funding became available only when the CIPM project was terminated in 1983 and approximately $3.5 million was redirected to regional IPM and biological-control grant programs. Federal support for IPM programs at land-grant universities was $0.5 million for CES pilot IPM projects in 1973 and $1.5 million for regional IPM and biological control research in 1981. For fiscal year (FY) 2000, federal funding for pest management programs included $10.8 million for IPM Extension implementation, $9.0 million for minor-use crop pest management (IR–4) (Figure 1.4), $4.0 million for the Food Quality Protection Act (FQPA) Risk Avoidance and Mitigation Program (RAMP) for cropping systems, $2.73 million for regional IPM and biological control research, $1.6 million for pest management alternatives research, $1.0 million for crops at risk (CAR) from implementation of the FQPA, and several million dollars in research funding from the National Competitive Grants Program in biologically based pest management and other program areas. Additional funding for IPM-related research and Extension came from the federal sustainable agriculture program. One-half to two-thirds of all projects funded by this program are estimated to have been related directly to IPM (Jacobsen, B. J. 2002. Personal communication).

Figure 1.4. The interagency IR–4 program ensures the safety of so-called minor-use chemicals before they are approved for commercial agricultural production. A California agronomist displays test-plot-grown broccoli that will be used to determine pesticide residue levels. Photo by Scott Bauer. Agricultural Research Service, U.S. Department of Agriculture, Beltsville, Maryland.

Regional grant programs are administered by the USDA–Cooperative State Research, Education, and Extension Service (CSREES) using grant managers in the northeastern, north central, southern, and western USDA administrative regions. These regional grant programs emphasize interdisciplinary research and preventive pest management tactics. The first competitive Extension funding was provided to the north central region in 1995 and to all regions of the United States in 1996. Starting in 1997, these regional IPM grant programs have focused on research and Extension priorities identified by state and regional IPM teams funded by the USDA's 1995 IPM Initiative. Still, concern has been expressed about the breadth of identified state priorities, which favor entomology heavily. Perhaps this funding emphasis should not be surprising: in 31 states, leadership in IPM programs rests with entomologists; in only 6 states, with pathologists; and in the remaining states, with nondepartmentally affiliated individuals (Gray 1995).

Clearly, the best IPM programs evolve from close interdisciplinary cooperation. It is crucial for weed scientists, plant pathologists, and nematologists to be involved with other stakeholders in the priority-setting process in both research and Extension. Since 1996, all regions have provided important incentives to weed scientists to apply for funding, and funding for weed management IPM research through the regional IPM programs increased from less than 10% of funded projects to approximately 25% in 2001. Although most funding still goes to entomology and plant pathology-related projects, funding of weed science is approaching parity with that of other pest management disciplines.

Funding from these regional programs was instrumental in the development of various computer-based decision support software products including the WISDOM computer software program for potatoes (Stevenson et al. 1995), AU-PNUTS for peanuts (Jacobi and Backman 1995), WeedSOFT (Mortenson et al. 1999), and many other IPM products. These decision support software programs have resulted in significant savings for growers and in decreased use of pesticides (Martin, Mortensen, and Lindquist 1998). For example, in 1998, implementation of WISDOM is estimated to have saved Wisconsin growers of Russet Burbank potatoes more than $168/acre (a.) (27.4 lb active ingredient pesticide, 123 lb nitrogen, and 5.2 inches of irrigation water/a.) (Connell et al. 1991).

Although funding for IPM research and implementation remained fairly static until 1983, in the mid-1980s IPM research and Extension scientists and practitioners began to discuss methods of increasing the biological component relative to the pesticide management component of IPM programs. Several developments stimulated this trend, including new federal programs emphasizing sustainable agriculture, increased resistance to pesticides of all types, increased public concerns regarding pesticides and food safety, increased difficulty of registering new pesticides or new use patterns (especially for minor crops), and increased pressure from growers and practitioners for new IPM tactics. These issues and factors continue to drive efforts to focus new resources on IPM development and implementation and were directly responsible for the Clinton Administration's IPM Initiative, an NAS call for investment in EBPM (Overton 1996), and a Consumers Union (CU) call for 100% implementation of biointensive IPM by the year 2020 (Benbrook et al. 1996).

The Clinton Administration's Integrated Pest Management Initiative

Several events led to the June 25, 1993 announcement by the Clinton Administration that it was committed to decreasing the use of high-risk pesticides and to promoting IPM. Key events included an EPA/USDA joint analysis of IPM's potential to address contemporary issues in agriculture, and the 1992 National IPM Forum. The recent CU book *Pest Management at the Crossroads* (Benbrook et al. 1996) suggests that the announcement was timed to head off the publicity expected to follow the release of the NAS report "Pesticides in the Diets of Infants and Children" (NRC 1993). That report received widespread coverage and was a major reason for the passage of the FQPA of 1996, which finally addressed the Delaney clause and differentially addressed the dietary risks of pesticide residues to infants and children. The Delaney clause in the 1954 Federal Insecticide, Fungicide, and Rodenticide Act prohibited the presence of any food additive (e.g., pesticide residue) that caused cancer in processed foods. Ultimately, increasing knowledge of carcinogen dose response data, and technologies allowing residue detection in the parts per billion or trillion range made the Delaney clause obsolete.

The calls for national leadership in IPM, initially issued by the Nixon and the Carter Administrations, were a consensus of the 1992 National IPM Forum (Sorenson 1992). That forum, sponsored by the USDA and the EPA, involved more than 600 individuals including scientists; regulators; food processors, marketers, and other agribusiness people; environmental and other public policy activists; and farmers from throughout the United States. Participants included Congressional staff and leading administrators from the USDA and the EPA. The success of this broad-based forum could be attributed largely to a 1991 workshop that contributed to a USDA/EPA joint analysis of the potential for IPM adoption beginning in 1990 in the United States. Sponsored by the Johnson Foundation, the Wingspread Conference focused on IPM analysis by private practitioners; university, federal, and industry-based scientists; farmers; environmental activists; and regulatory personnel for corn/soybean, cotton, tree fruits, and vegetables. Both the potential for IPM to address the economic and environmental issues of contemporary agriculture, and the constraints on implementation and adoption of IPM were reported in the 1992 book *Food, Crop Pests, and the Environment: The Need and Potential for Biologically Intensive Integrated Pest Management* (Zalom and Fry 1992).

As noted in the *Congressional Record*, on September 21, 1993, the Deputy Secretary of Agriculture, an EPA Administrator, and a U.S. Food and Drug Administration (FDA) Deputy Administrator, in joint testimony for the Clinton Administration, stated before Congress that "we are setting a goal of developing and implementing IPM programs for 75% of total crop acreage within the next seven years (yr). We believe Congress should endorse the goal." To that end, the USDA developed in 1994 a strategic plan for a department-wide IPM initiative with four objectives.

Objectives of the U.S. Department of Agriculture's Integrated Pest Management Initiative Strategic Plan

1. **To ensure that the USDA's IPM programs and policies are coordinated effectively across USDA agencies and that cooperation is facilitated with both public and private non-USDA entities, to meet national goals for IPM implementation**. The key coordinating mechanism was to be the USDA's IPM Program Subcommittee, chaired by the USDA's IPM Program Coordinator. The subcommittee had representation from the Agricultural Research Service (ARS); the Animal and Plant Health Inspection Service; the Forest Service; the Farm Services Administration; the Agricultural Marketing Service; the National Resources Conservation Service; the CSREES; the Economic Research Service (ERS); the National Agricultural Statistics Service (NASS); the Office of Budget and Policy Analysis; and the EPA. This broad-based working group was to be developed to ensure co-

ordination of federal research, education, and regulatory programs with the land grant university and locally based USDA programs in every state.

2. **To establish and to conduct a process for identifying the IPM implementation needs of producers and to provide necessary support and resources with which to conduct a coordinated program of research, development, and delivery of education and information in order to meet producers' IPM implementation needs.**
3. **To develop methods and to conduct programs accurately measuring progress toward the 75% IPM goal and assessing IPM implementation effects (as measured in terms of economic, environmental, public health, and other social factors) on both public and private sectors.**
4. **To implement a stakeholder-centered communication and information exchange program in order to increase public and policymaker understanding of the objectives, progress, and effects of the USDA's IPM Initiative** (USDA 1994, cited in and adapted from Fitzner 1996).

Unlike those European governments that mandated use-reduction strategies in the early 1990s, the United States adopted, through the USDA and the EPA, objectives regarding the decrease of risks from pesticide use and the development of more sustainable agricultural-production strategies. Evidence suggests that the European strategy resulted primarily in the substitution of lower-use-rate pesticides for higher-use-rate pesticides rather than in the adoption of a greater variety of pest management strategies.

In 1995, an increase of $25,000 in Smith-Lever 3(d) pest management education funding was provided to each state and territory to work with farmers and other stakeholders to identify and to prioritize needs for IPM implementation regarding key commodities in each state. More than 4,250 customers, including 3,205 farmers, were engaged in identifying priority research and Extension needs for IPM implementation at the state level. In addition to the state-level needs assessment process, 23 production-region IPM teams in 44 states identified IPM needs for crop production regions. These teams included 154 farmers or crop consultants, 36 food processors or marketers, state- and national-level commodity organizations, agribusinesses, USDA and EPA field personnel, and research and Extension faculty at cooperating land-grant universities (Jacobsen, B. J. 2002. Personal communication).

The Clinton Administration first requested increased budget support for the IPM Initiative in FY 1996. In that period, Congress appropriated increased funding of $1.6 million for one component of the IPM Initiative—the Pest Management Alternatives Program (PMAP)—but did not fund the other portions of the initiative. Funding for the USDA's complete IPM Initiative again was requested in the executive budget in FYs 1997, 1998, 1999, and 2000, but Congress appropriated only limited increases. These primarily addressed needs created by the FQPA legislation and included increases for the PMAP, the CAR from FQPA Implementation, and FQPA Risk Mitigation for Major Food Crops, the USDA–ARS Areawide IPM program, and the IPM activities of the ERS and the NASS. The total investment requested for IPM and related programs in FY 2000 was $366 million—$68 million more than was appropriated in the FY 1996 budget, the first year of the Clinton Administration IPM Initiative.

The Environmental Protection Agency Initiative

Whereas the USDA has focused on farmer development and implementation of IPM programs that identify stakeholder needs and priorities and develop new pest management tools to replace pesticides rendered obsolete due to regulation or to voluntary registrant cancellation, the EPA has focused on its Pesticide Use/Risk Reduction Initiative. This initiative has three basic components: (1) discouraging the use of high-risk pesticides, (2) providing incentives for development and commercialization of safer products, and (3) encouraging use of alternative control methods, thereby decreasing reliance on toxic and/or persistent chemicals. Designated the lead division for EPA, the Biopesticides and Pollution Prevention Division (BPPD) initiated two programs addressing the components of the initiative.

First, the Pesticide Environmental Stewardship Program (PESP) works with more than 114 commodity and pesticide user groups in a voluntary partnership promoting IPM adoption. The BPPD has provided grants to support on-farm research and demonstration projects, including funding of commodity groups to identify a descriptive set of IPM elements so that farmers can score their own levels of IPM implementation. Second, the PESP has expedited registration of biopesticides and other decreased-risk chemicals including pheromones. In 1995, the EPA set the following goals for the year 2005: (1) illegal residues detected in FDA monitoring will be decreased 25% from current levels, (2) a significant decrease in use of pesticides with the highest potential for carcinogenic effects on food crops will occur, (3) no instances will occur, in either adults or children, in which a pesticide reference dose is exceeded, (4) IPM will be used on all but 10% of crop acreage and pesticide usage and (5) biological pesticides deemed safe will be used on twice as many acres as in 1995 (EPA 1995).

The Clinton Administration, through the USDA/EPA initiatives, proposed a structure that, if funded by Congress, would provide for IPM systems development and implementation based on the identified needs of farmers and other stakeholders. This IPM initiative also would provide farmers and other pesticide users with biologically based pest management tools and a fast track for registration of biopesticides and biological control agents. Certain nongovernmental organizations criticized the federal strategy for being pest management strategy-neutral instead of emphasizing biological controls. It can be argued, however, that the strategy-neutral course has been adopted to reassure farmers and others that they will be provided with proven tools for pest management so that they can move along the continuum from pesticide intensive systems to biologically intensive systems while maintaining profitability and managing risk. The success of this strategy can be seen by the fact that the primary increases in the IPM budget portfolio enacted by Congress and supported by farmers since 1996 have been the FQPA-related programs PMAP, CAR, and RAMP.

As has been indicated, the concept of a pest management continuum has been described and refined many times in the last 35 yr. At the Third National IPM Symposium Workshop, held in Washington, D.C. in 1996, a representative of the World Wildlife Fund (WWF) presented the IPM continuum on which Figure 1.5. is based in part (Jacobsen 1997). Originally published by the CU, this continuum, in that it emphasizes shifting reliance from treatments based on scouting and ETs to prevention and biologically intensive interventions, differs from that used by the USDA and described by Vandeman and colleagues (1994), who studied practices indicative of an IPM approach for selected field and vegetable crops. Covering the period 1990 to 1993 and conducted by the ERS, the Vandeman et al. study determined for the USDA what the baseline was for the 75% goal. The baseline, constructed from survey data available from 1990 to 1993 and from decision rules based on a count of practices indicative of IPM was that approximately 50% of U.S. crop acreage was already under IPM. (Approximately 5 to 15% of surveyed acres were described as under "low-level IPM," e.g., scouting and pesticide applications based on ET; 25 to 35%, as under "medium-level IPM," e.g., low level plus one to two additional practices indicative of IPM; and 20 to 30%, as under "high-level IPM," e.g., low level plus three or more practices indicative of IPM.) The USDA report, although based primarily on practices related to pesticide use, was the first attempt to assess on a national basis the adoption of IPM across a wide range of crops.

In 1999, the NASS published a summary of pest management practices used by farmers in 1998. This summary used the PAMS paradigm for IPM practices, a paradigm helping researchers examine utilization of IPM practices across crops and regions. Using the PAMS descriptors, the survey showed that multiple tactics were used on more than 40% of barley and wheat acres, on more than 50% of corn acres, on more than 50% of cotton acres, and on more than 40% of soybean acres. In comparison, multiple IPM tactics were used on more than 70% of fruit and nut

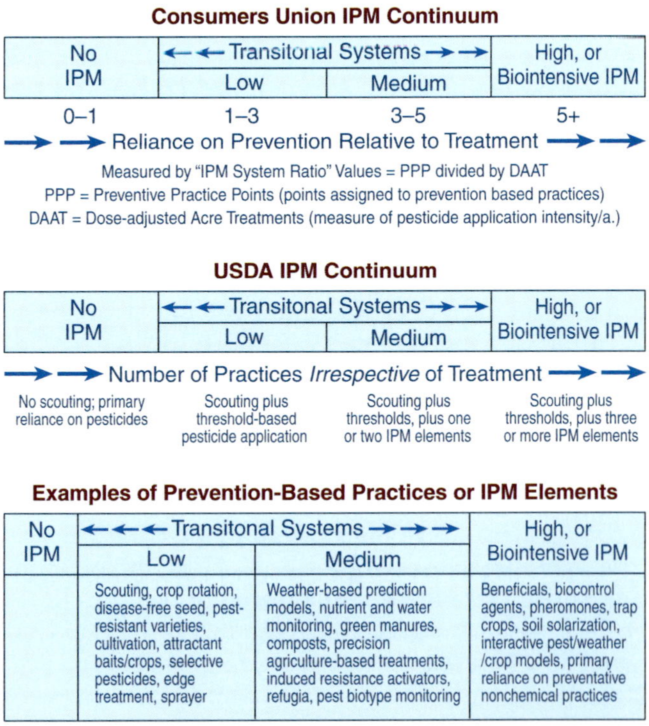

Figure 1.5. Comparison of Consumers Union and USDA IPM Continuums, and examples of IPM elements or prevention-based practices characteristic of low, medium, and high IPM.

acres and on 78% of vegetable acres. These findings are important because they concern the crops most dependent on pesticide use and thus the focus of FQPA calculations regarding potential pesticide-residues and risks to children. In 1998, acreage under medium- to high-level IPM implementation increased while acreage under low-level IPM implementation for most crops increased to levels approaching 70% or slightly greater. An assessment of IPM adoption is necessary if documentation of the effects of federal, state, and private investment in IPM is to be undertaken. The CU continuum is founded on an ecologically based set of factors accounting for both preventive and treatment-oriented IPM practices.

Call for Ecologically Based, or Biointensive, Pest Management

In 1996, the NAS published the book *Ecologically Based Pest Management—New Solutions for a New Century* (Overton 1996). Prepared at the request of the USDA, the EPA, and the Board on Agriculture, the book focused on the need for additional sustainable pest management strategies and emphasized that strategies should be based on ecological principles. The objective of the book, as of the CU's *Pest Management at the Crossroads* (Benbrook et al. 1996), was to shift the current IPM paradigm from focusing on pest management strategies relying all too often on pesticide management to a systems approach relying primarily on biological knowledge of pests and their interactions with crops. In EBPM, or biointensive, IPM systems, knowledge of the managed ecosystem and its natural processes suppressing pest populations is essential; intervention with pesticides or with physical or biological supplementation comes later. All parties calling for a new pest-management paradigm have called for increased investment in research to develop the required biological and ecological knowledge bases.

Differences in the NAS's and the CU's calls for IPM paradigms are more than nominal. *Pest Management at the Crossroads* calls for accelerated movement along the pest management continuum and for implementation of existing IPM knowledge and tools. The CU-sponsored book sets specific goals (e.g., 75% of crop acreage under medium- to high-level IPM by 2010 [Benbrook et al. 1996]) that are attainable in large part, with modest investments, given current estimates that 25 to 65% of crop acreage is at the medium-plus level of IPM (Benbrook et al. 1996; Vandeman et al. 1994). To attain EBPM as described by the NAS, massive investment will be required to build the ecological knowledge base needed for the multitude of cropping systems, environments, and pest complexes constituting U.S. agriculture. In all pest disciplines, this knowledge base is incomplete for most pests, and the development of interdisciplinary, agroeconomically and pest-complex-based system knowledge has just begun.

The authors of this Council for Agricultural Science and Technology report support the goals of EBPM but feel that they should be viewed as long-term strategic goals and thus as achievable only through collaboration with farmers and through encouragement of progression along the IPM continuum. Only through such movement will the necessary public/private financial and regulatory support be developed. Lack of success by the Clinton Administration in achieving fairly modest funding increases for the USDA/EPA IPM Initiative illustrated the new dynamics of federal funding for research and Extension education and demonstrated that the investments needed to achieve EBPM were unlikely to be made at that time. Pest disciplines, economists, ecologists, sociologists, public interest groups, and, most important, farmers would have to develop new dialogues and blueprints for interdisciplinary funding.

As farmers progress along the IPM continuum, they develop confidence in relatively basic IPM tactics and techniques before moving to more complex ones. This observation is supported by IPM adoption research done by Sorenson and Stewart (2000). Farmers also are leaders in calling for new research and technology transfer programs, as evidenced by the stakeholder inputs required by the USDA's IPM strategic plan and expressed at the five regional IPM workshops sponsored in 1993 by the National Foundation for IPM Education (Goodell and Zalom 2000). Clear examples are found among apple, potato, grape, and cotton growers, who are widely considered the most advanced implementers of IPM.

Integrated Pest Management in the Twenty-First Century

In March 1999, North Carolina State University (NCSU) hosted a second international IPM conference (Kennedy and Sutton 2000). Unlike the 1970 conference, which had focused on insect-pest management, the 1999 conference embraced multidisciplinary IPM concepts that had evolved out of the basic concepts and principles developed by entomologists in the 1960s. The 1999 NCSU conference identified IPM as the key contemporary crop-protection paradigm. Its impact on agriculture and the factors affecting its adoption were addressed, and its potential to increase profitability while helping to decrease both negative environmental agricultural effects and human risks from pesticides was demonstrated. The strategy succeeded by encouraging an array of crop-protection strategies. Information was presented in light of the assumption that ever-greater interdisciplinary involvement, including that of the social sciences, would be necessary to address future changes in agriculture—if IPM was to be addressed on an agroecosystem basis. Numerous speakers emphasized that greater progress in IPM implementation would require increasingly careful attention to applied aspects of ecology so that multiple trophic layers within agroecosystems could be understood. Further implementation of biological controls by means of microbes would require basic research in microecology of all plant surfaces, including the rhizosphere.

Speakers at the NCSU IPM Conference emphasized that the rate of change and the challenges for crop producers were increasing and that these factors would influence IPM. Factors in this increasingly dynamic economic and sociological flux included changes in federal farm-policy, decreased availability of economic safety nets for producers, increased farm size, increased consumer involvement, nongovernmental groups and other nontraditional groups' wishing to influence farm and pesticide policies, importance of the global marketplace and reliable information services, biotechnology, and public disinvestment in research and Extension education supplied by the land-grant university system.

Although these factors are global and IPM concepts general, speakers emphasized that applications were site specific and that producers needed solutions and reliable, unbiased information services that were applicable to local situations. Since the 1970 NCSU conference, successful establishment of a crop consultant industry throughout the United States has been an important advance in IPM implementation. Excellent ideas regarding future directions for the IPM paradigm are discussed by Goodell and Zalom (2000), Sorenson and Stewart (2000), and Merrigan (2000), all of whom address the importance of risk management as part of the IPM adoption equation.

Risk Management

Risk management currently is an emphasis of farm policy discussions. As a result of the 1995 Farm Bill, farmers could, in exchange for loss of price supports, grow almost any crop (with the exception of certain fruits and vegetables); for its part, the federal government worked aggressively in the international marketplace to ensure free and fair access for U.S. commodities. World surplus production in feed grains, cotton, and oilseeds, heavily subsidized production, lack of free and fair market access, and economic turmoil are major factors contributing to the lowest commodity prices in recent history. Thin to nonexistent profit margins have increased the importance of risk management for producers, especially in light of the availability of crop insurance.

Because insurance companies require production histories, the crop rotation/pest management opportunities made possible by the 1995 Farm Bill have been diminished or even eliminated for farmers with no production history for new crops. Additionally, for certain crop producers, the 50% insurance product provides inadequate risk management. But without actuarial data regarding the efficacy of specific production practices, insurers are uncertain about the financial viability of offering insurance products covering 65 to 85% of expected yields. Integrated Pest Management certification may enhance the marketability of such products. The relation between IPM and crop insurance is likely to be an important issue in the near future, and producers may be encouraged to increase IPM implementation as a result of the risk management afforded by insurance products.

The effect of biotechnology on agriculture and on IPM in particular was highlighted by many speakers at the 1999 NCSU conference. Significant progress was documented in terms of not only genetically engineered crops but also highly sensitive diagnostic technologies for detection or identification of pests, pesticide-resistant insects or pathogens, and insecticide residues. These new technologies will play an increasingly important role in IPM implementation and regulation.

The use of transgenic pest- or pesticide-resistant crops, a.k.a., *genetically modified organisms* (GMOs), has had a major effect on IPM. Acreage on which this technology has been implemented has expanded logarithmically in the past 5 yr, and environmental effects have been considerable. For example, cotton and corn plants expressing one or more insecticidal endotoxin proteins from *Bacillus thuringiensis* (*Bt*) were planted on 20 and 22% of acres, respectively, in 1998, compared with 5 and 13% of acres, respectively, in 1997. Crops modified for herbicide tolerance have implementation figures increasing even more dramatically from 1997 to 1998, with GMO corn acreage increasing from 2 to 11%; GMO soybean acreage, from 10 to 48%; and GMO cotton acreage from 5 to 34% (USDA–NASS 1999). These developments have brought about dramatically decreased pesticide use. For example, it is estimated that the use of *Bt* transgenic cotton resulted in the use of 300,000 fewer gallons of insecticide in 1997 than in 1995. The increased usages of these GMOs were even more striking for 2002, with GMO acreages increasing to 74% for soybean, 71% for cotton, and 32% for field corn (Cornell Cooperative Extension 2002).

Herbicide-tolerant crops and GMO crop plants expressing plant protection proteins constitute the first wave of transgenic crops; plants expressing quality factors such as improved protein, oil, starch, or processing characteristics will constitute the next wave, followed by plants that will be factories for specific products such as vaccines and proteins. More than 70% of USDA GMO permits issued are for herbicide-tolerant or insect-resistant crops (Bridges 2000). This emphasis on pest control through the use of GMOs puts in perspective clearly the importance of GMOs and their development to future IPM programs.

The effect of GMOs on IPM has come with real as well as perceived costs. Real costs have been associated with decreases in scouting by growers and in the use of multiple tactics, and thus with new problems arising from altered agroecosystems. Decreased scouting has been noted by growers using *Bt* transgenic cotton varieties. As growers come to rely on genetic resistance as a means of pest management, they often perceive resistance as a "silver bullet" and ignore the potential for problems associated with single-tactic pest management strategies. Use of herbicine-tolerant cotton cultivars has resulted in increased aphid activity associated, for instance, with early weed flushes and sudden elimination of habitat (Bridges 2000).

Although the term *IPM* is familiar to many growers, most still cannot define it confidently. Nevertheless, growers identify with the IPM tactics or elements described by the PAMS concept. Examples of tactics growers identify with are rotating; scouting; monitoring temperature/humidity or other environmental variables to predict disease, crop growth, irrigation needs, or weed emergence; predicting EIL by means of insect growth models; applying fungicides or bactericides; using certified seed, disease-resistant varieties, high-quality seeds or sanitation; controlling alternative or overwintering hosts; controlling pests biologically; managing plant fertility and water; and planting under favorable conditions for rapid germination and seedling emergence. To achieve IPM, EBPM, or biointensive IPM, each IPM discipline must consider such tactics to be elements in an IPM continuum that growers implement by applying information gained from other disciplines. The interdisciplinary research and Extension education needed to achieve IPM goals will require focusing on growers' implementation needs and not simply on high-quality, single-disciplinary science or Extension education. To be successful in implementing IPM, team members must come from areas outside the pest disciplines. Implementation will require partnerships among economists, sociologists, ecologists, horticulturalists, agronomists, agricultural engineers, geographic information specialists, food processors, crop consultants, pesticide applicators, regulatory agents, computer scientists, consumers, agribusinesses such as food providers, members of public-policy interest groups, and—most important—farmers.

One of the most compelling IPM success stories is that of the Campbell Soup Company's IPM implementation with growers/suppliers in Florida, California, Ohio, Michigan, Texas, New Jersey, and Mexico. In early 1989, the corporation made IPM implementation a priority. By diffusing among their growers IPM research from plant pathology and entomology, the company helped decrease pesticide use approximately 50% by 1994, without loss of yield or quality (Bolkan and Rienert 1994). Keys to success were the epidemiological models for early blight and anthracnose of tomato, for septoria and cercospora blight of celery, for use of resistant varieties, for the epidemiology of gemini viruses, and for the biology of the vector white fly populations. The company program addressed weeds, arthropods, and pathogens by emphasizing cultural practices, environmental and pest monitoring, and economically/ecologically based pesticide application. Similar IPM success stories have been documented for potato, cotton, sweetcorn, soybean, and most fruit and nut crops.

From their inceptions, the most successful IPM programs have emphasized interdisciplinary teams and grower involvement. The recent use of IPM labeling by the State of Massachusetts and by Wegman's grocery chain in New York state has depended on growers' defining how a product qualifies as *IPM grown*. The definition agreed to by Wegman contract growers was more restrictive than that proposed by the State of New York (Cornell University) IPM team of researchers and Extension educators (Tette, J. 1997. Personal communication). This fact suggests that growers will implement IPM if it helps them market products. A more recent example is the coalition between the WWF and Wisconsin potato growers, whereby growers meeting certain IPM standards are allowed to market potatoes with the WWF panda logo. Whether IPM labeling will contribute to IPM adoption is yet to be seen. Although consumer demand for organic foods has been increasing rapidly, Merrigan (2000) questions the potential impact of IPM labeling.

Appendix A. Abbreviations and Acronyms

a.	acre
ARS	Agricultural Research Service
BPPD	Biopesticides and Pollution Prevention Division
Bt	*Bacillus thuringiensis*
CAR	crops at risk (from implementation of the Food Quality Protection Act)
CEQ	Council on Environmental Quality
CES	Cooperative Extension Service
CIPM	Consortium for Integrated Pest Management
CSREES	Cooperative State Research, Education, and Extension Service
CU	Consumers Union
EBPM	ecologically based pest management
EIL	economic injury level
EPA	U.S. Environmental Protection Agency
ERS	Economic Research Service
ESCOP	Experiment Station Committee on Policy
ET	economic threshold
FDA	U.S. Food and Drug Administration
FQPA	Food Quality Protection Act
FY	fiscal year
GMO	genetically modified organism
IPM	integrated pest management
NAS	National Academy of Sciences
NASS	National Agricultural Statistics Service
NCSU	North Carolina State University
PAMS	prevention, avoidance, monitoring, and suppression
PESP	Pesticide Environmental Stewardship Program
PMAP	Pest Management Alternatives Program
RAMP	Risk Avoidance and Mitigation for Cropping Systems
USDA	U.S. Department of Agriculture
WWF	World Wildlife Fund
yr	year

Appendix B. Glossary

Avoidance. The practice of using cultural techniques to avoid losses to pests that could be present in a field.

Economic injury level. The pest population level capable of producing damage, the cost of which exceeds that of control.

Economic threshold. The pest population level (just below the EIL) at which control options are used to prevent the pest from reaching the EIL.

Integrated control. Applied pest control that combines and integrates biological and chemical controls.

Integrated pest management. "A strategy that uses various combinations of pest control methods, biological, cultural, and chemical in a compatible manner to achieve satisfactory control and ensure favorable economic and environmental consequences" (Bellinger and Rowe 2002).

Key pest. "A serious, perennially occurring species that dominates control practices because in the absence of deliberate control by man, the pest population usually remains above the economic injury levels" (Smith and van den Bosch, p. 295).

Monitoring. The practice of observing pest populations, weather, and nutrients.

Pests. Arthropods and rodents, as well as pathogens, weeds, and nematodes, that are economically or aesthetically damaging to other organisms.

Prevention. The practice of keeping a pest population from infesting a crop or field.

Preventive pest management. Practices that prevent or avoid pest outbreaks or maintain pest populations below the economic threshold.

Secondary pests. Those organisms whose populations normally are maintained below the economic threshold by biological agents but which may be disrupted by pesticide application aimed at key pests.

Suppression. The practice of using cultural, physical, biological, or pesticidal means to suppress pest populations.

Literature Cited

Apple, J. L., P. S. Benepal, R. Berger, G. Bird, and W. G. Ruesink. 1979. Integrated pest management: A program of research for the state agricultural experiment stations and colleges of 1890. Intersociety Consortium for Plant Protection for the Experiment Station Committee on Organization and Policy, Raleigh, North Carolina. 190 pp.

Bellinger, R. G. and R. C. Rowe. 2002. Integrated pest management (IPM), <http://www.savvygardener.com/Features/ipm.html> (22 May 2002)

Benbrook, C. M., E. Groth, J. M. Halloran, M. K. Hansen, and S. Marquardt. 1996. *Pest Management at the Crossroads*. Consumers Union, Yonkers, New York. 272 pp.

Bolkan, H. A. and W. R. Reinert. 1994. Developing and implementing IPM strategies to assist farmers: An industry approach. *Plant Dis* 78:545–550.

Bottrell, D. G. 1979. Integrated pest management. Council on Environmental Quality, U.S. Government Printing Office, Washington, D.C.

Bridges, D. C. 2000. Implications of pest-resistant/herbicide-tolerant plants on IPM. Pp. 141–153. In G. C. Kennedy and T. E. Sutton (eds.). *Emerging Technologies for IPM: Concepts, Research, and Implementation*. APS Press, St. Paul, Minnesota. 513 pp.

Carson, R. 1962. *Silent Spring*. Houghton Mifflin Company, Boston, Massachusetts.

Cate, J. R. and M. K. Hinkle. 1994. Integrated pest management: The path of a paradigm. U.S. Government Printing Office, Washington, D.C.

Coble, H. 1998. A new tool for measuring the resilience of IPM systems: The PAMS diversity index. In E. Day (ed.). *Integrated Pest Management Measurement Systems Workshop*, <http://www.farmlandinfo.org/cae/wp/sp98-1/ipmcov.htm> (22 May 2002)

Connell, T. R., J. P. Koenig, W. R. Stevenson, K. A. Kelling, D. Curwen, J. A. Wyman, and L. K. Binning. 1991. An integrated systems approach to potato crop management. *J Prod Agric* 4:453–460.

Cornell Cooperative Extension. 2002. Genetically engineered organisms—GE food in the market. Homepage, < http://www.geo-pie.cornell.edu > (23 January 2003)

Council on Environmental Quality (CEQ). 1972. Integrated pest management. U.S. Government Printing Office, Washington, D.C. 41 pp.

Fitzner, M. 1996. Integrated pest management: Backgrounder, <http://wwwnysaes.cornell.edu/ipmnet/ init.background.html#ob> (22 May 2002)

Goodell, P. B. and F. G. Zalom. 2000. Delivering IPM: Progress and challenges. In G. G. Kennedy and T. B. Sutton (eds.). *Emerging Technologies for Integrated Pest Management, Concepts, Research, and Implementation*. APS Press, St. Paul, Minnesota.

Gray, M. E. 1995. Status of CES–IPM programs: Results of a national IPM coordinators survey. *Am Entomol* 41:136–138.

Jacobi, J. C. and P. A. Backman. 1995. AU-Pnuts leafspot advisory 1: Development of a weather-based system to time leafspot fungicide applications. *Plant Dis* 79:666–671.

Jacobsen, B. J. 1997. Role of plant pathology and integrated pest management. *Annu Rev Phytopathol* 35:373–391.

Kennedy, G. C. and Sutton, T. E. (eds.). 2000. *Emerging Technologies for Integrated Pest Management: Concepts, Research, and Implementation*. APS Press, St. Paul, Minnesota. 513 pp.

Martin, A. R., D. A. Mortensen, and J. L. Lindquist. 1998. Decision support models for weed management: In-field management tools. Pp. 363–369. In J. L. Hatfield, D. D. Buhler, and B. A. Stewart (eds.). *Integrated Weed and Soil Management*. Ann Arbor Press, Inc., Chelsea, Michigan.

Merrigan, K. A. 2000. Politics, policy, and IPM. Pp. 497–504. In G. C. Kennedy and T. E. Sutton (eds.). *Emerging Technologies for Integrated Pest Management: Concepts, Research, and Implementation*. APS Press, St. Paul, Minnesota. 513 pp.

Mortensen, D. A., A. R. Martin, F. A. Roeth, T. E. Harvill, R. W. Klein, M. Miner, G. A. Wicks, R. G. Wilson, D. L. Holshouser, D. J. Lyon, P. J. Shea, and J. T. Rawlinson. 1999. weedSOFT$_{SM}$ (Version 4.1). Department of Agronomy, University of Nebraska, Lincoln.

National Academy of Sciences. 1969. *Principles of Plant and Animal Pest Control*. National Academy Press, Washington, D.C. 508 pp.

National Research Council (NRC). 1993. Pesticides in the diets of infants and children. National Academy Press, Washington, D.C.

Office of Technology Assessment. 1979. *Pest Management Strategies in Crop Protection*. U.S. Government Printing Office, Washington, D.C. 204 pp.

Overton, J. (ed.). 1996. *Ecologically Based Pest Management: New Solutions for a New Century*. National Academy Press, Washington, D.C. 144 pp.

Pedigo, L. P. 1972. Integrating preventive and therapeutic tactics in soybean insect management. Pp. 10–19. In L. C. Copping, M. B. Green, and R. T. Rees (eds.). *Pest Management in Soybeans*. Elsevier Science Publishers, London. 388 pp.

Rabb, R. L. and Guthrie, F. E. (eds.). 1970. Concepts of pest management. Proceedings. North Carolina State University, Raleigh. 242 pp.

Smith, R. F. and R. van den Bosch. 1967. Integrated control. Pp. 295–340. In W. Kilgore and R. Doutt (eds.). *Pest Control: Biological, Physical, and Selected Chemical Methods*. Academic Press, New York.

Sorenson, A. A. (ed.). 1992. *Proceedings of the National Integrated Pest Management Forum*. American Farmland Trust, Washington, D.C. 86 pp.

Sorenson, A. and P. A. Stewart. 2000. Factors affecting adoption of new technologies. Pp. 12–31. In G. C. Kennedy and T. E. Sutton (eds.). *Emerging Technologies for Integrated Pest Management: Concepts, Research, and Implementation*. APS Press, St. Paul, Minnesota. 513 pp.

Stern, V. M., R. F. Smith, R. van Den Bosch, and K. S. Hagen. 1959. The integrated control concept. *Hilgardia* 29:81–101.

Stevenson, W. R., L. K. Binning, J. A. Wyman, and D. Curwen. 1995. Integrated Pest Management WISDOM: Professional Software for Agricultural Systems. (Version 1.0). University of Wisconsin Integrated Pest Management Program, Madison. 101 pp.

U.S. Department of Agriculture–National Agricultural Statistics Service (USDA–NASS). 1999. Pest management practices: 1998 summary. National Agricultural Statistics Service, U.S. Department of Agriculture, Washington, D.C. 35 pp.

U.S. Environmental Protection Agency (EPA). 1995. Partners for pesticide stewardship. EPA–730–F–95–002. Division of Prevention, Pesticides and Toxic Substances, U.S. Environmental Protection Agency, Washington, D.C.

Vandeman, A., J. Fernandez-Cornejo, S. Jans, and B. H. Lin. 1994. Adoption of integrated pest management in U.S. agriculture.

Informational Bulletin Number 707. Economic Research Service, Washington, D.C. 26 pp.
Waddell, C. 2000. *And No Birds Sing: Rhetorical Analyses of Rachel Carson's* Silent Spring. Southern Illinois University Press, Carbondale and Edwardsville.
Zalom, F. G. and W. E. Fry. 1992. *Food, Crop Pests, and the Environment: The Need and Potential for Biologically Intensive Integrated Pest Management.* APS Press, St. Paul, Minnesota.

Related Web Sites

Gempler's Incorporated. 2002. IPM almanac, <http://www.ipmalmaac.com> (22 May 2002)
International Association for the Plant Protection Sciences. 2002. Homepage, <http://plantprotection. org> (22 May 2002)
National Academy Press. 2002. Ecologically based pest management, <http://www.nap.edu/readingroom/books/pest/summary.html> (22 May 2002)
North Carolina State University. 2002. Homepage, <http://www.ncsu.edu> (22 May 2002)
University of California. 2002. Sustainable Agriculture Research and Education Program, <http://www.sarep.ucdavis.edu.concep.htm> (22 May 2002)
University of California–Davis. 2002. IPM online, 6 May, <http://www.ipm.ucdavis.edu> (22 May 2002)
University of Nebraska–Lincoln. 2002. The internet center for wildlife damage management, <http://wildlifedamage.unl.edu> (22 May 2002)
U.S. Department of Agriculture–Cooperative State Research, Education, and Extension. 2002. Plant and animal systems, 24 January, <http://reeusda.gov/ipm> (22 May 2002)
U.S. Environmental Protection Agency. 2002. Rules and regulations, 15 May, <http://www.epa.gov /docs/ozone/mbr> (22 May 2002)

2 Integrated Pest Management Toolbox

The components of operational integrated pest management (IPM) include (1) biological and environmental monitoring, (2) decision-support systems, (3) decisions, (4) procedure implementation, and (5) outcome evaluation (Barker and Koenning 1998; Kennedy and Sutton 2000). Central to implementation is the availability of effective strategies and tactics for managing single and multiple pest infestations. Overall pest management plans, or *strategies*, to eliminate or to minimize pest problems include exclusion or avoidance techniques such as (1) quarantines, (2) eradication, (3) suppression through host resistance, pesticide, or biological control use, and (4) limitation or prevention by means of host tolerance or related tactics. The many *tactics*, or specific tools, used under one or more strategies continue to evolve and are the focus of this chapter. Typically, on a specific crop a complex set of pests needs to be controlled, and integration of specific tools from each of the five basic strategies is required for effective pest management. Determination of which set of tools to use in IPM depends on understanding of the pest life cycle; influence of both biotic and abiotic environments on pest population dynamics; the pest loss potential the crop-value relations; the pest presence or absence; and the availability of pest-resistant cultivars, registered pesticides, or biological controls and other tools.

Exclusion/Avoidance

Pests requiring no control inputs are the most economical to control; strategies excluding the pest from a production area are, therefore, important. As discussed in the last section of this chapter, many of the most damaging pathogens, insects, and weeds have been introduced into the U.S. from other countries (Waterworth 1993). Examples include Dutch elm disease and its vector, the European elm bark beetle; potato cyst nematode; witchweed (*Striga asiatica*); Russian knapweed, and many others. Several techniques are used to exclude pests. These techniques are effective in areas where the pest is absent; where conditions do not favor pest establishment, reproduction, infection, or spread; and where the pest is introduced with propagation material, soil, water, or plant debris.

Quarantines and inspections are tools used by international, national, state, or local governments to prevent introduction of pests not known to be present (Agrios 1997). Every country has quarantine laws to protect their agriculture and plants from invasive pests. When based on science, properly implemented quarantines and inspections are highly effective and economical. Additional introduced foreign plant diseases or pathogens include chestnut blight which devastated American chestnuts in the eastern and central United States, white pine blister rust, downy mildew of grape in Europe, coffee rust in South America, citrus canker in the United States and South America, soybean cyst nematode, and new aggressive strains of the potato late blight fungus in the United States and Europe. Examples of important introduced insect pests include European corn borer, Mediterranean fruit fly, Asian long-horned beetle, wheat stem sawfly, Russian wheat aphid, imported fire ant, Africanized honeybee, brown citrus aphid, soybean aphid, and gypsy moth. Many weeds, including diffuse and spotted knapweed and leafy spurge, are of foreign origin. Each of these disasters could have been prevented had the pest cycle been understood and timely, science-based quarantine and inspection procedures implemented.

In the United States, the Plant Quarantine Act of 1912 provides the legal framework for quarantine programs operated by the U.S. Department of Agriculture's (USDA) Animal and Plant Health Inspection Service (APHIS 2003). The Food and Agriculture Organization (FAO) of the United Nations provides technical services and a forum to resolve quarantine disputes between countries. Their respective web sites provide extensive information about FAO and USDA–APHIS programs (<www.fao.org> and <www.aphis.usda.gov>). APHIS has inspectors in other countries to pre-inspect shipments to the United States and to maintain inspections at all port of entry. When entering the United States, foreign travelers experience questionnaires and luggage inspection for contraband plant materials, soils, etc. Additionally, APHIS maintains quarantine facilities where imported plant materials are observed and tested for pathogens and insect pests over varying periods before release. Although quarantines and inspections are highly effective, they have been used as justifications for what are, essentially, trade barriers. APHIS (2003) provides an extensive list (and status) of the U.S. quarantines involving pests, plants, and animals. Currently, the World Trade Organization requires a scientific basis for application of both pest management tools.

In addition to quarantines and inspections, exclusion methods include production of certified disease-free planting materials, and methods of avoiding or evading pests or their vectors (Agrios 1997; Hollings 1965; Kahn 1991). The use of certified disease-free planting stock is the foundation disease control method for any plant that is propagated vegetatively. Generally any viral, viroid, phytoplasma, vascular bacterial or fungal pathogen, or nematode present in the mother plant can be expected to be transferred via cutting, rhizomes, tubers, buds, rootstocks etc. Certification programs have had tremendous impact on crops such as potato, tree and small fruits, citrus, and ornamentals, including roses, geranium, carnation and chrysanthemum. Use of pathogen-free propagation material is the foundation of pest management programs for these crops. Pathogen-free certification programs are often operated by state agencies, focus on specific pathogens, and may have precise tolerances for individual pathogens. For example, potato certification programs generally have a zero tolerance for ring rot or cyst nematode and 0.1–5% for certain virus and fungal diseases. Requirements to use zero tolerance foundation seed in production of certified seed, field inspections, serological tests for pathogens, storage inspections, winter growouts, and limited generation propagation from foundation stock have made the potato certification programs a critical part of potato-disease management programs.

Typically, pest infestations are avoided by growing propagation material under environmental conditions not favoring pest/pathogen spread/increase, or by planting crops under conditions allowing development free of infestation (Horn 1988). One example of this strategy can be seen in the production of bean or crucifer crop seed in areas with low rainfall. Such areas are unfavorable for pathogens such as bacteria or certain fungi depending on water splash for in-field dispersal. Moreover, in certain states routine soil assays are used to identify high-risk fields for potential losses to *Aphanomyces* and *Pythium* spp.-incited root rot on peas and beans; in this way, high-risk fields can be avoided by growers. Integrating certified disease-free foundation seed with unfavorable environments for pathogen dispersal allows producers to grow beans free of bacterial blight and fungal anthracnose pathogens. Although these diseases can be devastating in wet environments, the crop can be grown there if disease-free seed is used. Testing seed to ensure the absence of seedborne pathogens is a crucial step for many vegetable crops including crucifers (cabbage, broccoli, cauliflower, etc.), beans, peas, lettuce, and tomatoes. For certain diseases, only 1 infected seed in 10,000 seeds can result in an epidemic. Planting wheat after the Hessian fly-free period has dramatically decreased damage in winter wheat from this pest. Planting winter wheat late in the planting season decreases the likelihood for transmission of barley yellow dwarf virus by aphids, and wheat streak mosaic virus by wheat curl mites.

Other pest-avoidance strategies include removal of overwintering hosts, planting on "new" land not infested with soilborne pathogens, planting in soils with a pH unfavorable to disease development, and growing plants under conditions excluding vectors or otherwise unfavorable to pest development. For example, avoiding poorly drained soils where there are "water mold" or root-rot pathogens such as *Pythium*, *Phytophthoram*, or *Aphanomyces* spp. is a problem basic to controlling disease, as is maintaining humidities below 95% in greenhouses to control Botrytis gray mold. So that damping-off diseases can be measured, high-vigor seeds are planted routinely under conditions that favor rapid germination, emergence, and plant establishment. For these techniques to be used effectively, the germination and vigor potential of seeds must be known, as well as temperature and moisture requirements for germination.

Storage problems can be avoided by storage under conditions unfavorable for growth and reproduction of storage insects, fungi, and bacteria (Christensen and Meronuck 1986; Dennis 1983). For fruits and vegetables, this may mean refrigerated storage where free moisture is avoided. This technique is used commonly to prevent losses from soft rot bacteria, and fungi such as Sclerotinia and Botrytis. For fleshy fruits and vegetables, injuries to the epidermis, a common infection site, must be avoided (Dennis 1983). Seed storage at moisture levels below the minimum for growth of fungi such as *Aspergillus* and *Penicillium* species are used widely in grain storage. Thus, the interactions between temperature, seed moisture, and oil composition of seeds must be understood. At higher temperatures, storage fungi can grow faster and produce their own heat and moisture by means of metabolism once growth starts. Whereas starchy cereal seeds can be stored safely at 13 to 14% moisture, seeds with high oil contents need to be stored at 8 to 10% moisture for safe long-term storage. Short-term storage can be at higher moistures, and cold temperatures also can limit mold growth. Anaerobic or cold-temperature atmospheres commonly are used to control storage insects.

Eradication

Methods for decreasing or eradicating pests and pathogen inoculum include host eradication; removal or destruction of infected/infested plant tissue; sanitation; crop rotation; heat treatment; antagonistic plants; trap crops; soil or space fumigation; disinfection/disinfestation of equipment, storage facilities, pots, etc.; and many forms of biological control (Agrios 1997: Johansen 1962). Host plant eradication may provide control of initial infestations of introduced pathogens in many nursery situations. An example is eradication of citrus where citrus canker has been introduced into Florida (Schubert et al. 2001). Eradication of host plants so that pests cannot survive between crops or alternate hosts also has been quite effective, as in the use of crop-free periods, control of volunteer plants, and wild plant hosts near crop areas. Eradication and destruction of elm trees through Dutch elm disease has been an essential part of managing this disease. Control of Johnson grass also controls maize dwarf mosaic virus because the virus overwinters in that reservoir weed. For pathogens requiring two hosts to complete their life cycle, eradication of alternate hosts decreases both inoculum and genetic variability. The barberry eradication program, began in the 1930s in the United States, has done much to control wheat stem rust and to diminish genetic variability of rust so that stem rust resistance genes are effective over longer periods. Alternative host eradication also is effective for control of white pine blister rust, cedar-apple rust, and oat crown rust (Agrios 1997).

Pruning to remove infected or infested host tissue has been effective in the management of many insects and pathogens surviving in crop residues. Examples include fire blight and many fungal canker diseases of woody plants. Removal of dropped mummified peach fruit is an important component of brown rot control programs, and destruction of corn stalks is important in controlling the overwintering of European corn borer. Removal of prunings and dead plant material is important for control of apple black rot in orchards and of Botrytis gray mold in greenhouses (Agrios 1997). Extensive discussion of eradication programs for invasive species is found later in this chapter.

Crop rotation is a powerful pest-management tool, especially for pathogens attacking members of a single or a few plant families, and unable to survive after infected host tissue has decayed. Such pathogens are termed *soil invaders* because they cannot compete with saprophtyic organisms for foodbases and do not produce long-term survival structures such as fungal sclerotia or oospores. Rotation is of little value for controlling pathogens producing long-term survival structures, competing with saprophytes, or having broad host ranges (called *soil inhabitants*). Still, because of the effect of antagonistic organisms on long-term survival of pathogen propagules, rotations may aid in the control of pathogens in the soil inhabitant class. Examples include Verticillium, Sclerotinia, and certain cyst nematode species. Many foliar pathogens are soil invaders, and rotations will provide excellent control provided that infected crop residues are decayed before a susceptible host plant is replanted (Agrios 1997).

The use of *fallows* (i.e., bare soil or flooding) lasting for several weeks has been shown to eradicate or to decrease substantially populations of several nematodes and numerous other pests. Flood fallows are most effective in the tropics.

Heat commonly is used to eradicate or to decrease the numbers of various pests in soils, bulbs, corms, nursery stocks and seeds. Most plant pathogenic fungi, nematodes, and bacteria can be killed by a 30-minute exposure to temperatures in the 50 to 80°C range, with moist heat being more effective than dry. In warm, sunny environments, the use of clear polyethylene plastic mulches will allow soil temperatures to reach 50 to 55°C to a depth of more than 5 to 10 cm. This process called *soil solarization*, is used widely in tropical and subtropical regions (Agrios 1997). When the mulch is kept in place for 2 to 8 weeks and soil temperatures reach 50 to 55°C daily, most soil-borne bacteria, fungi, and nematodes in the affected zone will be killed. Also, there is evidence that populations of organisms antagonistic to plant pathogens will increase. Pasteurization of soils with steam (80°C for 30 minutes.) is used commonly in greenhouse and nursery situations. Soils must not be sterilized because this procedure will kill antagonistic organisms and create a biological vacuum such that reintroduced pathogens will cause more damage than in nonsterilized soil. High temperatures reached in composting can achieve results similar to those achieved in pasteurization, and there is evidence that pathogen-suppressive microbial communities are established and that antimicrobial compounds can be released from certain compost substrates (Katan and DeVay 1991).

Heat treatment of dormant plant parts and seeds is effective insofar as fungi, nematodes, and bacteria are killed at temperatures lower than those lethal to certain plant tissues. Plant parts and seeds generally are soaked in hot water at 50 to 52°C for 20 to 30 minutes although lower temperatures and longer durations are used to free bulbs and rootstocks from nematode infestation. Hot water treatment is used for crucifer seeds to eradicate *Xanthomonas campestris* (black rot) and *Plenodomus lingam* (blackleg) infection, in citrus rootstock to eradicate the burrowing nematode *Radolpholus similes*, and in bulbs for control of the bulb and stem nematode *Ditylenchus dipsaci* (Agrios 1997). Heat also is used to obtain meristems free of certain viruses that do not replicate at high temperatures. For example, potato plants grown for 4 to 6 weeks at 36°C will have meristems free of most common potato viruses. Virus-free planting stock can be multiplied from these meristems. Dry heat (72°C for 10 days) also can be used to free barley seed from *Xanthomonas campestris* pv. *translucens* (black chaff) infection without germination damage (Agrios 1997).

Antagonistic plants or trap crops can be used to control several nematodes and other pests (Barker and Koenning 1998; Magdoff and van Es 2000). Members of the marigold family produce root exudates toxic to several nematode species. Green manures and organic soil amendments (oil cakes, decomposition byproducts, etc.) of several different plants have been shown to decrease populations of plant pathogenic nematodes and fungi in soils. Oat green manures have been shown to decrease the severity of Aphanomyces black root rot of sugarbeets. Trap crops can be highly effective for control of cyst and root-knot nematodes. For example, oil seed radish cultivars resistant to *Heterodera schachtii* can be used to control this nematode. Crotalaria is a trap crop for root-knot nematode, and black night shade can be used as a trap crop for the potato golden nematode. The trap crop stimulates egg hatch when the nematode invades the trap-crop root, but the pest cannot complete its lifecycle because feeding sites cannot be established.

Disinfestation of tools, pots, storage containers, equipment, and soil with steam, chemicals such as bleach, quaternary ammonia compounds, iodine, formaldehyde, propylene oxide, methyl bromide, and chloropicrin is practiced widely. Soil treatment with fumigant biocides such as methyl bromide, chloropicrin, and methyl bromide-chloropicrin mixtures are used widely to control soil-inhabiting fungal pathogens, nematodes, and certain insects and weeds. (Methyl bromide is being phased out per international agreement; see Figure 2.1.) Several isothiocyanate-generating soil fumigant compounds (metam-sodium and dazomet) are used to decrease populations of both

Figure 2.1. **Ranchers in California set aside portions of their farms for collaborative studies on methyl bromide alternatives for strawberries. Photo by Scott Bauer, Agricultural Research Service, U.S. Department of Agriculture, Beltsville, Maryland.**

nematodes and soil-inhabiting fungi. Both fumigant (1,3 dichloropropene) and nonfumigant nematicides (carbamate and organophosphate) are used to decrease nematode populations (up to 99%) before planting. In the instance of nematodes and annual crops, population decreases below the economic threshold for 4 to 8 weeks after planting will provide economic control even though populations are larger at the end of the cropping season. But soil fumigation is quite expensive, and its use generally is limited to high-value crops. Fumigation with methyl bromide and phosphene gas also is used for space or cargo treatment for control of insects. These practices are widespread in grain storage and fruit export-import shipment.

Many biological control agents decrease pathogen populations by producing antibiotic compounds or by direct parasitism of the pathogen (Boland and Kuykendall 1998). *Trichoderma* spp. and *Gliocladium* spp., used commercially for biological control of *Rhizoctonia* and *Pythium* spp., do so by means of mycoparasitism. The fungus *Ampelomyces quiqualis* is used commercially for control of certain powdery mildew fungi. *Sporidesmium sclerotivorum* and *Coniothryium minitans* are effective parasites/antagonists of *Sclerotinia* spp. The use of fungal viruses has shown promise in controlling canker diseases of woody plants. Many other biological control microorganisms protect the plant by tying up available iron through siderophore sequestration or by inducing resistant reactions in plants. Biological control organisms typically function through several mechanisms of action.

Suppression of Pest Increase through Host Resistance

Host Resistance

One of the most efficient methods of pest control, host resistance, has been the focus of breeding programs for all types of plants. Plants are resistant to pests because they have specific resistance genes, they are not hosts of the pest, or they have other characteristics allowing them to escape pest damage. Genetic resistance can be complete (i.e., immunity) or incomplete (i.e., on the continuum from immunity to susceptibility). It can be governed by single genes providing resistance to specific races or biotypes of a pest (*vertical resistance*) or by multiple genes providing resistance to all pest races (*horizontal resistance*). Vertical resistance is controlled by a single gene and is a more easily achieved breeding objective. Vertical resistance, however, may be less durable than horizontal resistance because it is controlled by single genes. Crucial to an understanding of the durability of vertical resistance is the gene-for-gene concept (Boller and Meins 1992), which suggests that for each resistance gene in the host plant there exists a corresponding virulence gene in the pathogen. A small change in the virulence gene can defeat the resistance gene. This fact explains how new races can develop to attack previously resistant plants and why vertical resistance is considered unstable in environments where pathogens have sexual stages and where infection cycles occur in a given year. To combat the development of new pathogen races, breeders can (1) use several resistance genes in a single cultivar (*pyramiding resistance genes*), (2) use mixtures of cultivars, each with different resistance genes (*multilines*) that would require a pathogen race to have multiple virulence genes, or (3) use horizontal resistance.

Classification of host plant resistance also may be based on the mechanisms affecting the pest. This method, however, ignores the fact that the underlying genetic causes of resistance are myriad (Russell 1978). Deployment of host plant resistance obtained through standard breeding practices tends to enhance rather than to deplete the genetic diversity of agroecosystems.

Resistance exists to a broad spectrum of harmful organisms, including insects, bacteria, fungi, viruses, mites, nematodes, birds, and even mammalian herbivores or other plants (*allelopathy*) (Boller and Meins 1992; Russell 1978; Stalker and Murphy 1991). Plants also can possess resistance to a wide range of abiotic maladies or stresses, including salt, drought, ozone, and aluminum (Ayers 1948; Musgrave and Ding 1998). Resistance in noncultivated plants often develops during the evolutionary selection process, when plant populations are exposed to pests.

Crop resistance is considered a fundamental component of IPM because the technology is in the seed. Management intensity (suitable for conventional or other management systems) is one feature differentiating host plant resistance from other IPM tools, many of which are management intensive (Kennedy and Sutton 2000). For example, pest populations, frequently monitored in IPM, are not usually measured when host-plant resistance is deployed. Like other tools, however, host plant resistance must be deployed in a manner ensuring longevity of use or durability. This situation may require monitoring the crop for new pathogen or pest biotypes, rotating resistance genes, or developing multiline varieties composed of lines with differing resistance genes.

Wherever genetic uniformity exists within a crop, devastating epidemics likely will occur (NAS 1972). When growers plant a single genotype over large areas, breeders have used a single source of resistance, or the host evolved separately from the pest, such a negative result is quite possible. Examples of disasters relating to widespread use of a single type of resistance in the United States include the southern corn leaf blight epidemic, brown plant hopper resistance in rice, greenbug resistance in sorghum, and Hessian fly resistance in winter wheat. Devastation of the American chestnut by the chestnut blight pathogen that evolved in Asia is an example of separate evolution. The question of genetic vulnerability of crop plants due to narrow germplasm bases has been addressed several times in the past 30 years (Duvick 1984; Fermin-Muñoz et al. 2000; NAS 1972).

Breeding for disease resistance is perhaps the dominant disease management tool for crops with low unit values such as small grains, corn, soybean, and rice for which growers cannot afford to use multiple fungicide sprays or other tactics such as long-term rotation. Breeders have developed cultivars with resistance to multiple pathogens. For example, wheat with resistance to one or more viruses and to one or more rust and smut pathogens, as well as resistance to Septoria or Helminthosporium leaf spot, is available. Corn varieties with resistance to multiple diseases are the rule. Vine crops with resistance to bacterial wilt, Fusarium wilt, cucumber mosaic, anthracnose, downy mildew, and powdery mildew are available. Breeding for disease resistance is done by private companies, universities, government agencies, and international research centers such as the International Maize and Wheat Improvement Center (CIMMYT) (wheat and maize), the International Rice Research Institute (IRRI), the International Center for Tropical Agriculture (CIAT) (beans), the International Center for Agricultural Research in Dryland Areas (ICARDA) (barley, wheat, and pulse crops), and the International Potato Center (CIP).

How Is Plant Resistance Obtained?

Pest-resistant varieties usually are developed by means of crossing parents having desirable horticultural or agronomic characteristics with parents having genetic characteristics for pest resistance, and then by selecting offspring with resistance and other desirable characteristics. This process may take several years and multiple generations to select for highly performing hybrids (Stalker and Murphy 1991). Donors of resistance traits may be of the same species as the recipient or may be of a related species whose target genes are transferred by traditional plant breeding technologies. Target genes from unrelated species must be transferred by molecular or other modern technologies (Russell 1978; Kennedy and Sutton 2000). Pest resistance may be identified in plants that have co-evolved with the pest(s) for long periods. Thus, resistance genes often are found at the center of origin of the host plant, where evolution with the pathogen has provided selection pressure for development of resistant individuals. Examples of centers of origin are the Andes Mountains of South America for potato, Eurasia and the Middle East for wheat and barley, and Central America and Mexico for maize. Other sources of resistance genes are land race cultivars or close plant relatives used by indigenous peoples for centuries. The importance of preserving genetic diversity and germplasm collections has been identified as a national priority (NAS 1972). Through molecular technology, identified resistance genes can be transferred with specificity between plants and unrelated plants or other life forms, a process that would be impossible with traditional breeding techniques (Kennedy and Sutton 2000; Stalker and Murphy 1991). Genes transferred from unrelated species include *Bacillus thuringensis* (*Bt*), a bacterium, transferred to field corn, cotton, or vegetables for insect control; and plant-virus coat protein, transferred into tomato, various curcubits, papaya, and certain ornamentals for control of plant viruses (see Chapters 3 and 14). Herbicide-tolerant (HT) transgenic crops such as cotton and soybean now are accepted widely in the United States (See Chapters 3 and 14).

With regard to the development of resistant varieties, two main steps are taken. First, a source of resistance, the donor, must be identified. Second, the trait must be transferred from the donor to a well-adapted, desirable crop variety (Russell 1978; Stalker and Murphy 1991). Several years and multiple generations usually are required to complete the transfer. A feature common to both activities is a technique for identifying plants carrying resistance genes. This technique often requires challenging plants with pests and determining response (e.g., pest departure, or death, or plant damage). As knowledge is gained about the nature of resistance, other techniques may be used to aid in identification of resistant plants (Boller and Meins 1992). For example, the phytochemical(s) responsible for resistance may be measured or the presence of either the genes responsible for resistance or the genes closely linked to those responsible may be verified. Regardless of the technique used to identify resistant plants, procedures usually involve a considerable commitment of time and resources by the breeder and his or her collaborators in the pest disciplines.

The degree of difficulty involved in identifying resistance traits varies. If quite strong in the donor species, resistance usually is evident. If weak, it may be difficult to identify. Resistance traits also may be masked by other traits. For example, host resistance conferred by repellence may mask resistance conferred by toxicity. The resistance trait often occurs at a low frequency in the general population, and thus evaluation of many individual plants is required to determine the presence of resistance.

Genetic transfer of resistance can be straightforward or complex. Traits present in distantly related species usually are relatively difficult to transfer, as are traits complex in themselves (i.e., controlled by several or many genes). For complex traits and for traits present in distant relatives, the intermediate step of gene pool enrichment often is necessary.

Breeding and introduction of crop varieties with high levels of disease resistance have been notably successful, especially for certain high-value crops such as tomato, as illustrated in Table 2.1. These varieties often are introduced into existing production systems with little modification of traditional or IPM operations. This success can be attributed in part to identification, in the same crop species, of high resistance levels controlled by single genes. Transfer of single genes from a donor of the same species usually is a straightforward task for plant breeders. But transfers of multigenic traits and traits from other species are more difficult and time-consuming. Varieties of major crops resistant to disease are common. Success in breeding for host plant resistance has been attributed to the presence in primary gene pools of a high frequency of strong resistance traits controlled by one or a few genes. Crop varieties resistant to nematodes and insects also have been developed, but these successes are fewer in number than those for disease resistance involving fungi, bacteria, and viruses. This difference is due, in part, to (1) the greater complexity of the attacking organisms, especially of insects compared with bacterial and fungal pathogens, and (2) the resulting complexity of interactions between insects and plants. Because of this complexity, traits contributing to host plant resistance in insects are increasingly likely to be multigenic and thus increasingly difficult to transfer. Compared with disease resistance, which often has been uncovered in conspecific or closely related species, insect and nematode resistance often is present only in more distantly related relatives, making transfer difficult.

Host Plant Resistance, Value, and Successes

Comprehensive estimates of the value of host plant resistance do not exist, but a number of striking examples are available. In 1976, Schalk and Rattcliffe estimated that 319,000 tons of pesticides were saved annually as a result of the use of insect-resistant varieties of corn, barley, grain sorghum, and alfalfa. Ruesink (1980) estimated that the use of a resistant variety of alfalfa would result in 40% larval mortality of alfalfa weevil and that, combined with the use of a weevil parasite, such a variety would save producers $73,000,000 in the eastern United States. In 1982, Bradley and Duffy stated that a single nematode-resistant variety of soybean increased U.S. farmers' profits by more than $66,000,000/year (yr). Buntin and Raymer (1989), concentrating on the forage value of winter wheat grown in Georgia, estimated the value of host plant resistance to Hessian fly at between $25 and $60/acre (a.). The development and deployment of hybrid corn in the United States with increased resistance to diseases and pests resulted in augmented yields from less than 30 bushels (bu)/a. in 1900 to more than 120 bu/a. in 1999 (USDA–NASS 1999). Additional benefits such as those to the ecology or to public health were largely ignored in these estimates. But the estimated percentages of crop losses caused by insects, pathogens, and weeds continue to increase (currently 42%) worldwide and amount to $500 billion in damage (Fermin-Muñoz et al. 2000).

Two of the most striking examples of success in disease resistance have occurred in the elimination of plant disease epidemics, namely of wheat stem rust and southern corn leaf blight. Since the 1950s, wheat stem rust has been controlled by means of resistant varieties; without such varieties, farmers' losses probably would have exceeded $1 billion/yr (Agriculture and Agri-Food Canada 1998). The southern corn leaf blight epidemic devastated U.S. farmers in 1970, causing a $1 billion loss to corn growers. Use of resistant varieties since has eliminated the threat of another epidemic (Ullstrup 1972).

Host Plant Resistance: What Has Limited Its Development and Use?

The main limitation to deployment of resistant varieties is the lack of varieties possessing useful agronomic or horticultural characteristics combined with effective resistance characteristics. There are many reasons for this deficit; they can be classified, however, as one of three main types. Funds necessary for successful breeding of host-resistant varieties are unavailable, host resistance is difficult or impossible to develop, or breeders have focused on yield and quality and overlooked pest resistance. The presence of acceptable solutions alternative to host resistance, such as effective and economical pesticides, tends to diminish the urgency of solving pest problems through host resistance. Strategic difficulties in breeding for host plant resistance occur because (1)

Table 2.1. Economically important disease, insect, and mite pests of tomato to which tomato breeders have successfully bred host-plant resistance (Compiled by J. Snyder, University of Kentucky)

	Pest	
	Common name	Scientific name
Fungal diseases	Collar rot	*Alternaria solani*
	Early blight	*Alternaria solani*
	Leaf mold	*Cladosporium fulvum*
	Anthracnose	*Colletotrichum coccodes*
	Target leaf spot	*Corynespora cassiicola*
	Didymella canker	*Didymella lycopersici*
	Fusarium wilt	*Fusarium oxysporum* f. sp. *lycopersici*
	Late blight	*Phytophthora infestans*
	Phytopthora fruit rot	*Phytophthora parasitica*
	Phytopthora root rot	*Phytophthora parasitica*
	Corky root	*Pyrenochaeta lycopersici*
	Septoria leaf spot	*Septoria lycopersici*
	Gray leaf spot	*Stemphylium solani*
	Verticillium wilt	*Verticillium albo-atrum; Verticillium dahliae*
Bacterial diseases	Bacterial canker	*Clavibacter michiganensis*
	Bacterial speck	*Pseudomonas syringae* pv. *tomato*
	Bacterial wilt	*Ralstonia solanacearum*
	Bacterial spot	*Xanthomonas campestris* pv. *vesicatoria*
Nematode parasites	Potato cyst nematode	*Globodera* spp.
	Sugarbeet nematode	*Heterodera schactii*
	Root-knot nematodes	*Meloidogyne arenaria, M. hapla, M. incognita, M. javinica*
Viral diseases	Cucumber mosaic virus	Cucumber mosaic virus
	Curly top	Beet curly top virus
	Potato virus Y	Potato virus Y
	Spotted wilt	Tomato spotted wilt virus
	Tobacco mosaic	Tobacco mosaic virus
	Tomato yellow leaf curl	Tomato yellow leaf curl virus
	Tomato leaf curl	Tomato leaf curl virus
Insect pests	Cotton bollworm/Tomato fruitworm	*Heliothis armigera*
	Tomato pinworm	*Keiferia lycopersicella*
	Egyptian cotton leaf worm	*Spodoptera littoralis*
	Tomato fruit worm	*Heliothis zea*
	Vegetable leafminer	*Liriomyza sativae*
	American serpentine leafminer	*Liriomyza trifolii*
	Potato aphid	*Macrosiphum euphorbiae*
	South American tomato pinworm	*Scrobipapula absoluta*
	Southern armyworm	*Spodoptera eridania*
	Greenhouse whitefly	*Trialeurodes vaporarium*
	Beet armyworm	*Spodopter exigua*
	Tobacco fleabeetle	*Epitrix hirtipennis*
	Potato moth	*Phthorimaea opercullela*
	Tomato looper	*Plusia calcites*
	Cotton aphid	*Aphis gossypii*
	Colorado potato beetle	*Leptinotarsa decemlineata*
	Tomato hornworm	*Manduca sexta*
	Green peach aphid	*Myzus persicae*
	Silverleaf whitefly	*Bemisia argentifollii*
Mite pests	Carmine spider mite	*Tetranychus cinnabarinus*
	Two-spotted spider mite	*Tetranychus urticae*
	Mariana mite	*Tetranychus marianae*

the resistance characteristics are unavailable, (2) the resistance characteristic is complex and difficult to transfer, or (3) the science and/or technology needed to effect transfer is lacking (Frey 1996, 1997, 1998).

Indirect Effects of Plant Characteristics on Pests, Especially Insect Pests

Host plant resistance often is considered a direct effect of plant characteristics on attacking organisms. But plant characteristics can have indirect effects, as well. These interactions may be unrelated or related directly to host resistance. For example, leaf characteristics alterable by plant breeders can improve leaf ability to retain pesticides (Antonious and Snyder 1993). In certain instances, presence of host plant resistance has been shown to increase pest susceptibility to pesticides (Muid 1993; Nicol et al. 1993). Host plant resistance also may decrease the rate at which pests develop resistance to pesticides (Gould 1998). Other characteristics may synergize, act additively upon, or hamper the use of pest-control tactics.

Plants produce chemicals that can be sensed or used by other organisms, including pests. For example, plants produce chemical signals that attract searching enemies (Turlings, Tumlinson, and Lewis 1990). Such characteristics always may be present in the crop, may be induced by attack of a pest, or may be a characteristic of an associated plant in a multiple-species cropping system. For example, aphids, a pest of cabbage, are parasitized to a greater degree when the cabbage crop is interplanted with beets, which attract aphid parasites (Read, Feeny, and Root 1970). Plant chemicals can be used as *kairomones*, or plant-produced compounds that become incorporated into the body odor of herbivores. These odors are sensed by searching enemies, who use them to locate prey (Lewis, Jones, and Sparks 1972). Plant chemicals also can mask other chemicals that attract enemies (Moneith 1960). Insect pests can sequester chemicals from their host plants, and the presence of these chemicals in insects may alter their interactions with natural enemies (Campbell and Duffey 1979; Thorpe and Barbosa 1986).

Resistance or nutritional status of host plants can alter interactions between pests and enemies. Pests with slower growth rates are exposed to natural enemies for a longer period and consequently may suffer greater mortality (Price 1986; Price et al. 1980). The efficacy of *Bacillus thuringensis* (controls several insects) and polyhedrosis viruses (targeted for corn earworm [*Helicoverpa zea*], tobacco budworm [*Heliothis virescens*] on cotton, beet armyworm [*Spodoptera exigua*] on vegetables, and others) may be greater when insect pests feed or are present on insect-resistant than on noninsect-resistant plant varieties (Hamm and Wiseman 1986; Schuster, Calvin, and Langston 1983). For example, the nuclear polyhedrosis virus-related mortality of *Helicoverpa zea* larvae was dramatically potentiated by feeding them insect-resistant corn silk, compared with feeding non-resistant silks. Nutrients available to a crop also can have profound effects on a pest population (Gutierrez et al. 1988a,b; Tamo 1991).

Morphology and anatomy of host plants can influence interactions between insects and natural enemies. Insects on plants with an open architecture may be discovered more easily by natural enemies (Pimentel 1961). Additionally, plant features may be present that prolong the time that insects are exposed to natural enemies or decrease the ability of natural enemies to find and to exploit prey (Natarajan 1990).

Current Trends in Breeding for Host Resistance

Recent reports of trends in plant breeding research indicate a decline in public plant breeding programs and an increase in private plant breeding programs (Brooks and Vest 1985; Collins and Phillips 1991; Frey 1996, 1998). According to Frey (1996), most plant breeding research in the public sector is directed toward basic plant breeding research or genetic evaluation and enhancement, also called *gene pool enrichment*. Gene pool enrichment is the public research area underpinning improvement of host plant resistance. Evidence suggests that the research efforts needed to improve gene pool enrichment are not increasing but decreasing.

Searches of the Current Research Information System (CRIS) database (USDA 1999) can provide a snapshot of related scientific efforts in plant breeding for resistance. A recent search of the database revealed approximately 2,700 citations that include plant breeding as an activity. Approximately 40% of these citations included disease resistance as an objective; 17%, insect resistance; 7%, nematode resistance; and 1%, weed control, or allelopathy. Only 3% of the 2,700 citations included IPM as an activity. Frey (1998) presented data on success in plant breeding that support a similar division of effort among attempts to breed for resistance. These limited surveys suggest that plant-breeding efforts directed to host resistance probably are most frequently undertaken to study disease resistance, followed by insect, nematode, and weed resistance. Efforts dedicated to breeding specifically for IPM systems seem minimal.

Future Challenges and Directions

As has been mentioned, host plant resistance is a fundamental tool or component of an IPM system. Characteristics required of a successful variety carrying host plant resistance are constrained doubly, requiring pest resistance and high, acceptable yields. Because this constraint will remain in the future, successful breeding for host resistance is unlikely to occur independently of breeding for other traits.

Successful varieties with host plant resistance will arise from active breeding programs, including those with host plant resistance and improved yield and quality as breeding objectives. Increasing the frequency of resistant varieties will require improvement along the following lines: (1) increasing the number of public research programs directed at improving quality, yield, and resistance; (2) improving the efficiency or success rates of these programs; and (3) increasing the priority of breeding for host resistance to pests.

Production of varieties adapted specifically to IPM systems provides additional opportunities and challenges for plant breeders and for IPM specialists. Improved understanding of the role played by the host in tritrophic interactions likely will underpin production of varieties specifically for IPM systems. Genetically controlled plant characteristics can differ greatly. For example, certain varieties of flax produce oil; others produce fiber. Different varieties of mint have very different tastes and uses. Field corn, popcorn, and sweet corn are the same species but exist in these various forms because of selection, the primary tool of the plant breeder. Unquestionably, genetic variation exists for characteristics that could enhance other IPM tools. But exploiting such variation will require longer-term collaboration between plant breeders and IPM specialists.

Frey (1998) has stated that gene pool enrichment is conducted primarily by public-sector scientists, and this activity is largely uncoordinated, underfunded, and absent for many crops. In short, the traditional genetic research leading to the development of host plant resistance is inadequate. Frey (1998) also has stated that a dramatic increase in germplasm evaluation and gene pool enrichment will be required if plant breeding is to meet the goal of decreased agrochemical use without compromising production while meeting public demand. This goal is consistent with the goals of IPM.

Trends in public breeding programs, including coordination with genetic engineering programs, must be maintained or enhanced if host plant resistance is to remain a tool for IPM. Plans have been offered for reinvigorating traditional plant breeding research in the United States (Frey 1997, 1998). These plans call for added national support of gene pool enrichment and include goals of producing crops sustaining or improving environmental quality, improving economic viability, and decreasing agrochemical use. To the extent possible, all stakeholders, and not only plant breeders, should support such improvements.

To enhance the availability of resistant varieties, the IPM scientific community must continue to support both traditional and molecular research programs targeted to improve host plant resistance. The community also must recognize the opportunities and the unique challenges that IPM systems present in regard to host plant resistance. Finally, these diverse scientific communities must explore new ways of generating meaningful interchange.

Cultural Practices for Minimizing Crop Pests

Many cultural practices have proved successful in the management of a wide range of crop pests. Agricultural systems developed before the era of pesticides used a variety of cultural practices, which often were based on empirical research (Thurston 1990). Fundamental mechanisms frequently have not been understood, but efforts in IPM and sustainable agricultural research have provided a basis for understanding the biology behind many cultural practices.

There is societal pressure to minimize or to eliminate pesticides. In certain locations, there also is political pressure to promote organically grown agricultural food products. For example, in Berkeley, California, Mayor Shirley Dean announced on June 9, 1999 that she would like the entire city to drink only organically grown coffee. One goal of sustainable agriculture is to develop biologically based management systems requiring limited levels of material inputs (See Chapter 4).

Developing cultural control methods within a crop management system requires extensive information. Long-term planning is essential and must take into account the characteristics of the agroecosystem, including vegetative diversity, crop, and climatic stability (Hall and Kuepper 1997; Southwood and Way 1970). The cropping system must include features that can be manipulated to enhance biological control or that can discourage establishment of pests and diseases. The underlying principle of cultural controls, therefore, is to modify management practices so that the environment becomes less vulnerable to pest invasion, reproduction, survival, and dispersal, and so that a decrease in pest numbers can be achieved either to avoid economic injury or to allow biological control to take effect (Altieri 1994; Barker and Koenning 1998; Takahashi 1964).

An array of management strategies and tactics can be considered cultural controls. Some of the most common, including some of the oldest practices, involve cultivation, manuring, crop rotation and selection, planting date, sanitation, intercropping, residual root destruction, antagonistic cover crops, and trap crops (Barker and Koenning 1998; Bridge 1996; Sullivan 1998). Most of these cultural control measures are primarily preventive and need to be applied well in advance of the pest attack or disease development. With soilborne pests, long-term strategies are required to minimize the effects of causal organisms.

From the 1950s through the 1980s, with focus often on the use of highly effective pesticides and high-input agriculture, cultural practices tended to receive limited attention. The concepts of IPM and sustainable production systems, however, have generated renewed interest in cultural control methods (Kennedy and Sutton 2000; Liebman, Mohler, and Staver 2001; NRC 1996). Key factors in cultural control include diversification or polyculture, and environmental and biological manipulation leading to biological diversity (Altieri 1994).

Knowledge of crop and pest biology, ecology, and phenology is essential for the design and implementation of cultural control measures, with attention given to recognizing the "weak links" in the pest life cycle (Barker and Koenning 1998; Bridge 1996; Coaker 1987). Such knowledge should allow for the development of a system making the crop less prone to damage by pests and diseases and simultaneously should diminish pest survival and enhance biological control.

Specific cultural controls addressed herein are crop rotation, tillage and no-till, trap crops, green manure and cover crops, multiple cropping or polyculture (intercropping), and refugia.

Crop Rotation

Crop rotation, one of the oldest pest-management tactics, ideally involves changing plant species and/or cultivars and planting dates so that host crops of prevalent pests are grown only when target pests are near or below damaging levels (Altieri 1994; Bridge 1996; Sullivan 1998). Rotation of a host and a nonhost crop is very effective against pests with a narrow host range and low dispersal ability. Crop rotation is especially well suited for pests and pathogens that cannot survive for more than one or two seasons in the absence of a host crop (Table 2.2). Selection of the "nonhost" crop must take into account potential effects on other pathogens and pests (Altieri 1994).

Rotation of crops enables rotation of pesticides and may slow the increase of pesticide-degrading microorganisms, thus allowing chemical effectiveness to be prolonged. A herbicide program to control weeds in one crop that may not be controlled in another because of crop sensitivity to the pesticide is made possible with crop rotation. For example, in Australia, a great brome weed was controlled easily by simazine in the lupin phase of a lupin-wheat rotation (Heenan, Taylor, and Leys 1990). This weed was difficult to control in wheat. Rotating wheat with sorghum was very effective for controlling wild oats (Martin and Felton 1993). This rotation required two winters to control the wild oats with cultivation or herbicides, and an exponential decline in seed numbers resulted.

Crop species differ considerably in terms of their overall competitive ability (Boerboom 1999). This fact may affect initial requirements for herbicide use. A crop's ability to shade out weeds can affect the frequency and/or the rate of herbicide application. Management practices such as planting in narrow rows or at higher crop population densities can enhance crop competition and decrease the herbicide rate required or the need for sequential herbicide applications (Boerboom 1999). For example, seed production

Table 2.2. Crop response in pest- or pathogen-infested fields following selected cropping systems

Target crop	Crop response		Reference
	Monoculture (or high annual frequency)	Host–Nonhost	
Soybean	1,345 kg/ha	2,776 kg/ha	Weaver et al. 1998
Soybean	2,437[a] kg/ha	3,110[b] kg/ha	Young 1998
Turnip	70% marketable	98% marketable	Sumner et al. 1978

[a]These means were derived by averaging all continuous soybean yields.
[b]These means were derived by averaging all soybean yields for the year following corn.

and survival of eastern black nightshade was much lower in drilled soybean than in soybean planted in 76-cm rows (Quakenbush and Andersen 1984). Corn root worms (*Diabrotica* spp.) commonly are controlled in corn areas by rotating corn crops with soybeans (Pedigo 1989, 1999). Adult beetles lay their eggs in mature cornfields in which their eggs overwinter. The next spring, root worm eggs hatch, but the highly specific root worm larvae, which feed only on corn roots, cannot disperse from their habitat to find other sources of corn, and die within the soybean crop. Widespread use of a corn/soybean rotation, however, has led to the corn root worm's acquiring a behavioral resistance with both an extended diapause in one population to survive the rotation and the ability to survive and lay eggs in soybeans in another population (Levine and Oloumi-Sadeghi 1991; Onstad et al. 1999).

Rotating pesticides may be as important as rotating crops, especially for certain insecticides and fungicides. This practice also is important for certain nematicides; for example, the nematicide fenamiphos is degraded rapidly by soil microflora after several years of continual use (Figure 2.2) (Davis, Johnson, and Wauchope 1993; Johnson 1998). Research in The Netherlands and in Australia indicates that metam-sodium also is subject to biodegradation (See Chapter 14).

Altering planting dates along with crop rotation is a cultural practice that may help alleviate pest problems in certain environments. The soybean cyst nematode has a winter dormancy that begins to break with the onset of warming soils, and its population densities begin to decline rapidly when dormancy ends (Koenning and Anand 1991; Koenning, Schmitt, and Barker 1993). Thus, a delay in planting often exposes the crop to fewer pests. For pests such as the soybean cyst nematode, which has an exponential growth phase late in the season, an early maturing crop may limit late-season increases (Figure 2.3). Limited nematode reproduction is beneficial, for their population densities are inversely related to crop yield (Figure 2.4). Altering planting date along with crop rotation should be effective with any pest-host combination whose biology is similar to that of the soybean cyst nematode. Control of cheat (*Bromus secalinus* L.) in winter wheat is achieved best by cultural methods (Baker and Peeper 1990). Seeding rate, seeding date, and row spacing of wheat influence plant density of cheat (Koscelny et al. 1991). For example, plots with wheat seeded in 3- and 9-inch rows in September at 120 pounds/a. had fewer cheat plants present in April than plots seeded at 60 pounds/a. Fewer cheat plants

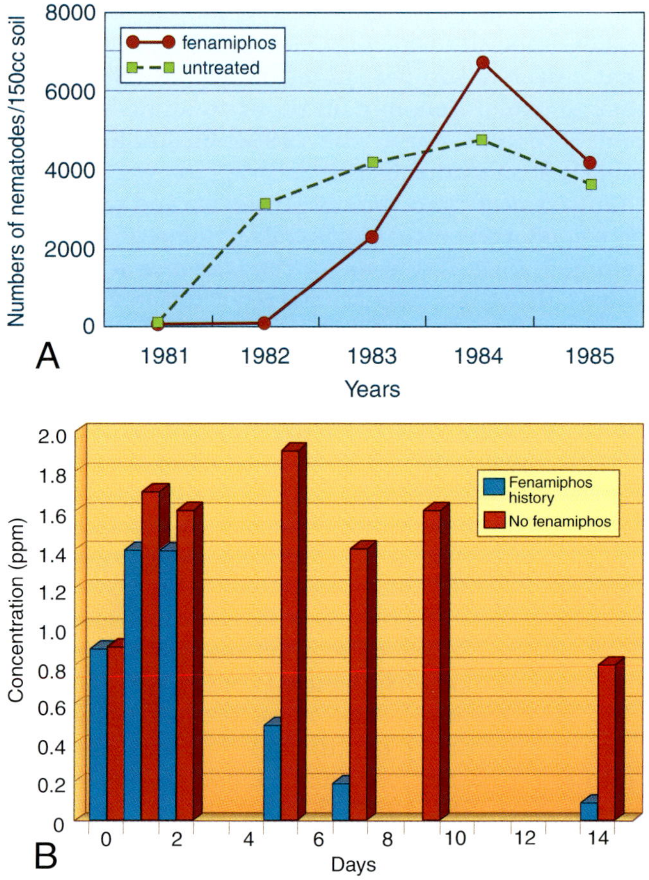

Figure 2.2. Evidence for microbial breakdown of fenamiphos resulting from continuous usage. A. Population densities of *Meloidogyne incognita* second-stage juveniles in a sweetcorn-sweet potato-vetch cropping system (Johnson et al. 1992). B. Concentration of fenamiphos sulfoxide as affected by fenamiphos history (Davis et al. 1993).

were present in wheat seeded in 3-inch rows in October at 120 pounds/a. than at 60 pounds/a. Cheat also was affected by seeding date. In plots seeded in September, October, and November, cheat density averaged 84, 71, and 11 plants/m^2, respectively. This effect of planting date, however, may have been confounded with tillage.

Tillage and No-Till

Tillage affects the physical characteristics of soil and crop residues. These effects can influence survival and movement of pests harbored by soil and/or residue. The positive aspects of no-till on soil quality are documented (Valenzuela 2000) and will not be discussed in depth here. Instead, the focus will be on the beneficial effects arising from pest control.

Effective management of several pests is readily

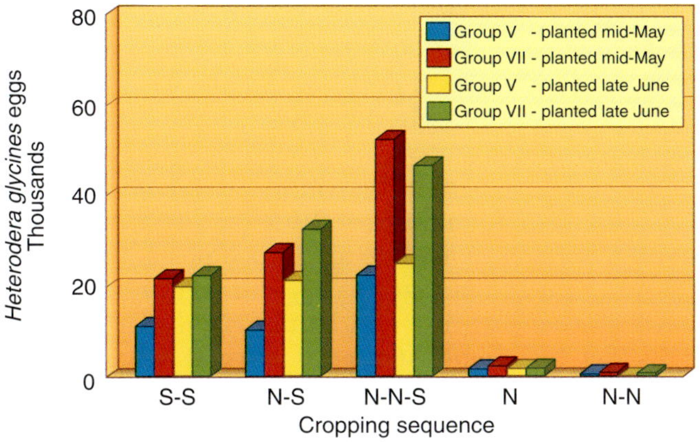

Figure 2.3. Mean population density of eggs of *Heterodera glycines* (soybean cyst nematode) at soybean harvest on susceptible cultivars grown in rotation with nonhosts for 0, 1, or 2 yr (Koenning et al. 1993).

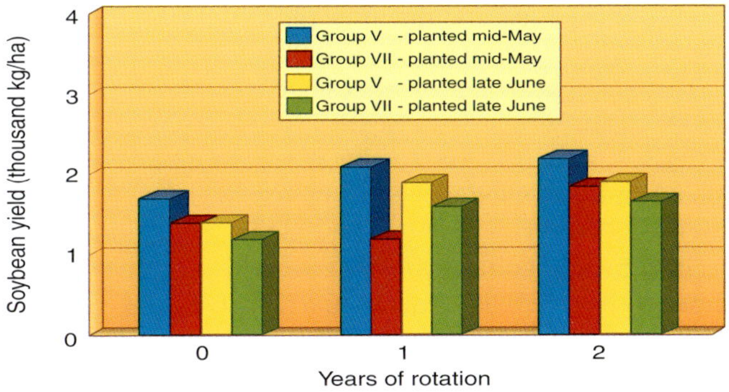

Figure 2.4. Mean soybean seed yield (kg/ha) of soybean cultivars susceptible to *Heterodera glycines* (soybean cyst nematode) planted in mid-May or late June and grown in rotation with nonhosts for 0, 1, or 2 yr (Koenning et al. 1993).

achieved by complete destruction of the residual stalks followed immediately by incorporation of the shredded material into the soil by plowing. Numbers of pink bollworm (*Pectinophora gossypiella*) on cotton, tobacco hornworm (*Manduca sexta*) on tobacco, and root-knot nematode (*Meloidogyne* spp.) on both crops are dramatically decreased by this procedure (Allen 1994; Barker, K. R. 2003. Personal communication; Godfrey et al. 2001). This stalk destruction program is legislated for cotton in several U.S. states, with owners and tenants jointly and/or severally being responsible for compliance (Pink bollworm regulation 5.179[c]).

Overall, total population densities of weeds, insects, and pathogens often are related directly to tillage amount. That is, as tillage increases, the population density of associated pest organisms decreases. A study of weeds in southwestern Ontario, Canada, demonstrated that the total population density of weeds/field was greatest in no-tillage fields and least in conventionally tilled fields (Table 2.3) (Frick and Thomas 1992). Certain grasses such as rye and wheat decreased populations of certain broadleaf weeds (Worsham 1989).

It is common for tillage to affect certain pathogens adversely, but seemingly to affect others beneficially. In a regional assessment of selected pathogens in the Midwestern United States, the incidence of brown stem rot was greater in fields in which tillage was minimized than in fields in which tillage was sufficient to decrease residue to less than 15% (Workneh et al. 1999). In the same study, population densities of the soybean cyst nematode were greater in tilled than in no-till fields. In other studies such as one in Iowa, tillage had little impact on this nematode (Gavassoni, Tylka, and Munkvold 2001; Table 2.3).

With recent developments in precision agriculture, site-specific applications of herbicides hold much promise for greatly decreasing the amounts of herbicides and pesticides used in IPM programs (See Chapters 3 and 14). The selection of a weed-control program also may affect insects and pathogens. For example, poor weed control in soybean often results in increased insect and soybean cyst nematode problems (Alston et al. 1991).

Herbicides have enabled growers to decrease the amount of tillage for the majority of acreage of row crops being treated. Nevertheless, mechanical weed control still is a viable option and was used on 73% of the corn crop in Iowa and 68% of that crop in Wisconsin (Hanna et al. 1996; Owen, M. D. 2002. Personal communication). Cultivation to control weeds

Table 2.3. Effects of tillage on pest population densities

Pest/Pathogen density	Tillage			Reference
	Conventional	Reduced	No-till	
Total weed (no./m^2)	17.0	24.7	31.7	Frick and Thomas 1992
Early-season weeds (% suppression)	9–30		76–81	Worsham 1989
Soybean cyst nematode eggs	5,445	3,823	5,838	Gavassoni, Tylka, and Munkvold 2001

(ca $6.00/a.) is considerably more economical than chemical weed control (ca $30.00/a. per application). Cultivation allows growers to use band application of herbicides, which lowers the chemical cost by 50%. Further, mechanical weed control may be beneficial through decreased selection pressure related to herbicides. Today, with new designs in rotary hoes and cultivators, mechanical weed control used in conventionally tilled fields can be used in reduced till and no-till systems (Boerboom 1999). A major advance in mechanical weed control is the development of the row guidance (vision guided) systems (Åstrand 2000; Baerveldt and Åstrand 1997). This equipment allows the cultivator to follow a row of plants accurately.

Trap Crops

Trap crops can be used to decrease pest populations in various ways. A susceptible variety can be planted and destroyed before the nematode has the opportunity to complete its life cycle (Barker and Koenning 1998; Bridge 1996). A more practical alternative is to use a crop that attracts the nematode and allows penetration but does not support completion of the life cycle. Crops more attractive to insect pests than the main crop, or specific stages of crop growth, may function as trap crops. Thus, one IPM strategy involves planting a row or several rows of insect-attractive alfalfa between several rows of cotton (El-Zik and Frisbie 1991). The effectiveness of such a program can be demonstrated with a leaf hopper-vectored viral disease of rice called *tungro* (Table 2.4) (Saxena, Justo, and Palanginan 1988), whose negative effect was decreased by a trap crop program. This situation most likely was related to a lower population density of the insect vector. More nymphs and adults were recorded in the untreated control, and yield was significantly higher in trapped and in insecticide-treated rice fields than in the control. Growing smaller trap crop plantings earlier than the main planting diverts insect attacks away from the crop at risk by providing insects with a more attractive food source (Coaker 1987). This system works where the pest has a narrow host range.

The effectiveness of trap crops will be affected by the population dynamics of the target pest (Valenzuela 2000). Highly mobile arthropod pests, for example, may not be managed easily with trap crops; pests with limited mobility can be managed more easily. This fact was demonstrated in Hawaii, where a diamondback moth population failed to become established in cabbage fields with a canola/cabbage border (Luther, Valenzuela, and DeFrank 1996). Economics, as well as planting and harvesting equipment, may discourage growers from using trap crops.

Green Manure and Cover Crops

Green manure crops have a wide range of beneficial effects on crop production. They improve soil-moisture holding capacity and increase the availability of other plant nutrients such as phosphorus (Lewis and Hunter 1940; Pieters and McKee 1929; Rodgers and Giddens 1957; Sullivan 1998). Winter annual legumes are excellent green manure crops, providing an important nutrition source and being associated with high yields (Lewis and Hunter 1940; Pieters and McKee 1929, 1938; Rodgers and Giddens 1957). Undoubtedly, the effects on water holding capacity and on nutrients are the major benefits provided by a green manure crop. Cover crops and green manure incorporation into soil also have been used to break disease cycles. In Idaho, sudangrass and corn green-manure treatment resulted in control of verticillium wilt and in the highest yield of potatoes in the trial. Disease was most destructive in the fallow treatment. Responses in other green manures such as rape, Austrian winter pea, oat, and rye were intermediate be-

Table 2.4. Incidence of rice tungro virus (RTV) disease on rice as influenced by various numbers of rows of a trap crop in the Philippines (Saxena et al. 1988)

Trap crop[a]		Main crop[a]	
Number of border rows	Insecticide sprayed weekly	Insecticide sprayed weekly	Incidence of RTV (%)
0	–	–	36.8
0	+	+	6.4
2	+	–	5.2
3	+	–	7.5
4	+	–	8.5

[a]Rice IR42 planted 15 days before the main crop (also IR42 rice) was used as the trap crop.

tween fallow and sudangrass (Davis et al. 1996).

The effects of green manures are not well understood, but probably involve certain breakdown products giving rise to detrimental effects in microorganisms, and a favorable environment for enhancing microbial activity resulting in biological control (Davis et al. 1994). Certain microorganisms, especially a number of plant-growth promoting rhizobacteria, have been shown to be antagonistic to pests or even to induce systemic resistance to crop pathogens (Barker and Koenning 1998; Kloepper et al. 1992; Wei, Kloepper, and Tuzun 1996).

Multiple Cropping or Polyculture (Intercropping)

Polyculture, a cultural system common in the developing world, especially in the tropics (Valenzuela 2000), can result in higher yields for several reasons. These include decreased weed competition due to soil conservation; dense crop cover; more effective use of incident radiation, water, and soil nutrients; low to moderate pest pressure; and diminished economic risks (Altieri 1994; Norton and Conway 1977; Valenzuela 2000; Willey 1979). If intercropping or multiple cropping culture is to be understood and used as a widespread means of decreasing pest abundance in the United States, farmers must become aware of the mechanisms affecting pest populations. An understanding of mechanisms also would allow improved agronomic use of crop mixtures with respect to spatial arrangements, planting times, and relative crop growth-rates, thus making the cultural control system maximally effective (Altieri 1994; Coaker 1987).

Intercropping can decrease phytophagus pest levels in two ways: (1) natural enemies are favored in intercropping systems because of wider temporal and spatial distribution, increased ground cover, and prey numbers; and (2) associated plant species have a direct effect on the ability of insect herbivores to find and to use host plants (Coaker 1987).

Polycropping likely would have its greatest effect on mobile pests such as flying insects. A definitive conclusion is elusive, however, so long as the literature remains inconsistent. In a review of some 150 papers, 53% reported population declines of insects as a result of intracrop diversity (Risch, Andrew, and Altieri 1983). For soilborne organisms, the primary impact of intercropping would be decreased dispersal and overall populations in the field because of a nonhost or a poor host. Control of the northern root-knot nematode with intercropping was associated with decreased spread under field conditions that included lower soil populations, egg hatch, and juvenile penetration and galling (Heald 1987).

Polyculture using multiline cropping systems has decreased the disease incidence of several foliar pathogens. A mixture of disease-resistant and susceptible cultivars seems a promising method for controlling *Uromyces* bean rust on *Phaseolus* in Colombia (Panse, Davis, and Fischbeck 1997). In Germany and India, multilines enable growers to decrease fungicide applications (Wolfe 1990).

Although in its infancy, crop breeding for polyculture is directed at increasing crop yields and economic returns to the grower (Valenzuela 2000). Research in the area of improving a crop's ability to perform in mixed cultures is difficult to conduct because of the multitude of interactions; pests will be affected, however, and research will be needed to identify the impacts.

Refugia

The refugia concept involves the conservation of both pesticide-susceptible individuals within a pest population and natural enemies (Altieri 1994). The purpose is to prevent populations or subpopulations from being subjected to insecticide selection pressure and to decrease exposure so as to minimize selection pressure (DeSouza, Holt, and Colvin 1995). Refugia, as a system for resistance management, can be instituted in various ways: site-specific IPM tended to limit the need for an insecticide to control the Colorado potato beetle (Midgarden et al. 1997). Site-specific IPM conditions created temporarily dynamic unsprayed refuges within fields and resulted in spatial variations in insect phenotypes. Where selection pressure was low on a whole-field basis, susceptible phenotypes of the insect were conserved. In low-spray levels of the field, insects were more susceptible to insecticide at the end of the season than at the beginning.

The high dose/refugia strategy has been the focus for resistance management strategies for *Bt* crops since they were first commercialized in 1995 (EPA 1998, 1999, 2001; Gould 1998). This strategy focuses on the transgenic *Bt* crop producing a high dose of the *Bt* toxin(s) to kill greater than 99% of the susceptible individuals. The structured refugia of non-*Bt* fields is planted in close enough proximity to the *Bt* fields so that the rare resistant individuals will randomly mate with susceptible individuals produced in the refugia fields and produce fully susceptible heterozygous individuals that will be killed by the *Bt* crop, thus diluting resistance. The current resistance management strategies for *Bt* are briefly discussed in the section on "Resistance Management Tactics and Tools–Insects and Mites."

Site-specific IPM also has the potential to conserve natural enemies (Altieri 1994). In fact, any type of management practice maintaining unsprayed areas, whether unsprayed strips or unsprayed blocks, should help maintain populations of natural enemies and, so, decrease selection pressure on pests.

Integrating Cultural Management Programs

Pest management clearly must be integrated with the cropping system. Through such integration, all aspects can be manipulated to optimize the benefits of all practices required for crop production. Areawide implementation of practices can (1) help farmers achieve the goal of breaking pest and pathogens cycles or help make microclimates unfavorable for pest and pathogen development and (2) avoid or restrict outbreaks and epidemics, thus minimizing crop damage. In the production of cotton, a program aimed at decreased pesticide use has been implemented successfully (Summy and King 1992). In addition, after the introduction of transgenic *Bt* cotton into the field in 1996, pesticide use dramatically decreased (Carpenter et al. 2002; EPA 2001; Gianessi and Carpenter 1999).

A new era in pest management calls for a new focus on maintaining natural ecological balances (NRC 1996). The challenge is to shift from managing components or individual organisms to an approach examining processes, flows, and relationships among organisms. The goal is to achieve ecologically based pest management solutions that are safe, profitable, and durable.

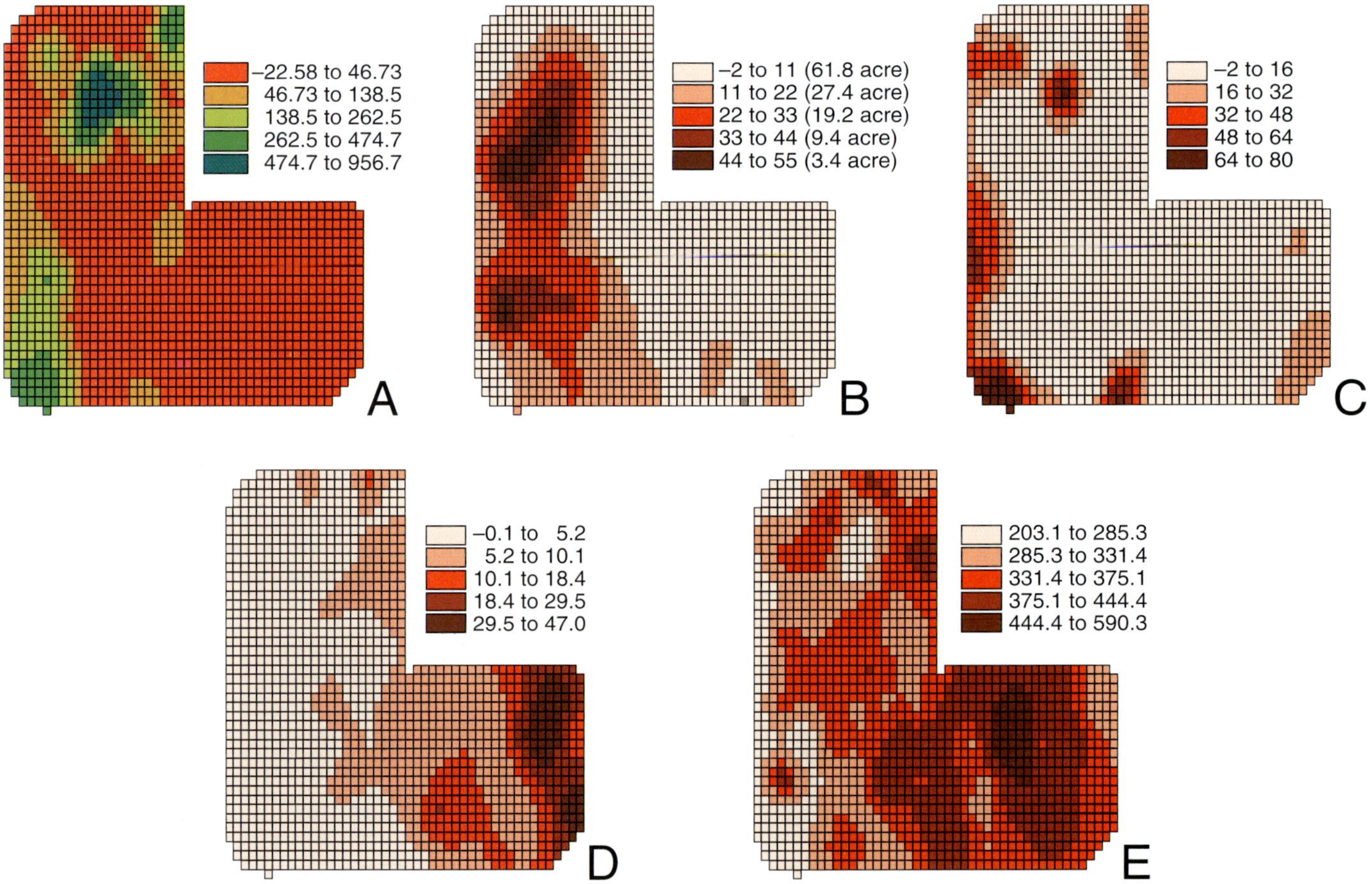

Figure 2.5. **Geo-referenced information on the field distribution of plant pathogens on potatoes and yield that could be useful in designing control strategies while minimizing pesticide use. A. Plant-parasitic nematodes; B. Verticillium; C. Early dying of potato (caused by Verticillium and lesion nematodes); D. Early blight surface; E. Yield of Russett Burbank (> 4 0z) CWT per acre. (Courtesy of W. Stevenson, A. E. MacGuidwin, G.D. Morgan, and L. K. Binning, University of Wisconsin–Madison.)**

Decision Support Aids and Diagnostic Systems

Field and Region

Effective IPM programs require accurate and timely information that is readily available in a usable form to the grower. Key types of information include accurate diagnosis of pest problems, quantification of pests including current populations and distribution, measurement of pest movement over time, preseason soil sampling for pests and nutrients, crop monitoring during the growing season for pests and nutrient states, collection of weather data and application of these data to pest prediction models, presampling before harvest to determine yield potential, and postharvest monitoring of stored products. Traditionally, this information has been acquired by trained personnel engaging in labor-intensive scouting of individual fields. Depending on cropping system and information requirements, fields may be sampled before planting to determine levels of soil fertility, weeds, insects, fungal pathogens, and plant parasitic nematodes.

Knowledge of the overwintering populations of important soilborne insect pests assists growers in key management decisions before susceptible crops are planted. Likewise, quantitative information about the soil populations of plant pathogens such as plant parasitic nematodes on several crops (Barker and Davis 1996) and soilborne fungi on potato (Nicot and Rouse 1987) provides growers with information crucial to the management of potential pest problems (Figure 2.5). Early dying of potatoes, caused largely by the combined activity of *Pratylenchus penetrans* (root lesion nematode) and *Verticillium dahliae*, leads to premature vine death and decreased potato yield. By sampling field soils in late summer before the subsequent spring planting of potatoes, growers can determine the soil populations of both pathogens. In fields with populations that exceed economic thresholds, growers can decide to extend the nonpotato rotations or to fumigate as basic methods for decreasing levels of these pathogens.

Although these activities are often tedious and time consuming, the economic return more than pays for grower investments in IPM and results in improved pest control with decreased pesticide use. Expansion of the IPM industry has facilitated greatly the adoption of IPM programs on a national scale. If adoption of IPM technology is to increase, additional partnerships must be forged between IPM consultants, universities, state agencies, growers, and agribusinesses.

In the upper Midwest, where bean and pea root rot can lead to total crop loss, bioassay of soil samples to assess the risk of root rot on pea and snap beans provides information essential to decisions regarding which fields can be planted safely to these crops. Avoiding high-risk fields has saved the industry millions of dollars in the past 30 yr. During the growing season, fields are sampled periodically to determine plant nutrient status, populations of harmful and beneficial insects and mites, distribution of weeds, and appearance and spread of plant diseases (Figure 2.6). These data subsequently serve as the basis for selecting fields for planting with specific crops or crop cultivars, adjusting fertility programs, timing pesticide applications, and scheduling harvests.

Reliable weather information is a key feature of many IPM programs. The interpretation of this information often provides a basis on which to predict crop development, the appearance of important crop pests, and the development of those pests. The interpretation also may influence decisions made by management. Numerous crop and pest models useful to implementation of IPM programs have been developed (Textbox 2.1). Some models are designed simply and incorporate only a few parameters, such as maximum and minimum temperatures for the calculation of growing degree days (GDD) and physiological days (P-Days). Other models are much more complex, requiring multiple parameters and sophisticated computer simulations (Russo 2000). The availability of reliable and inexpensive weather information to drive these models continues to improve.

Figure 2.6. Field scouting with "backpack" that allows for georeferencing of information. (Courtesy of W. Stevenson and associates, University of Wisconsin–Madison.)

Whereas the federal cooperative weather network traditionally has provided data for early pest models, advances in technology have made it possible for individual growers to own and to operate their own

Textbox 2.1. A model IPM recommendation document for vegetables (Source: M. Haining Cowles, Cornell University)

Integrated Crop and Pest Management Recommendations for Commercial Vegetable Production, a 305-page book published in 1999, is the result of a two-year grant funded by the Northeast IPM Grants program. Editors Stephen Reiners, Department of Horticultural Sciences, NYSAES; Curt Petzoldt, IPM Program; Mike Hoffmann, Department of Entomology, Cornell University; and Christine Cefalu Schoenfeld, IPM Program, worked with 19 other discipline editors from Cornell University to create this comprehensive volume. The new *Recommendations* is a major revision of what was the *Cornell University Pest Management Recommendations for Vegetable and Potato Production*, a document that focused solely on pesticides. The purpose of the revision was to create a document that provides all pest management options in a concise, reader-friendly format and that serves as a model for revision of similar pest management recommendations produced in other states and regions.

The new vegetable recommendations book is written for more than 20 different vegetable cropping systems in New York. It includes cultural, biological, mechanical, and chemical pest management options and information on cultural and fertility practices and crop varieties. Pest complexes covered are weeds, diseases, insects, and wildlife. The book is organized to allow for easy transition from the "Elements of IPM" (a list of the latest and best IPM practices) for each crop to corresponding recommendations. It is available both in print and on the World Wide Web at http://www.nysaes.cornell.edu/recommends/.

The revision and the upload onto the Internet were funded by a grant from the USDA Northeast IPM Grants Program #97-EPMP-1-0127.

automated weather stations. Growers needing in-canopy environmental weather data can purchase reliable and relatively inexpensive instrumentation with which to collect data for use in their IPM programs. Many states have developed statewide weather networks to provide regional and local data for IPM programs. In California, for example, a weather data network of approximately 350 stations located throughout the state provides a database of current and historical daily weather. Growers with access to the Internet can download these data for operation of IPM models (Textbox 2.2) on their farms.

Textbox 2.2. Wisconsin vegetable growers use WISDOM to manage pests (Source: B. Jensen, IPM Program Manager, University of Wisconsin)

WISDOM, the culmination of long-term cooperation between Wisconsin vegetable growers, the University of Wisconsin researchers, and the University of Wisconsin Integrated Pest Management (IPM) Program, is decision-oriented software. WISDOM helps growers manage vegetable crop pests using field scouting data and environmental crop conditions. The software includes modules to manage insect, disease, and weed control practices as well as to schedule irrigation on potato. The program also includes modules for disease and insect management on snap beans, an important rotational crop in Wisconsin's vegetable industry.

Estimated pesticide and irrigation savings from WISDOM use on potatoes are approximately $75 /acre. These savings were achieved by

- eliminating two fungicide sprays and reducing rates of application ($27.50/acre saving),
- reducing pre-emergence applications and one postemergence herbicide application ($23/acre saving),
- eliminating a systemic insecticide application at planting and two to three additional foliar applications ($20/acre saving), and
- reducing irrigation costs ($4.50/acre savings).

WISDOM includes profiles of more than 150 key vegetable pests with important information on each pest, including

- a description of the pest and life cycle,
- the environmental factors influencing pest development,
- economic importance and treatment thresholds,
- photographs of each pest, key life stages, symptoms/damage to assist in pest identification,
- nonchemical and chemical management recommendations, and
- scouting and assessment techniques.

WISDOM currently influences 90% of Wisconsin's potato crop that is grown on 86,000 acres and is valued at approximately $170 million. The program is also used successfully in several other potato production areas of the United States and serves as an educational tool in the classroom, demonstrating the value of IPM programs in potato production.

More recently, commercial and university efforts are leading to site-specific (latitude/longitude specific) forecasts of environmental information that will allow IPM programs to make accurate real-time predictions of pest and crop developments. Providing this real-time and site-specific information to growers and to IPM consultants has the potential to improve the reliability of IPM programs. This level of precision is likely to foster adoption of IPM practices and to decrease further the use of areawide applications of prophylactic pesticides.

Site-Specific (Precision) Agriculture

Precision agriculture using geographic information systems (GIS) to reference field information has demonstrated its potential in terms of management efficiencies (Ellsbury et al. 2000; See Chapter 3). While early applications dealing with precise application of plant nutrients have helped refine the technology, the potential benefit for pest management applications is only beginning to be realized in the field. By the linking of existing pest and crop knowledge bases with GIS technologies, increased efficiencies in pest management can be introduced into agricultural production. Recognizing the precise in-field location of individual pest problems (spacial distribution) and viewing this information in the context of precise information related to fertility, irrigation, and other crop management inputs over time (temporal changes) can lead to more precise and efficient use of pest and crop management tools. Use of georeferenced sampling protocols and site-specific applications of control measures will assist growers in the implementation of IPM programs in specific fields (Figures 2.5–2.6). The greatest gains, however, likely will be in the use of this technology on whole farms and on an areawide basis where georeferenced information can promote IPM efficiencies within the grower community. In Wisconsin, georeferenced risk maps are being developed to describe potato pest pressures on an areawide basis. These maps are proving useful to the grower community in planning rotations, allocating areas for insect refugia, implementing strategies for management of resistance to pesticides, and collectively planning pest management strategies.

Use of Diagnostics

Although concepts and approaches related to pest management depend on the plant protection disciplines, diagnosis is crucial to minimizing crop losses due to insects, pathogens, and weeds. As concerns have increased regarding production costs and unnecessary pesticide use, precise assessment and pest monitoring challenges have become increasingly important. Even though approaches to characterizing infestations of different pest groups differ, the primary goal always is to facilitate production of healthy plants and crops. This discussion focuses on key classical and newly developed molecular- and computer-based diagnostic pest technologies.

Advanced DNA sensors and sensor arrays for direct genetic analysis of pathogens are being developed, primarily for human pathogens (Henkens et al. 2000). In addition to molecular diagnostics, precision agriculture, which can incorporate global positioning systems (GPS) and GIS, has great potential to facilitate identification of site-specific pest-management needs. Geographic information systems also can facilitate delivery of ideal doses of crop-protection agents and other production inputs for subunits of the target area (Hall 2000).

In recent years, the range of diagnostic procedures within agriculture has increased greatly. Many immunological diagnostic tests now are available for viral and bacterial pathogens, and a number of biochemical tests are becoming available for fungal pathogens (Stewart 2000). Laboratory-oriented tests, including DNA analyses, also have been developed for plant parasitic nematodes. The many questions that must be addressed in choosing a diagnostic procedure and determining how to use it were addressed recently by Stewart (2000). Field tests are essential for determining the utility of a diagnostic procedure for IPM. Studies conducted jointly by university and industry personnel often address these issues.

In addition to GPS/GIS, remote sensing is an emerging technology that can be used to help predict pest pressure and trigger more precise targeting of pest control measures. Through interpretation of the spectral analysis of reflective light wavelengths, one can indirectly discern the level of pest infestation in a field or fields. This technology is likely to be very useful as a diagnostic tool for pest management decisions in IPM programs. An example of using hyperspectral analysis approaches to quantify chlorphylls and carotenoids at a leaf and canopy scale is found in Blackburn (1998).

Assessment of Insect Infestations

Assessments of insect infestations usually are linked closely to economic thresholds and regulatory programs (Pedigo 1999). Because of the dynamic nature of insect populations, available sampling procedures for these pests are more advanced than those for other pest groups. Today, insect-infestation assessment frequently goes beyond traditional scouting and diagnosis by a skilled specialist, and this is especially true where resistance to a given insecticide is encountered (Figure 2.7). Approaches for resistance detection can be biochemical, immunochemical, molecular, or bioassay (Roe et al. 2000). This information is important for maintaining the durability of certain insecticides.

Figure 2.7. Using a gasoline-powered insect vacuum, a technician samples the number of spiders at various points in an Oklahoma wheat field. Photo by Scott Bauer, Agricultural Research Service, U.S. Department of Agriculture, Beltsville, Maryland.

Data on insect activity, including population shifts, are fundamental to pest management. Key types of information include long-distance movement, crop invasion, local movement and feeding, and reproduction (Barbosa and Schultz 1987; Carlson et al. 1991; Pedigo 1999; Price 1997; Southwood 1978a,b). Local, state, and national agencies often have interests in insect and other pest activities. As has been indicated, APHIS provides an extensive regulatory and inspection system for insect and other plant and animal pests in the United States.

Pathogen Detection and Diagnoses

Detection refers to establishing the presence of a certain target organism in a sample whereas diagnosis refers to identification of the nature and the cause of an observed disease problem (Louws, Rademaker, and deBruijn 1999). Depending on the timeframe, detection and diagnosis can be crucial aspects of an IPM program.

One of the primary issues concerning agricultural pest detection and diagnosis is the sampling method. Because of the clustered dispersal patterns of many plant pests, sampling procedures require careful consideration. Once the physical sampling area of interest is defined, other factors such as crop, pathogen(s) of interest, environmental conditions, crop growth stage, and desired testing level are considered (Stewart 2000). The second step in sampling is to select a random or a systematic sampling method, such as walking a "W" pattern, or another conventional sampling procedure. Variables include sample number and sample size. The specific procedures for pest sampling depend on target organisms and diagnostic methods. So that changes in pathogen numbers can be avoided, samples should be processed as soon as possible after collection. Some pathogens, including certain fungi, may grow throughout the sample whereas others, including plant parasitic nematodes, may decline sharply in number, especially when exposed to adverse conditions. Thus, tissue or soil samples should be treated as perishable products and kept in a cool, dark environment at all times.

Virus Tests

On numerous crops, and especially on greenhouse ornamentals, an enzyme-linked immunosorbent assay (ELISA) is the primary method used to detect and to diagnose viral problems (Dinesen and van Zaayen 1996). This assay has proved to be a highly reliable and sensitive, as well as a relatively inexpensive, procedure. By means of serological and enzymatic reactions, ELISA detects the presence of an antigen such as a plant virus. Commercially available kits for detecting groups of viruses including potyviruses aid in the testing of plant samples for unknown viral infection (Dinesen and van Zaayen 1996). Bioassays using susceptible hosts often are required, however, to confirm negative ELISA results.

Bacteria Tests

Numerous polymerase chain reaction (PCR)-based protocols such as rDNA-based PCR and others have been devised and adapted to enhance both the detection and the identification of plant pathogenic bacteria (Louws, Rademaker, and deBruijn 1999). Such protocols have been used widely in research programs and are resulting in the development of fundamental information on the ecology and the population dynamics of bacterial pathogens. They remain to be applied in IPM programs as related to this group of pathogens, however.

Additional protocols used for identifying bacteria include immunofluorescence and staining, ELISAs, and other immunobinding assays (e.g., Lateral Flow Strips) (Dinesen and van Zaayen 1996). Currently, use of these methods is limited largely to research programs and to diagnostic laboratories.

Fungi Tests

Although diagnosis of fungal-induced maladies still relies heavily on traditional laboratory-oriented isolation and identification, much progress has been made in developing immunology-based tests and PCR-based protocols (Dinesen and van Zaayen 1996; Stewart 2000). Commercial ELISA kits now are available for fungi such as *Septoria tritici* and *S. nodorum* (Stewart 2000), as well as for other fungi, such as *Fusarium* species and *Rhizoctonia solani*.

Nematode Assessments

Protection of plant parasitic nematodes and diagnoses of related problems rely on soil and tissue sampling, a range of extraction procedures, and identification by means of various methods. Traditionally, nematode morphology and differential hosts have been the primary bases on which to identify nematode species and host races. Today, protocols involving immunology, protein electrophoresis, isoelectric focusing, and a range of DNA-based protocols are used in laboratories and in research programs (Barker and Davis 1996; Fleming and Powers 1998). For IPM purposes, the presence of nematode-specific symptoms (galled roots for root-knot nematodes or root lesions for lesion nematodes) and signs of nematodes (the pear-shaped or round cysts of the cyst nematodes), as well as the typical spotty growth patterns of a given crop in an infested field, are reliable indicators of nematode problems. Precise identification of nematode species, however, continues to be done mainly by nematode advisory programs and by research laboratories. Nevertheless, protein electrophoresis, immunology, and DNA-based protocols are beginning to have diagnostic applications. For example, an immunological procedure has been used to detect the pinewood nematode, *Bursaphelenchus xylophilus*, in wood products such as those intended for international shipment (Lawler and Harmey 1993).

Weed Assessments

In contrast to other plant pests, weed infestations usually develop in most fields in the absence of effective control measures (Kennedy 2000). Most agricultural fields harbor great numbers of seeds of a great diversity of weed species (Table 2.5) (Schweizer, Westra, and Lybecker 1998). Because weeds represent an annual threat to crop yield and quality, most growers use some combination of cultural and herbicidal tools to minimize yield loss from weeds. But in certain instances, growers apply more herbicide than the biology of the system warrants. Assessment of the weed-seed bank in target fields is an effective but costly tool in weed diagnosis and management (Wiles and Schweizer 1999). The cost of counting and identify-

Table 2.5. Number of weed seeds/m^2 extracted from 50 eastern Colorado corn fields (Courtesy of P. Westra, Colorado State University)

Weed	Average number of seeds/sq yard	Maximum number of seeds/sq yard
Redroot pigweed	10,700	89,500
Lambsquarter	1,160	7,580
Kochia	10,360	109,200
Velvetleaf	1,270	3,430
Foxtail spp.	7,350	46,000
Barnyard grass	1,990	13,710
Wild-proso millet	2,700	7,160
Sandbur	1,220	4,230

ing 36 seeds of 4 species ranges from \$3 to \$11/soil core, and the cost of counting and identifying 37 seedlings of 6 species in a 7-inch by 5-foot quadrat (a sampling subunit or cell of land) averages \$0.26/quadrat.

The time-density-equivalent concept offers a new approach to weed assessments over a growing season, but its potential remains to be elucidated (Berti et al. 1996). Scouting fields for weed seedlings in most crops is not done systematically. Using standard-sized quadrats on a grid system to count seedlings by species in a field is a precise method but requires an ability both to identify weeds by species and, at times, to count hundreds of seedlings per quadrat. When such data are generated, weed-management decision models provide robust, reliable recommendations for weed control (Schweizer et al. 1993). Attempts to automate weed-seedling counts by means of remote sensing with specialized cameras have met with limited success because most fields contain five or more dominant weeds, which typically have image signatures different from the crop signature. Many excellent weed-identification guides (e.g., *Weeds of the West*) have been published detailing the identifying characteristics of weed seedlings and full-grown plants. As color guides are integrated into web sites, scouts may be able to access diagnostic materials on-line in the field, to identify weeds or crop injuries due to herbicides. Site-specific application of herbicides for the control of weeds detected by soil sensors also is under evaluation (Hall 2000).

Biological Control

Although widespread interest in biological control has developed only recently, the first success story for biological control in the United States occurred more than 100 yr ago. In 1888, the Vedalia lady beetle, *Rodolia cardimalis*, and the endoparasitic fly, *Cryptochetum iceryae*, were introduced and deployed effectively to control the cottony cushion scale, *Icerya purchasi*, in California citrus groves. Except where disrupted by pesticide applications, these natural enemies of the cottony cushion scale have been used effectively ever since in the United States and have been distributed to more than 25 additional countries (Hoy 2000). Even with this more than century-long success story in biological control, biopesticides still constitute only about \$1 to 2 billion/yr in a \$31 billion/yr industry (Meneley 2000).

Just what is the "biological control" of pests? Definitions differ considerably. Hoy (2000) states that "biological control of arthropod pests and weeds is a method of pest management employing parasitoids, predators, pathogens, and entomophyllic nematodes (natural enemies) to reduce pest populations." The National Academy of Sciences defined *biological control* more inclusively, as the use of natural or modified organisms, genes, or gene products, to reduce the effects of undesirable organisms (pests) and to favor desirable organisms such as crops, trees, animals and other beneficial insects, and microorganisms (Cook 1987). *Biological-control organisms* may be differentiated from *biological-control products* as follows: the former are living organisms that can be used to manage arthropod (mites and insects), weed, and plant (bacteria, fungi, viruses, and nematodes) pests and pathogens; the latter are genes or gene products derived from living organisms that kill, disable, or otherwise regulate the behavior of plant pests or biological-control products (NRC 1996).

Many insects and plant parasites may remain below damaging numbers as a result of being suppressed by natural enemies, without inputs from humans (Hoy 2000). Applied technologies used to enhance biological control include three basic approaches: classical biological control, augmentation biocontrol, and conservation (Hoy 2000). Under classical biological control, natural enemies are imported and established to provide long-term control of nonnative and, occasionally, native pests. The success story discussed earlier for controlling cottony cushion scale of citrus is an example of classical biological control. Efforts at augmentation focus on activities fostering build-up of antagonist populations or favoring the beneficial effects of natural enemies, including the periodic release of natural enemies. Additional augmentation activities include the use of cropping practices favoring natural enemies, providing hosts or prey that are alternative or fastidious (i.e., have exceptional growth requirements), and providing nests, sites, and/or food. Conservation approaches, which may overlap augmentation activities, focus on practices maintaining existing natural enemies by modifying crop management practices and other environmental components. Specific methods for conservation include strip cropping, changing planting and/or harvesting times, and altering pesticide-use patterns to preserve natural enemies and to enhance their effectiveness (Hoy 2000).

Today, continuing loss of registered pesticides offers a challenge and an opportunity for developing and deploying biological controls (Hoy 2000). Implementation of the Federal Food Quality Protection Act (FQPA) of 1996 in the United States likely will result in the loss of pesticides or pesticide uses, especially for minor-use crops (See Chapter 14). For these crops, IPM programs may shift from a pesticide-centered to a more biologically based system. Compared with traditional synthetic pesticides, biological control or biopesticides offer a number of advantages, including safety and relative environmental benignity. Moreover, related government regulations usually are less cumbersome for biological control or biopesticide strategies; there is a less risk that pest-resistance will develop; biotechnology offers methods for genetically improving strains, as well as new and improved methods for identifying and processing. Biologically based IPM interfaces well with sustainable agriculture and organic production methods (Meneley 2000). The disadvantages of biological controls or biopesticides compared with synthetic pesticides are that the former have a slower rate of kill; they are less efficacious and may not provide complete control; they usually control a narrower range of pests; their storage and shelf lives are briefer; they may require specialized application equipment; they may be incompatible with fertilizers and chemical pesticides; they often are more costly than chemical pesticides (Meneley 2000); and growers may be required to obtain specialized education in their use.

Development of Biopesticide Products

Meneley (2000) states that the biopesticide industry still is rather immature, for most companies have inadequate resources to develop large, intensive discovery programs. Still, the basic cost of $250,000 to $500,000 for the toxicology testing to meet TIER 1 data requirements for registering biopesticides with the U. S. Environmental Protection Agency (EPA) is much less than the estimated $8 million cost for registering chemical pesticides (Meneley 2000). A number of useful biocontrol products are not subject to EPA regulations, including entomophilic nematodes, insect predators and parasites, and microscopic parasites (Hoy 2000). Growth in the biopesticide market continues to be slow, with an increase from 0.2% of total pesticide sales in 1988 to approximately 0.4% in 2000 (Meneley 2000).

Although the total volume of biopesticide sales is still rather small, the number of biopesticides has increased sharply over the last decade. Today, hundreds of biocontrol agents are available, through many commercial suppliers (Tables 2.6–2.10) (EPA and USDA 2003; Garcia 2000; Julien and Griffiths 1998). A comprehensive list of microbial pesticides, including trade names, manufacturers, and suppliers, is provided by Dufour and Bachmann (1998). As discussed in the following sections, much progress has been made in developing bioinsecticides, bioherbicides, microbial fungicides, and bacteriocides, but only a few bionematicides are in use.

Insect Biocontrol

With pests continuing to invade new regions of the United States, a situation due partly to globalization of agriculture and commerce, the need for classical biological controls is growing (Hoy 2000). In addition to the examples listed earlier, citrus insects, including the leaf miner, round citrus aphid, and psylla, are targets of biological control activities. The successful establishment of one or more foreign, natural enemies of a target insect in the new environment is crucial to classical biological control. Generally, rates of successful establishment of these pests range from 16 to 34% (Hoy 2000). The relatively low efficacy of biocontrols often is a focus of debate. Although current biocontrol efficacy still needs to be improved, additional challenges involve characterization of the effects of these natural enemies on nontarget species in classical biological control programs. Despite these challenges, a great number of bioinsecticide active ingredients and bioinsecticide products have been developed (Tables 2.6, 2.7) (EPA and USDA 2003).

Currently, releases of natural enemies of insects for augmenting biocontrol are directed against insect pests of greenhouse crops, field-grown strawberries or other crops with high case-values, and nurseries. The associated high cost and at times low quality of augmentative biological control are responsible for the relatively low level of use of this form of control in agronomic crops (Hoy 2000). In contrast, as described by Stanton, Clement, and Dutky (1999), the very diverse array of natural enemies of greenhouse crops, provides an attractive arsenal for insect management of greenhouse crops. Hoy (2000) concludes that the conservation of natural enemies of insects is limited by insufficient information on the compatibility of traditional pesticides with natural enemies. With this understanding, chemical pesticide use may be decreased by 50 to 80% without affecting crop quality or yield. For certain insect pests, pesticide use probably should not be eliminated completely. Improved understanding of how to integrate appropriate pesticide use and conservation biocontrol is especially important in such situations (Hoy 2000). The integration of conservation biocontrol with transgenic crops also warrants careful consideration. For example, using natural enemies of insects in a system centered on the use of *Bt* transgenic crops could delay the de-

Table 2.6. Commercial bioinsecticides (In part, after Dufour and Bachmann [1998] and Meneley [2000])

Active ingredient	Trade names	Pests controlled	Type of action
Abamectin	Agrimee, Avid	Broad spectrum	Toxin
Autographa californica	VFN80	Alfalfa looper (*Autographa californica*)	Causes disease in larvae
Azadivachtin	Azatin, Ecozin, Neemax	Numerous insects and nematodes	IER, repellent
Bacillus thuringiensis var. *aizawai*	Agree, Design, Xintari, Mattch	Lepidoptera in vegetables and corn	Endotoxin
B. thuringiensis var. *israelensi*	Vectobac, Gnaterol, Technar, Bactimos, Aquabac, BMP144, Aquabac, primary powder? Bactis, Teknar	Larvae of mosquitoes, black flies, and midges	Endotoxin
B. thuringiensis var. *kurstaxi*	Dipel Larvo-BT Thuricide, Javelin WG, Vault, Raptor, Bactec, Bernan, BMP123, Condor, Cutlass, Foil BFC, Raven, Forwabit, MVP, MVPII, Biobit HP, Biobit 16K WP, Biobit 320B FC, Foray, Foray 48B, Foray 68 B, Futura, Bactosisk, Agrobac, Troy-BT, Biocot	Most lepidoptera Larvae with high gut pH	Endotoxin
B. thuringiensis var. *tenebrionis*	Norodor, M-Trak	Colorado potato beetle and some other leaf beetles	Stomach poison
B. papilliae	Doom, Japademic	Larvae of Japanese beetles, chafers, some May and June beetles	Stomach poison
B. sphaericus	Vectolex		Bacterium
Beauveria bassiana	Naturalis-L, Naturalis-O, (ornamentals), Naturalis-T (turf), Ostrinil, Mycotrol, Botanigard	Mole cricket, chiggers, white grubs, fire ants, flea beetle, boll weevil, white flies, plant bug, grasshoppers, thrips, aphids, mites, and others	Insect-specific fungus
Granulosis virus (Baculoviruses)	Capex Cyd-X	Leafroller Codling moth	Disease-causing virus
Heliosis zeae Nuclear polyhedrosis virus (NPV)	Gemstar LC	Lepidoptera	Disease-causing virus
Heterorhabditis spp.	Larvanem	Lepidoptera larvae	Parasitic nematodes
Heterorhabditis bacteriophora	Otinem, Cruiser, Lawn patrol	Blackvine weevil, turf grubs (including Japanese beetle) and other soil insects	Parasitic nematodes
Logenidium giganteum	Laginex	Mosquitoes	Fungal
Mamestra brassecae Nuclear polyhedrosis virus (NPV)	Mamestrin	*Trichoplusia, Heliothis, Diparopsis, Phthorimaca operculella* (potato tuber moth)	Disease-causing virus
Metharizium anisophae	Bay Bio1020, BioBlast	Soil-inhabiting beetle, termites	Pathogenic fungus
Nosome locustae	Nolo-bait, Semaspore	NA[a]	Protozoan
Paecilomyces tumosoroseus	PFR97	White flies, aphids, and thrips	Fungal parasite/antagonist
Syngrapha falcifera, Nuclear polyhedrosis virus (NPV)	Celery looper virus	Lapidoptera	Pathogenic on larvae
Spodoptera exigua NPV	Otienem-5, Spod-X	Beet army worm (*Spodoptera exigua*)	Pathogenic on larvae
Steinernema carpocapsae	Bio-Safe, BioVector, Ecomask, Seanmask, Guardian	Black vine weevil, strawberry root weevil, cranberry girdler, and other insects	Pathogenic nematode
Steinernema feltiae	Nemasys, Nemasys M, Otienem-S, Entonem	Larvae of vine weevils and fungus gnats	Pathogenic nematode

[a]NA = not available

Table 2.7. Additional bioinsecticides/biocides (In large part, after Meneley [2000])

Active ingredient	Trade name	Type
Abamectin	Agrimec, Avid	Toxin
Azadirechtin	Azatin, Ecozin Neemen	IGR repellent
Capsaicin	Hot pepper wax	Plant extract
Cinnamaldehyde	Cinnamite, Cinnacure, Hefty/Sergeant repellent	Toxin, repellent
Cyromozine	Citation, Trigard, Larvadex	IGR
Diatomecceous earth	Various formulations	Abrasive
Diflubenzuron	Dimilin	IGR
Fexoxy carb	Precision, Comply	IGR
Horticultural oils	Sunspray UF, Stylet oil	Light oils
Hydropene	Gentrol, Mator	IGR
Kinopene	Enstar II	IGR
Methopene	Altosid	IGR
Neem oil extract	Triact	Toxin
Orange oil	Power Plant, Scant-off	Repellent
Potassium bicarbonate	Armicarb, Remedy	Toxin
Pyrethrum	Various formulations	Toxin
Pyriproxyfan	Distance, Pyrigro, Knack	IGR
Sabadilla	Red Devil, Natural Guard M-Pede	Alkaloid Insecticide soap
Spinosad	Conserve, Spintor, Success	Toxin
Sulfur/lime sulfur	Various formulations	Toxin

velopment of insect resistance to *Bt* in transgenic plants (Hoy 2000).

Weed Biocontrol

Insects and mites, microbial plant pathogens, nematodes, and fish have been used successfully as biological control agents of weeds in the United States and elsewhere (Charudattan 2000). Insects have been deployed successfully as biological control agents of weeds for many years (Mortensen 1998). Much research also has been aimed at exploiting microbial-plant pathogens as classical and inundative biocontrol agents of weeds (Charudattan 2000). Plant parasitic nematodes also have been used in classical and augmentative biocontrol of weeds, and grass carp (*Ctenopharyngodon idlla*) can be effective as a nonselective herbivore for managing submerged aquatic vegetation. Thus, biological approaches for weed management have focused on introduced weeds because such pests usually have escaped their natural enemies (Mortensen 1998). In conjunction with a strategy greatly decreasing the competitive advantage of introduced weeds, numerous bioherbicides have had excellent results, controlling as much as 90 to 100% of the target weed (Charudattan 2000). Bioherbicides often involve use of a plant pathogen as a weed-control agent by means of inundative and repeated applications of inoculum or through augmentation of natural, seasonal disease levels, with smaller releases of inoculum. Because plant pathogens applied as prescriptive weed controls are considered a form of pesticide (Charudattan 2000), these organisms must be registered and approved as biopesticides in the United States and in many other countries.

Some 250 bioherbicidal agents with proven efficacies have been reported, but only a small number of related commercial products have reached the mar-

Table 2.8. Promising bioherbicides (In large part, after Charudattan [2000], Dufour and Bachmann [1998], and McFadyen [1998])

Active ingredient	Trade names	Pests controlled	Type of action
Cerosporella ageratinae	NA[a]	Pamakani weed	Fungal pathogen
Chonodrostereum purpureum	Biochon	Stump-treatment product	Prevents resprouting
Colleotrichumn gloesporioides f. sp. *aeschynomene*	Collego	Northern joint vetch	Pathogenic fungus
C. gleosporioides f. sp. *malvae*	Biomal	Round-leared mallow	Pathogenic fungus
Cylindrobasidium laeve	Stumpout	Stump-treatment	Prevents resprouting
Phytophthora palmivora	Devine	Strangler vine *Morrenia odorata* (citrus in Florida)	Root pathogen
Puccinia canaliculata	Dr. BioSedge	Yellow nut-sedge	Obligate parasite++
Xanthomonas campestris	X-PO	Turf weeds	Pathogenic bacterium
X. campestris pv. *poae*	CAMPERICO	Annual bluegrass	Bacterial pathogens

[a]NA = not available

Table 2.9. Commercial microbial bacteriocides (In part, after Dufour and Bachmann [1998] and Loper and Stockwell [2000])

Active ingredient	Trade names	Pests controlled	Type of action
Agrobacterium radiobacter	Norbac 84, Galltrol, Nogall, Diegall	Crown gall caused by *Agrobacterium tumefaciens*	Antagonist
Pseudomonas fluorescens	Conquer	*Pseudomonas tolassí* on mushrooms	Antagonist
P. fluorescens	Blight Ban A506	*Erwinia amylovora* on apple, cherry, almond, peach, pear, potato, strawberry, tomato	Antagonist
P. fluorescens or strain NCIB12089	Victus	*P. tolassí* bacterial blotch on mushrooms	Antagonist
Pseudomonas syringae	Bio-Save 10, 11, 100, 110, 1000	Postharvest pathogens on apples, pear, citrus	Antagonist

ketplace (Charudattan 2000; Liebman, Mohler, and Staver 2001). Factors hindering development include the narrow host-specificity of weed pathogens, early hurdles in developing formulations that would yield a shelf life of 1 to 2 yr for the product at room temperature, unknown interactions with conventional pesticides, and limited support from industry (Charudattan 2000; Green et al. 1998).

But with public pressures to eliminate pesticides in IPM programs increasing, extensive efforts are being made to control weeds through biological means. In addition to the organisms listed in Table 2.8, many weed pathogens are under evaluation as potential biocontrol agents in the United States and other countries. Some of the pathogen-based weed control combinations under precommercial study include *Alternaria destruens* on dodder (*Cuscuta* spp.), *Ascochyta caulina*, common lambsquarters, (*Chenopodium album*), *Colletotrichum truncatum*, Sesbania, *Sesbania exaltata*, and others (Charudattan 2000).

Rhizobacteria also are receiving considerable study as potential weed controls, especially for grassy weeds and cereal crops (Mortensen 1998). For example, application of *Pseudomonas fluorescens* for control of downy brome (*Bromus tectorum*) in winter wheat has led to increased wheat yields and to suppressed overall growth and seed production of downy brome. The very rich and dynamic microfauna and microflora of the rhizosphere can be affected greatly by cultural practices, including crop rotations. Improved understanding of the biology and the ecology of the rhizosphere is essential if the efficacy of these types of biocontrol agents is to be enhanced (Mortensen 1998).

Postemergent and pre-emergent applications of multiple-pathogen inocula offer promise for controlling multiple-weed infestations (Charudattan 2000). For example, the combination of three fungal pathogens, *Brechslera gigantea*, *Exserohilum londiristratum*, and *E. rostratum*, when applied as an emulsion, was effective in killing seven weedy grasses. Specific formulations of the inocula are crucial for improving level of control, consistency of performance, and host-range on multiple-weed infestations (Charudattan 2000). This multiple-pathogens strategy has great potential for greenhouse operations, where few conventional herbicides can be used. Still, the fact that only five bioherbicides are available currently for commercial use suggests the challenges that this endeavor will meet. Although the successes for biological control of weeds have been limited, greater understanding of the ecology of weeds and their antagonists should advance this tactic as well as contribute to the ecological management of weeds (Charudattan 2000; Liebman, Mohler, and Staver 2001).

Biocontrol of Plant Pathogens, Including Nematodes

As with conventional plant-disease control, much effort has been directed at the development of effective strategies and tactics for biological control of plant pathogens. Although the goal has been to develop effective biocontrols based on ecological principles, many highly complex ecological interactions of soilborne pathogens with resident fauna and microflora remain to be elucidated. Much progress has been made, nevertheless, in developing efficacious biocontrols of numerous plant pathogens. Currently, at least 37 biocontrol agents are available commercially (Loper and Stockwell 2000). The discovery of *Agrobacterium radiobacter* K84, a highly effective biocontrol of the crown-gall pathogen, *Agrobacterium tumefaciens*, was a major breakthrough in both plant pathology and plant disease biocontrol (Loper and Stockwell 2000). As a result of this landmark research, much of the subsequent research on biological controls for plant disease has focused on single, empirically selected biocontrol agents to suppress a given target pathogen. Examples of available products that are effective in controlling diseases induced by bacteria or fungi appear in Tables 2.9 and 2.10. A comprehensive list of registered biopesticides in the United States is available (EPA and USDA 2003).

In recent years, because of the variable efficacy

Table 2.10. Commercial microbial fungicides (In part, after Dufour and Bachmann [1998], Meneley [2000], and Loper and Stockwell [2000])

Active ingredient	Trade names	Pests/Disease controlled	Type of action
Ampelomyces quisqualis	AQ-10	Powdery mildew	Hyperparasite
Bacillus cereus	Pix Plus	Leaf spot of peanut	Antagonist
B. subtilis	Epic, Kodiak, System 3, Kodiak HB, Quantum 4000	*Rhizoctonia, Fusarium, Alternaria* + *Aspergillus* that cause root rots and seedling diseases	Antagonist
Burkholderia cepacia	Deny, Blue Circle, Intercept	*Fusarium, Pythium, Phytophthora*	Bacterial antagonist
Candida oleophila	Aspire	Postharvest pathogens *Botrytis, Penicillium*	Fungal antagonist
Gliocladium spp.	Gliomix	Soilborne fungal pathogens	Antagonist
Gliocladium virens	Soil Guard	Soilborne fungal pathogens, especially *Rhizoctonia solani* + *Pythium* spp.	Antagonist
Phelbia gigantea	Rotstop	Rust fungus, *Heterobasidíon annosum* on pine and spruce	Fungal parasite
Pseudomonas syringae	Bio-Save 10, 11, 100, 110, 1000	Postharvest pathogens on apple, pear, citrus	Antagonist
Pseudomonas sp. plus *Azospirillum*	BioJet	Soilborne pathogens that cause brown patch and dollar spot	Antagonist
Pythium oligandrum	Polygandron	*Pythium ultimum* on sugarbeet	Antagonist
Streptomyces griseoviridis	Mycostop	Soilborne pathogens: *Alternia, Botrytis, Fusarium, Phomopsis, Pythium, Phytophthora, Rhizoctonia*	Antagonist
Streptomyces hydicus	Acinovete	Soilborne pathogens as listed above	Antagonist
Trichoderma harzianum	RootShield, Bio Trek 22G, Supresivit, T-22G, T-22 HB, Trichodex	Soilborne pathogens: *Botrytis, Pythium, Rhizoctonia, Sclerotium, Verticillium,* and others	Parasite/Antagonist?
T. harzianum + *T. polysporum*	Binab	Tree-wound pathogens	Mycoparasites
T. harzianum + *T. viride*	Trichopel, Trichojet, Trichodowels, Trichoseal	*Armillaria, Botryosphaeria,* and other fungi	Antagonist/Parasite

encountered over space and time with available control products, considerable progress has been forthcoming in the development of more innovative approaches. Recently developed strategies include the formulations of mixtures of biocontrol agents and their integration in field and greenhouse situations; the exploitation of rhizobacterially mediated systemic induced host resistance; and the integration of biological and conventional pesticides (Wilson 1997). Additional approaches to enhancing efficacy of biological controls include genetic manipulation of biological control agents and improved fermentation and formulation procedures favoring their longer-term survival and activity (Loper and Stockwell 2000).

Formulation of mixtures of pathogen antagonists has considerable potential. For example, a mixture of *Trichoderma harzianum* T-95 and *Pythium nunn* proved much more effective in controlling *Pythium* damping-off of cucumber in different soil preparations than either organism alone did (Loper and Stockwell 2000). Use of the product GREY GOLD, which consists of the fungi *Trichoderma hamatum* and *Rhodotorula glutinis* and the bacterium *Bacillus megaterium* offers promise for controlling the pathogenic fungus *Botrytis cinerea* (Wilson 1997). The bacterium *Pseudomonas fluorescens* A506 and the streptomycin-resistant derivative of *Pantoea agglomerans* can be combined with the antibiotic streptomycin for controlling fire blight in affected orchards (Wilson 1997).

As has been suggested, certain pathogen antagonists may have both direct suppressive effects on pathogens and indirect effects by inducing the host to acquire systemic resistance (Loper and Stockwell 2000; Wilson 1997). Certain growth-promoting rhizobacteria, in fact, may induce certain host plants to become resistant to different types of plant pathogens. For example, rhizobacteria may induce systemic resistance to cucumber foliar pathogens such as *Pseudomonas syringae* pv. *lacrymans* and *Colletotrichum orbiculare* (Wei, Kloepper, and Tuzun 1996). The bacteria *Bacillus sphaericus* B43 or *Agrobacterium radiobacter* G12 may induce resistance to the potato cyst nematode *Globodera pallida* (Hoffman-Hergarten et al. 1997). The development of cropping systems favoring or augmenting the build-up of rhizobacteria has much promise for exploiting these biocontrol agents for large farm operations (Barker and Koenning 1998; Kloepper et al. 1992).

A wide range of other antagonistic bacteria, nematode trapping fungi, predaceous nematode species, and other soil fauna has considerable promise in the biological control of plant parasitic nematodes (Kerry and Evans 1996; Stirling 1991). Numerous species of fungi such as *Paecilomyces lilacinus* and *Verticillium chlamydosporium* have provided nematode control under greenhouse and certain field conditions, but have not been used effectively on a large-scale basis. More fundamental information on soil ecology is needed before biological control of plant parasitic nematodes can be implemented.

Conventional biocontrol agents, including mixtures of antagonists, have much promise for highly controlled environments such as greenhouse operations. For example, *Burholderia cepacia* and binucleate *Rhizoctonia* strains provide excellent biological control of root-rot diseases of azalea and other woody ornamentals in soilless greenhouses (Burns and Benson 2000). In contrast, these same organisms provide only limited control of *Rhizoctonia solani* on cotton and standard soil mixes even in the greenhouse. Other antagonists, namely *Pseudomonas fluorescens* and *Gliocladium virens*, provide acceptable levels of pathogen biocontrol on greenhouse plants (Loper and Stockwell 2000).

Biocontrol also has potential in the management of postharvest disease. Certain high-value fresh foods are stored under controlled environmental conditions and usually are monitored for associated diseases. As illustrated in Tables 2.9 and 2.10, commercial biocontrol products are becoming available for managing postharvest disease, but this tactic to date has had only limited application.

A very different type of biocontrol involves the use of strain-specific bacterial viruses for suppressing *ineffective* nitrogen-fixing *Bradyrhizobium* or *Rhizobium*. Bacteriaphages or rhizobiophages commonly occur in the rhizosphere of legumes, in which they may affect the nitrogen-fixing capacity of *Bradyrhizobium* or *Rhizobium* in root nodules (Kuykendall and Hashem 1998). To enhance nitrogen-fixing capacity in fields, simultaneous introductions of effective rhizobiophage-resistant Rhizobia strains and phages virulent to ineffective strains in the soil will increase productivity and sustainability of the target crop, especially in soils where nitrogen is limiting (Kuykendall and Hashem 1998).

Additional Methods for Enhancing Biocontrols

Agricultural crops, which occupy approximately 25 to 30% of land worldwide, greatly affect biological diversity (Altieri 1994). Especially in Europe, widely used practice favoring the build-up of beneficial organisms such as insects, involves the maintenance of an abundant food supply in the form of polycultures (refugia) or farmscaping (Altieri 1994; Dufour and Greer 1995).

Farmscaping has been defined as the use of hedgerows, insectary plants, cover crops, and water reservoirs to attract and to support populations of beneficial organisms such as insects, bats, and birds of prey (Dufour and Greer 1995). The nectar, pollen, and honeydew (from aphids feeding on plants) on flowering plants, as well as the herbivorous insects and mites, serve as food for natural enemies of pests. The concepts of farmscaping or refugia refer to the development of cropping systems enhancing plant diversity and disrupting associated pest life cycles and other pest insect activities.

The goal is to provide a highly species-diverse environment, or extensive biodiversity, so that a variety of habitats and niches are available for organisms to exploit (Altieri 1994; Dufour and Greer 1995). This approach recaptures some of the natural ecosystem diversity lost in monocultures (Altieri 1994). Another benefit is decreased pesticide application. In the selection of cover crops and other plants as tools for this purpose, care must be paid to favor sustainable soil fertility while avoiding plants that would serve as plant virus reservoirs or favor soilborne pests such as nematodes and other root pathogens. Extensive lists of candidate plant species, including certain weeds and associated beneficials that mesh with polycultures, are available (Altieri 1994).

A range of new technologies offers much promise for enhancing the biological control of many crop pests. In addition to the widely deployed insect-resistant and herbicide-tolerant transgenic crop cultivars, technologies include the potential to genetically improve natural enemies (Hoy 2000) (e.g., transgenic entomopathogenic nematodes and other entomopathogens, parasitoids, and predators). Through identification and use of appropriate genes, transgenic natural enemies surviving even in adverse environments may become available for biocontrol.

Undoubtedly, molecular approaches to taxonomy and ecological studies will facilitate the arrival of effective biological controls for a range of pest. In addition to rapid, precise identifications of the most effective strains of growth-promoting rhizobacteria, the exploitation of nitrogen-fixing bacteria, mycorrihizal fungi, and beneficial nematodes will be enhanced as a result of their integration into well-planned cropping systems. In this way, crop productivity and biological control of targeted pests will be enhanced.

Pesticides

Ecologically based IPM is a promising option for decreasing negative pesticide impacts and agricultural production in the twenty-first century (Liebman, Mohler, and Staver 2001; NRC 1996). Many early IPM proponents have predicted that pesticide inputs in agriculture eventually would be displaced by biologically intensive IPM tools, tactics, and strategies. Their optimism has diminished somewhat in light of findings that pesticide use in the United States has stabilized since the mid-1980s, at around 1.0 billion pounds of active ingredient annually. Pesticide use in countries with rapidly developing economies, especially in the Pacific Rim nations, has grown at rates sometimes exceeding 8% annually (EPA 1999). Pesticides likely will continue to be among the primary tools of IPM (Jacobsen and Backman 1993).

Why Do Pesticides Remain a Critical Component?

Answering this question requires a review of the biological, legal, sociological, trade, and, ultimately, economic factors driving pesticide use. In the end, pesticides remain the least complex, most effective, and least expensive short-term solution to pest suppression in many, if not most, pest management systems. Additionally, pest management is complex because pests themselves are complex, and their effects on society are pervasive. Pesticides are used in virtually every theater of human existence. They are used in all aspects of producing, processing, storing, packaging, transporting, and marketing food. They are used in manufacturing, right-of-way management, forestry, schools, parks, golf courses, and municipalities, and in more than 90 million U.S. households (FCH 2001; Hall and Menn 1999).

Today's food production systems operate in a complex global market characterized by extreme competition. Producers in the United States are increasingly bimodal: either they are large commodity-producers with large production units that remain profitable through high-volume sales with low per-unit margins; or they are small, diversified niche or direct-market operators close to major population centers.

Large commodity-producers have been forced to increase efficiency so as to survive on smaller and smaller margins. Farm-gate returns for commodities such as small grains and potatoes have grown less than 3 times on a per-unit basis in the last 45 yr (Smith 1992), and net returns for certain commodities have actually declined over this period. Meanwhile, value-added, vertically integrated food companies and retailers have experienced an increase in per-unit food price, which has grown 400% over the same period. For the most part, agricultural producers have not shared in this expansion. Many large-scale producers have integrated chemically intensive IPM systems skillfully with sustainable agricultural tactics practical for large-scale farms. Pesticides remain essential because their consistent performance provides stability for businesses operating on small margins.

If there is growth in the number of agriculture producers, it is in the number of small, diversified production systems focused on niche or direct marketing. These producers often target horticultural crops and may provide recreation for consumers wishing to "pick-their-own" vegetables, bedding plants, or ornamentals on visits to roadside produce establishments or farmers' markets. These producers also may represent an alternative production system or lifestyle that suburban and/or urban consumers value (e.g., organic, biodynamic, or low-input sustainable agricultural systems). Often these producers use aspects of IPM, although they may not identify them as such. Pesticides are essential in many of these operations, and not only because effective nonpesticide alternatives are lacking for certain crops. Time, education, and economic constraints also may prevent producers from implementing complex biologically based pest-control systems.

Farm-related regulations have become more numerous and complex since the early days of IPM, and these regulations sometimes inhibit implementation of IPM (Kennedy and Sutton 2000; see Chapter 14). Hundreds of laws regulate all aspects of farm-related businesses, including finances, pesticide use, pesticide storage, pesticide application equipment, equipment operation, occupational safety, transportation, land use, crop storage, pest damage and contamination, commodity grades and standards, labeling, export market access, genetically modified organisms (GMOs), seed and production license or contract arrangements, surface water, ground water, odor generation, chemical trespass, taxes, inheritance and ownership, right to farm, labor, open land, riparian areas, wetlands, and wildlife habitat for endangered species.

Understandably, many producers cannot comply fully with the many regulations confronting their operations, and the time and energy required for mandated paperwork decreases time available for actively managing crop production. Furthermore, product grades and standards, contractual arrangements, farm credit and market access all may require, directly or indirectly, low levels of pest damage achievable only with pesticides.

Pesticides also are essential tools for helping farmers meet market grade and phytosanitary standards imposed by brokers, food processors, the USDA, the Food and Drug Administration (FDA), or export markets. For example, sooty blotch and flyspeck diseases cause only superficial blemishes on apple fruit, but the USDA grade standards and consumer expectations limit the size of blemishes acceptable in top-grade fruit. Assuming that the dark fungal blemishes represent pesticide residues on fruit, consumers at farmers' markets sometimes have refused to buy apples with sooty blotch. In other instances, pesticide applications are required to minimize insect contamination in processed food products. Baby food manufacturers, for example, reject loads of pears if the fruit contain Comstock mealy bugs in the calyx. These insects do not cause significant damage to the fruit but do contribute to detectable insect parts in the processed product. Many export markets demand a certain level of pesticide input to guarantee elimination of exotic pests threatening their own agricultural systems. These phytosanitary demands can function as nontariff trade barriers if the costs of additional required inputs make the resulting product noncompetitive in the international marketplace.

Another significant factor driving pesticide use is the introduction of new species, biotypes, and races of pests. More than half of all economically important arthropod pests in the United States have been introduced from foreign countries (OTA 1993). These pests have arrived by numerous routes. Certainly, global trade and mass transportation account for many new pest species introductions annually, but trade is not the only source. Long-range transport on weather systems also is a documented method of pest introduction. The issue is more complex than the introduction of new pest species, however. When a new pest is introduced, it usually comes without its natural enemies; there are, therefore, few or no biological controls operating in its new habitat. As a result, new pest populations sometimes grow to crisis proportions. Pesticides often are the only short-term pest control alternative available, especially when local, state, or federal laws mandate eradication. This issue is exacerbated when the "new pest" actually is the same species as an indigenous pest, but with additional features such as new or different capacities to transmit diseases (*vector competence*), pesticide resistance, or expanded host plant range. These latter capacities are almost intractable, as evidenced by the recent introduction of a new biotype or species of *Bemesia* white fly throughout the Southwest and the soybean aphid in the Midwest.

Economics and temporal/spatial scale issues often force growers to attempt to produce the highest returns even if variety, geographical location, or other factors make achieving this goal unlikely. For instance, there are four grades of apples recognized by the USDA: Extra Fancy, Fancy, US #1, and Utility. The difference in return between Extra Fancy, and Fancy and Utility often exceeds tenfold. Thus, many producers of high-value horticulture crops are risk-averse: rather than risk a damaged crop and a decreased grade, they apply pesticides.

The dynamics of weather, regional climates, and pest geographical and host range often encourage pesticide use. Certain regions have diverse pest and climatic conditions that exacerbate pest problems. The humid central and eastern United States, for example, uses more fungicide and bactericide than the irrigated, arid West does. Apple maggots appear primarily in the Northeast, and fire ants in the Southeast. Pink bollworms appear in the Southwest, and orange tortrix in the Northwest. Weather differences and pest vagaries constitute the backdrop influencing regional differences in pesticide use in IPM programs.

Although often requiring greater volumes of pesticide inputs than conventional pest management programs do, alternative production systems such as organic agriculture have escaped pesticide residue scrutiny because the pesticides used have tended to be less toxic, analytical methods for certain organic pesticides (e.g., sprayable compost teas) have not been developed, and many policymakers and consumers believe that organic fruits and vegetables are produced without the use of any pesticides (see Chapter 4). The EPA under the Federal Insecticide, Fungicide, and Rodenticide Act (FIFRA) defines a pesticide as anything that kills, repels, or mitigates a pest. Under this definition, elemental sulfur is a pesticide that is used heavily by organic and conventional farmers to control plant diseases. Sulfur must be used at rates of 10 to 20 lb/a./application, whereas other fungicides can provide equivalent or better disease control when used at rates of as low as 2 oz./a. Despite societal concerns about pesticide use, the use of sulfur has exploded in Europe and the United States. In California, for instance, sulfur use has increased dramatically in the last 5 yr, and it is one of the most-cited products in medical injury-complaints generated by pesticide use (Reigart and Roberts 1999). This paradox is just one outcome in an array of perceptions and societal valuations that make pesticide use so complex and controversial.

Pesticides are likely to remain important components of IPM because they are relatively easy, effective, and inexpensive to use. Historically, their negative effects have been relatively long term and have escaped detection for long periods. Pesticides developed over the past 10 to 15 yr, however, have been screened more carefully for adverse nontarget effects, and these newer products can provide safe, effective, and efficient pest control for many crops.

The Role of Pesticides

The role of pesticides in IPM systems depends on the specifics of crop, pest, region, site, weather, and economics. Furthermore, pesticide usage changes constantly as research and field experience lead to alternative pest protection systems, improved monitoring methods, and improved understanding of pest biology. Because they decrease risks of major crop losses, which otherwise would make a production system economically nonviable, pesticides are crucial IPM components for most crops. Farmers often are forced by simple economics to avoid high-risk endeavors such as those involving crops subject to uncontrollable pest infestations. Today, effective and practical nonpesticidal alternatives are lacking for many pests in high-value crops. Even where pest populations are managed by means of biological or cultural methods, farmers often view pesticides as essential tools for rescuing crops threatened by unusual pest infestations or unexpected failures of biologically based IPM strategies.

Pesticides can be used proactively to decrease densities of pests before they become established in crops. Alternatively, pesticides can be used as repellents, causing pests to avoid stored food, packaging, and crops. For example, pyrethroid insecticides applied to apple trees provide protection against the 17-yr cicada because cicadas quickly migrate back out of orchards sprayed with even very low dosages of the synthesized natural insecticide. The repellent effects of pyrethroids also may decrease viral transmission by repelling aphid vectors in certain vegetable crops.

Pesticides can be used as tools for adjusting ecological predator-prey balances within crops and as backstops to arrest pest populations exceeding established economic thresholds (Kennedy and Sutton 2000). For example, miticides frequently are used to adjust ecological balances between predatory and plant damaging species in various cropping systems. In apples, populations of European red mites sometimes reach economic thresholds before predators reach populations necessary for biological controls to work. When selective miticides kill more pests than beneficials, populations of European red mites can be suppressed until populations of predatory mites reach densities at which biological control will be successful. Thus, a pesticide can serve various functions depending on pest, crop, crop phenology, and geographic region.

Protectant fungicides often are used to prevent initial pest infestations and thereby decrease the total number of sprays needed to protect a crop during secondary infection periods. For example, four or five timely fungicide applications during spring can prevent establishment of apple scab in apple orchards in the northeastern United States. If primary scab infections are not prevented, then five to eight additional fungicide sprays may be needed to prevent secondary infections on apple fruit during summer.

In many crop/pest systems, pest populations are monitored, and pesticides are used only when pest populations exceed an economic threshold, which refers to the pest population density at which crop loss exceeds the cost of intervention (Kennedy 2000). Producers especially value pesticides that can provide rapid knock down as a population approaches a critical level. Availability of pesticides that can arrest pest development at critical thresholds adds flexibility to IPM programs because farmers have more opportunities to incorporate biological, cultural, sanitation, and preventive strategies if they know pesticides are available to rescue a crop, should alternative management strategies fail.

Effects of Changing Fungicide, Herbicide, and Insecticide Chemistries

Trends in New Chemistry

In the last 10 to 20 yr, the pesticide industry has undergone a remarkable transformation. In the 1960s to 1970s, as the petrochemical industry matured, pesticide companies that for the most part were subsidiaries were spun off into stand-alone companies. With decreased profitability and a slowdown in discovery and registration, fewer new pesticide products were brought to the market in the 1980s. Increasing registration restrictions in the United States, Canada, and Europe further focused industry discovery efforts toward safer, narrower spectrum, environmentally friendlier chemistries. These trends, together with the horizontal and vertical integration of the pharmaceutical industry, led to an aggregation of smaller pesticide companies into larger and larger conglomerates in the 1990s.

In the waning years of the twentieth century, life science companies emerged. These multinational conglomerates often merged pesticide companies into companies that previously had focused on pharmaceuticals, information science, and molecular biology. The evolution of conglomerates is continuing, but, more recently, certain conglomerates have sold off agriculture-related or pesticide units because of low profit margins.

Passage of the FQPA in the United States galvanized the move toward safer, more ecologically and environmentally friendly pesticides. Today, there is an array of new synthetic chemistries, fermentation products, *semiochemicals* (compounds involved in communication among organisms, such as pheromones in insects), natural chemistries, and transgenic crops or microorganisms that are transforming the face of IPM in nearly every pest management area (Kennedy and Sutton 2000).

Trends in Fungicides and Bactericides

Trends in fungicide development over the past 30 yr have had important effects on IPM programs.

1. **New fungicides often control a more limited range of pathogens than old broad-spectrum fungicides controlled**. As a result, the process of identifying pathogens to be controlled and selecting the appropriate fungicide has become more complex. Producers and farmers are more willing to hire IPM experts to provide professional services.
2. **New fungicides are expensive**. Integrated pest management programs allow farmers to optimize spray timing and fungicide selection. Farmers are becoming aware that fungicides are investments that may not produce a positive economic return if applied incorrectly. This awareness has caused many farmers to embrace IPM as the best strategy for optimizing investments in new fungicides.
3. **New fungicides often have both postinfection and protectant activities, whereas old fungicides were mainly protectants**. Fungicides with postinfection activity can be applied after an infection period has occurred, thereby decreasing the need for prophylactic spraying based on weather forecasts lacking precision or reliability. The combined protectant and postinfection activities of new fungicides allow for increasingly flexible spray timing and improved integration of disease-, insect-, and mite-control strategies.
4. **New fungicides usually are labeled at rates providing little margin for error in sprayer calibration**. Labeling products at the minimal effective rate allows manufacturers to address environmental and consumer concerns while improving profitability to the manufacturer. Manufacturers can charge competitive rates per acre for a product while minimizing the amount of product that must be applied. When products are labeled at minimum effective rates, sprayer calibration and optimal spray timing become increasingly important; both issues are addressed within IPM programs.
5. **New fungicides are single-site inhibitors arresting fungal development by interfering with a single biochemical pathway**. Such products must be managed carefully to avoid selection of fungicide- resistant strains. Integrated pest management programs incorporate resistance management.

The strobilurin fungicide group is an example of fungicide chemistry developed during the 1990s (Tedford and Brown 2000; Ypema and Gold 1999). Development of this group resulted from the discovery in the 1970s that a certain wood-decaying fungus growing in Europe contained powerful antifungal compounds. The natural compound was modified by various agrochemical corporations to produce the photostable azoxystrobin, trifloxystrobin, pyraclostrobin, and kresoxim-methyl chemistries. These fungicides have low toxicities to mammals and, because they break down quickly in soil and water, negligible environmental risk. In addition to being effective against many pathogens at very low rates (approximately an ounce active ingredient/a.), they have activity against a broad range of fungal pathogens, including both downy mildew and powdery mildew on grape. They also have postinfection and protectant activities and work by inhibiting mitochondrial respiration in fungi. But because these fungicides could be compromised quickly by the selection of resistant pathogen strains, label restrictions require that they never be applied more than two or three times in succession. Alternation of these fungicides with older fungicides having alternative modes of action, will, it is hoped, prevent or delay for many years development of resistance.

Biocontrol agents controlling or decreasing plant diseases have received considerable attention during the past several decades (Tedford and Brown 2000). Strains of *Trichoderma* species, for instance, have been commercialized as seed treatments preventing root diseases and wood decay. Several strains of the bacterium *Bacillus subtillis* have been registered for control of both fungal and bacterial diseases on fruit and vegetable crops. Strains of the bacterium *Pseudomonas syringae* and a strain of the yeast *Candida oleophila* have been registered for control of post-harvest diseases on apples and pears. With a few exceptions, however, performance of biocontrols against plant diseases has been inconsistent, and biocontrols have not been used widely yet as replacements for traditional fungicides. (See previous section of this Chapter.)

Development of bactericides for the control of plant pathogens plateaued after registration of the antibiotic streptomycin and oxytetracycline more than 30 yr ago. Bacteriologists expressed increasing concern that antibiotics applied to plants in the field might select for antibiotic resistance in nontarget bacteria. The possibility existed that antibiotic resistance might later be transferred to human pathogens, thereby impeding the effectiveness of antibiotic therapy for important human diseases. The dim future for the registration of new antibiotics for plants left selection of disease-resistant varieties as the most effective strategy for managing bacterial diseases.

Within the past 5 years, strategies have evolved for using chemical sprays that, instead of attacking bacterial pathogens directly, work indirectly by making host plants less susceptible to infection (Sticher, MauchMani, and Metraux, 1997). These products are known as systemic acquired resistance (SAR) inducers. When applied to plant foliage, they stimulate plant cells to produce biochemical products making plants more resistant to attack. Usually the resistance generated by SAR inducers is ephemeral (3 to 7 days in duration) and provides incomplete control of disease on a highly susceptible crop. Furthermore, crop response to SAR inducers has been quite variable and presumably is affected by other crop-stress factors, environmental conditions, and pathogen biologies. Nevertheless, SAR inducers represent an exciting new technology for managing recalcitrant bacterial diseases and certain fungal diseases. Integrated pest management should provide the tools needed to optimize effectiveness of this technology.

Another new strategy for managing bacterial diseases involves use of a plant growth regulator, prohexadione calcium, which, by altering plant physiology, makes apples and pears less susceptible to fire blight. Prohexadione calcium inhibits biosynthesis of gibberellin, a natural plant hormone essential for cell elongation. Plants treated with prohexadione calcium show decreased vegetative shoot growth, and this decrease evidently makes the growing shoot tips less susceptible to invasion by *Erwinia amylovora*, the bacterium causing fire blight. To use this product effectively, IPM practitioners need to consider both pest control and horticultural variables because, in certain instances, the growth suppression caused by prohexadione calcium can limit productivity, whereas failure to apply the product can result in significant losses to fire blight.

Trends in Acaricides and Insecticides

The introduction of dichloro-diphenyl-trichloroethane (DDT) and other chlorinated hydrocarbons in the 1940s revolutionized pest control and ushered in the modern synthetic pesticide era. The global adoption of these chemistries led to induced pest outbreaks (secondary pests), resurgence of target pests (loss of natural enemies), resistance evolution, and residue problems giving rise to acute and chronic environmental and ecological effects. In 1972, the use of DDT was essentially banned in the United States. Organophosphate and carbamate insecticides were popularized from the late 1960s through the 1980s. They dominated many insect management strategies until recently, when implementation of the FQPA threatened their long-term use (See Chapter 14). Photostabilization of the natural insecticidal pyrethrins led to the synthetic pyrethroids, which have paved the way for major inroads into pest management in the last 20 yr. All these chemistries have evidenced certain of the problems experienced by the chlorinated hydrocarbons, with the possible exception of long-term chronic environmental and ecological effects.

The chemistry of insect-juvenile hormone was elucidated in 1967 (Granett and Retnakaran 1977). Initially, there was much interest, and many pest managers thought that these chemistries would provide adequate arthropod control without the major drawbacks of the synthetic organic chemicals. Only six or seven of the juvenile hormone mimics, chitin synthesis inhibitors, or molt inhibitors are registered in the United States today, however, and they account for only marginal insecticide sales. Formamidine insecticides also have been disappointing in that only two (chlordimeform and amitraz) have been successful commercially. Both are able to control arthropods by means of a combination of lethal and sublethal effects but have been plagued with chronic human-toxicity concerns. Fermentation products such as *Bacillus thuringiensis* insecticidal proteins have great promise but have accounted for less than 2% of total pesticide sales in the United States (Nester et al. 2002). They suffer from comparatively low efficacy and slow action. The development of GMOs through the incorporation of these insect active proteins, by means of molecular biological technology, into plants also has significant promise but may be stalled as a result of negative public perceptions (See Chapter 14).

Novel Insecticidal Chemicals

The culmination of two decades of research and investment by agrochemical companies has netted an array of relatively safe, narrowly targeted, and environmentally friendly insecticides (Kennedy and Sutton 2000). Avermectins disrupt both gamma amino butyric acid (GABA) receptors and chlorine channels in the arthropod nervous system. Some of the most potent new chemicals are active against a range of insects, mites, and helminths and break down rapidly in the environment. Because they also have few nontarget effects, these chemicals have been viewed as environmentally friendly by the EPA. Several companies are developing neonicotinyl chemistries (e.g., imidacloprid, thiamethoxam, acetamiprid) to have activity against soft-bodied sucking insects and certain beetles. These compounds have both contact and systemic activity but can be long-lived in the environment. Development of resistance may be a problem, especially in aphids and Colorado potato beetle (See following section). Few or no nontarget effects have been detected with the use of neonicotinyl chemistries.

Tebufenozide, a novel insect-growth regulator, may have promise in controlling many lepidopteran larvae. It is an ecdyzone agonist (mimicking insect horomone, causing premature molting and death), recently registered by the EPA, on a number of crops. It has few or no nontarget effects and is short lived. Another GABA chlorine channel blocker—phenylpyrazole, or Fipronil—has promise in a number of insecticide markets. Fipronil has contact, ingestion, and systemic activity. It may be the best potential resistance-breaking compound on the horizon for organophosphate, carbamate, and synthetic pyrethroid-resistant insects. Because pyrroles are uncouplers of oxidative phosphorylation, they, too, may be good candidates for breaking resistance in many pests. These compounds have some contact as well as major ingestion toxicity. They exhibit a moderate spectrum of activity, as well as comparatively low nontarget activity.

The spinosyns are derived from Actinomycetes (filamentous bacteria) (Sparks et al. 1999). They exhibit both contact and ingestion toxicity against beetles, flies, termites, worms, and lice. Like the neonicotinyl chemistries, they bind to arthropod nicotinic acetylcholine receptors in the nervous system but have no known synergism, antagonism, or crossresistance with the former chemistries. Additionally, they have few nontarget effects, are short lived, and exhibit few or no mammalian acute or chronic effects. The EPA also has registered an array of behaviorally active semiochemicals, principally attractants, pheromones, and repellents. These compounds have some promise in IPM systems as mating dispruptors, monitoring tools, and "attract and kill" stratagems.

The EPA has registered these environmentally friendlier and safer compounds in almost a 2:1 ratio over other chemistries since 1996. For the most part, these chemistries have been on the EPA's registration fast track as the FQPA has created a near-crisis need for organophosphate and carbamate alternatives. These new chemistries also represent the outcome of both global regulatory pressure toward safer, more environmentally friendly pesticides and insecticide industry foresight, capital investment, and targeted research in the 1980s and 1990s. Since 2000, the division of EPA responsible for fast-tracking the registration of biopesticides has been struggling with decreased funding, fewer staff, and increased demands to shift focus to transgenic crops that produce pesticides (plant-incorporated protectants) rather than registering microbial and biochemical pesticides.

All these new compounds point to a trend toward narrower-spectrum insecticides with more environmentally friendly, safer attributes (Kennedy and Sutton 2000). These compounds will be put to various uses in IPM systems and will improve the development and adoption of biological (predator and parasite) integration in pest management greatly. The compounds also will accompany and augment the need for information regarding delivery, monitoring, and timing of treatment to ensure effective IPM. Additionally, pest management is likely to be more expensive with the use of these newer compounds. Not only are the individual chemistries more expensive than their predecessors, but they also have much narrower spectrums of control. More applications may be necessary, therefore, to achieve control similar to that achieved with organophosphate, carbamate, and pyrethroid insecticides.

Trends in Herbicides

One of the most striking achievements in herbicide chemistry development since 1980 has been the discovery and development of herbicides effective at very low use-rates (See Chapter 3). Many older herbicides required use rates of 0.5 to 3 lb active ingredient/a. New herbicides, typically applied at ≥ 1 oz/a., provide equivalent or better weed control. Herbicides in the sulfonylurea and imidazolinone families are examples of low use-rate products with very low mammalian-toxicities. An additional restriction placed on new herbicides is the leaching characteristics of the herbicide or its major metabolites. Herbicide candidates showing a propensity to move through soil to surface water or groundwater are not developed, because of environmental concerns. Certain of the newer herbicides, however, persist in the soil at levels undetectable by normal methods. Planting sensitive crop species in treated soil as much as 2 to 3 yr after application can result in severe damage. Thus, the use of these new chemistries tends to limit rotation options to relatively tolerant crop species.

A novel herbicide strategy used since 1996 has been the development of crops genetically modified to express tolerance to herbicides that ordinarily would have killed them. Crops with tolerance to glyphosate and glufosinate are two examples of new uses for safe herbicides developed through biotechnology (Bridges 2000) (See Chapter 3). Use of cropping systems based on such technology can decrease the amount of herbicide used in a field dramatically and can allow farmers increased flexibility to practice decreased tillage, which in turn decreases soil erosion. Still, repeated use of glyphosates and other chemistries is likely to result in the development of resistant weed populations, further compounding problems in adopting IPM practices. In fact, increased use of glyphosate because of the use of glyphosate-tolerant crops has led to the development of glyphosate-resistant weeds (Carpenter et al. 2002; Lyon et al. 2002). Glyphosate-resistant horseweed (*Conyza canidensis* L.) occurred in only three years of glyphosate use in a continuous glyphosate-tolerant soybean system (Van Gessel 2001).

Because of the rising costs of developing new herbicides and the significant investments required to develop genetically engineered crops, acquisitions and mergers in the herbicide chemical industry have become common. In this way, the power of technology development is concentrated in fewer and fewer companies that must be included in dialogues related to furthering IPM goals and adoption.

The FQPA requirements are driving companies to abandon herbicide labels for minor crops and farming systems in which their product profits are low. New mechanisms and incentives are needed to ensure that minor crop production remains competitive and profitable in the United States.

Trends in Nematicides

Although the first fumigation test for nematode control was conducted more than 130 yr ago, the first effective fumigant (dichloropropane-dichloropropene mixture [D-D]) was discovered much later—1943—and was invaluable as a nematode management tool and as a means of characterizing crop losses caused by plant parasitic nematodes. The subsequent development of additional fumigant and nonfumigant nematicides (organophosphates and organocarbamates) provided a limited but effective array of nematicides, which were used widely from the 1950s through the 1970s for control of a wide range of crops. Another fumigant, methyl bromide, has proved an especially effective biocide and has been used worldwide for control of nematodes, insects, fungi, bacteria, and weeds.

Because of environmental and health concerns such as contamination of ground water, numerous fumigant nematicides have been removed from the market, and others such as 1,3-dichloropropene and methyl bromide were under intense review in the 1990s. In fact, methyl bromide is being phased out by 2005 (EPA 2000). Certain other nematicides such as 1,3-dichloropropene and aldicarb are restricted to certain uses. The manufacturer of 1,3-dichloropropene indicates that this product is likely to remain cleared for use in the United States. In the November 21, 2001 issue of the *Federal Register* (Vol. 66, No. 225), the EPA announced the termination of the Telone Special Review because the Agency believes that the benefits of Telone use continue to outweigh its risks. Microbial decomposition of organophosphate and organocarbamate nematicides recently has become a serious problem, especially when a given product is used repeatedly (Jones and Norris 1998). Thus, the long-term utility of a number of the currently available nematicides is uncertain.

Considerable research has been directed toward discovery and development of new types of nematicides and biocides, including biologically based products. These efforts involve products formulated from chitinatious materials, fungi and fungal by-products, and plant and plant by-products such as Neem (NRC 1992). Compounds from the Neem tree have potential as nematicides as well as insecticides (NRC 1992). Neem oil and azadirachtin have both been registered as nematicides in the United States (EPA and USDA 2003). An example of a fungal by-product receiving extensive evaluation is DiTera, a complex, unpurified material from *Myrothecium* spp. (Warrior et al. 1999), but related results to date are rather variable. Most current research efforts are focused on the development of alternatives to methyl bromide. Promising replacement compounds include methyl iodide, metam-sodium, and dichloropropene-chloropicrin (Hutchinson et al. 1999; Locascio et al. 1997). For root-knot nematode control on carrots, methyl iodide was as effective as methyl bromide (Hutchinson et al. 1999). Unfortunately, the current costs and uncertain environmental effects of methyl iodide place this compound under question. Locascio et al. (1997) concluded that no single product can provide the broad-spectrum pest control offered by methyl bromide. Physical treatments such as solarization and irradiation have promise on dried plant parts during storage for soil pest control and for insect control, respectively.

Future nematode management probably will rely less on traditional nematicides unless significant breakthroughs occur. Current trends in nematode control rely on integrated cropping systems encompassing the use of resistant cultivars or crops, rotations including cover crops, soil amendments, and biological controls, especially those favoring soilborne antagonists/parasites (Barker and Koenning 1998). Future nematicide formulations will need to meet efficacy, environmental and health protection, and food safety standards.

Trends in Pesticide Delivery Technology

Pesticide application precision remains one of the most important environmental and ecological factors in all of pest management. Most pesticides applied today are either preemergent herbicides for row crops or postemergent herbicides for weed canopies. The standard wisdom is to spray large droplets that do not drift. High-ground clearance sprayers work with nozzles designed to produce a flat fan of nonoverlapping spray. Improved formulations and new active ingredients require lowered inputs to achieve similar or improved weed control. Additionally, application of herbicide bands to the row instead of to the entire field has decreased herbicide use in certain row crops.

Application equipment has become much more sophisticated, with many new technological changes such as improved nozzle types (Figure 2.8) and computer systems linked to pumping systems (Kennedy and Sutton 2000). Computer systems linked to satellites allow aerial applicators the opportunity to apply pesticides very accurately by means of global positioning application systems. These systems also eliminate the need for human flaggers for aerial application. Computer systems linked to spraying sys-

Figure 2.8. Hooded sprayers operated by a field technician direct herbicide just to areas between rows of grain sorghum. Photo by Jack Dykinga, Agricultural Research Service, U.S. Department of Agriculture, Beltsville, Maryland.

tems mean that mixtures can be injected while applicators are spraying. This process has eliminated the need to mix large tanks of spray solution before application or to fill tanks with mixed pesticides. These tanks can result in major spills on the highway and in costly clean-up procedures. Because the injection system allows applicators an opportunity to spray and to mix simultaneously, no unneeded pesticide mixtures are mixed that cannot be applied because of adverse wind conditions or a change of plans. Additionally, with injection systems, the possibility of pesticide breakdown through hydrolysis is eliminated because pesticide solutions are sprayed shortly after being mixed.

Precision agriculture, as discussed earlier and in Chapter 3, has great promise in further decreasing the quantity of herbicides as well as other pesticides and fertilizers used in row crop agriculture. With decreases in inputs, and increases in the "precision" of targeting, negative effects on the applicator, non-target organisms, and the environment also should be decreased. Tree fruit, brambles, vineyards, hops, certain vegetables, forestry, human habitation sites, and certain woody ornamental applications are especially challenging, and there is an array of alternative delivery systems available for them.

With an IPM system in place, pesticides should be used only on an "as needed" basis. There are two parts to a spray decision that involves sprayer engineering: timing and active ingredient. Appropriate timing requires efficient, relatively high-speed equipment to cover the necessary crop area within the narrow window of time before economic loss occurs. The characteristics of certain active ingredients also have influenced pesticide delivery apparatus design. For example, plant disease managers have not had the postinfection activity needed to eradicate or to arrest development of infectious diseases until very recently. Therefore, protectant residues of antimicrobial pesticides have been necessary in many plant protection situations. Delivery of these sprays with sufficient coverage for control has required dilute sprays with very high-volumes sometimes exceeding 428 gallons/a. Development of ultra-low-volume (ULV) sprayers and an array of new fungicides and bactericides has diminished the need for high-volume sprayers in many pest management systems.

The mode of action and the environmental attenuation characteristics of the active ingredient as well as the behavior of the target pest also may dictate the type of equipment, volume of water, and frequency of treatment necessary for control (MSU 2002; NRC 2000). Surfactants (spreader stickers) and ultraviolet light protectants have increased the efficacy of a number of active ingredients. Yet the most significant development has been the engineering of spray equipment that can deliver pesticides over large areas with good coverage, on an "as needed" basis. Spray drift remains the most intractable problem. Numerous mitigation inventions have been deployed, including covered and tunnel sprayers, droplets charged to adhere to leaf surfaces, boom delivery systems, horizontal flow supplemental air movement fans, and spray droplets with air bubbles inside to increase droplet size.

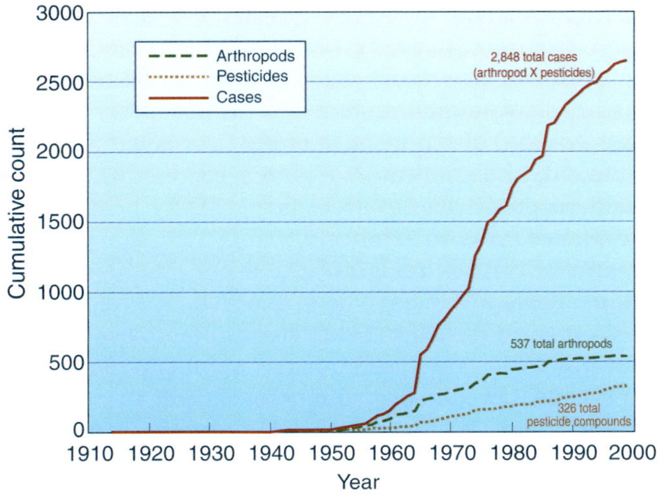

Figure 2.9. Timeline for the development of insect resistance in arthropods. (Source: M. Whalon, Michigan State University.)

Pesticide Resistance Management

The frequent occurrence of pesticide resistance in various pests emphasizes the importance of this phenomenon and the need for deployment of multiple-tactic IPM programs. More than 533 species of arthropods (Figure 2.9), 100 species of plant pathogens, and 200 biotypes of weeds (Figure 2.10) have developed resistance to pesticides. This resistance within a pest is an inherited genetic change allowing individuals in a population to survive exposure to

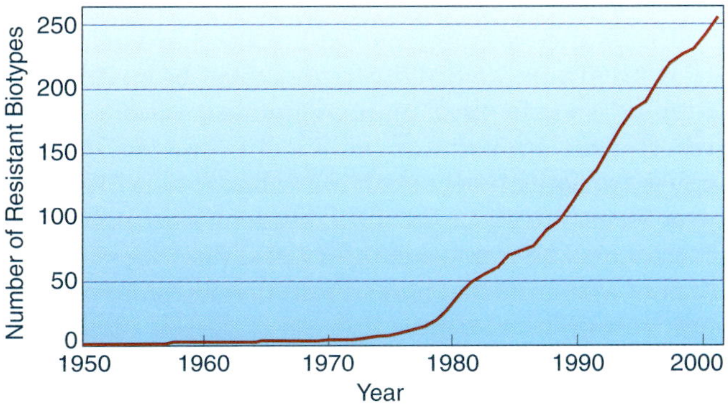

Figure 2.10. Worldwide development of herbicide-resistant weeds. (Source: I. Heap, <www.weedscience.com>)

pesticides. The evolution to resistance may require increased treatment rates or application frequencies, or a shift to other pest management strategies. Resistance development often is very disruptive of IPM systems and also may be harmful to ecosystem, environment, and applicators if increased amounts of pesticide inputs become necessary. Many pests have developed cross-resistance, whereby resistance to one pesticide confers resistance to other chemistries to which pests never have been exposed. Archetype resistant species such as the Colorado potato beetle may develop resistance mechanisms enabling populations to escape a broad spectrum of insecticides.

Resistance management (RM) is a set of strategies and tactics within IPM programs that seeks to ameliorate pest resistance. Today, most IPM programs include RM objectives for arthropod, pathogen, and weed pests. Resistance management may include: (1) decreasing pesticide-related selection pressure on pest populations; (2) monitoring for shifts in susceptibility of pest populations; (3) using alternative population-suppression mechanisms; (4) conserving susceptible genes in a pest population through refugia or enhanced immigration; and (5) developing models to assist in designing resistance management programs (Denholm and Rowland 1992).

Tactics for pesticide RM within IPM systems often include applying nonpersistent pesticides, rotating or sequencing pesticides with different modes of action, targeting a certain life-stage, and using different pesticides or IPM strategies for successive generations. Pest managers often alter the dosage applied to achieve high, moderate, or low dosage exposures. In other instances, a geographic mosaic may be achieved as a result of adjusting timing, application rates, and products used in fields or regions.

Resistance Management in an Integrated Pest Management Context

Resistance management adds another variable to IPM programs, so RM strategies must be customized to meet specific pest management situations. As a result, development and implementation of RM strategies are continually evolving information intensive processes. Resistance management strategies and tactics may be implemented by pest managers or producers with minimal investment in additional personnel, research, or time. But effective RM strategies usually require detailed knowledge of RM genetics, biologies of both pest and crop, and interactions among pest, crop, and environment. Pesticides and GMOs often become available and are used commercially before scientists can evaluate fully all the important factors potentially contributing to effective RM. Thus, most RM strategies evolve over time and, almost always, after resistance has developed. RM strategies for *Bt* crops, however, were developed in a proactive fashion before their commercialization and these strategies have been modified over time (EPA 2001). No *Bt* resistance has been detected in the field since *Bt* crops were commercialized.

Resistance management strategies frequently must be constructed from a combination of general principles, pest biology and ecology, population genetics, resistance management models, laboratory bioassays and field monitoring data, informal observations, and anecdotal experiences from IPM practitioners in the field. But there are usually a number of uncertainties in the available information that prevent a definitive, proscriptive RM strategy. Yet even in the absence of sophisticated RM understanding, practitioners often can delay resistance development by decreasing the number of unwarranted applications, rotating pesticides with different modes of action, and using pest preventative cropping practices such as sanitation, rotation, and early or late planting to avoid peak activity.

Factors contributing to the rapid development of resistance include repeated application of the same or related pesticides, multiple pest generations/yr, large pest populations present at treatment time, and lack of pest immigrants to dilute the genetics of pesticide-resistant individuals developing within treated areas (Roush 1989). In designing RM strategies that minimize selection pressure for resistance while maximizing the long-term benefits of available pesticides and GMOs for both farmer and pesticide/GMS producers, effective IPM programming will take into account all these factors.

The EPA and the Pest Management Regulatory Agency of Canada (PMRA) have developed and published guidelines for voluntary pesticide resistance management labeling based on rotation of mode of action for implementation in North America under the auspices of the North American Free Trade Agreement (NAFTA) (EPA and USDA 2001; NAFTA 2000). These voluntary labeling guidelines provide acceptable schemes of classification of pesticides according to their mode/target site of action and RM recommendations in the "Use Directions" section of the pesticide label. Implementation of these voluntary guidelines will help improve the likelihood of adoption by users of good RM as part of an IPM program and, potentially, reduce the likelihood of pest/pesticide resistance (Matten, S. R. 2003. Personal communication).

An illustration of the impact of differences in pest biology on resistance development is that of benzimidazole resistance in several apple pathogens. For example, the apple scab fungus developed resistance to benzimidazole fungicides within 3 yr after the fungicides were introduced. In contrast, the benzimidazoles still provide excellent control of flyspeck on apples after more than 25 yr. The difference is that *Venturia inaequalis*, the fungus causing apple scab, has more generations per year and often has higher populations within the orchard than *Zygophiala jamaicensis*, the fungus causing flyspeck, docs. Furthermore, *V. inaequalis* survives from year to year within the orchard whereas most of the inoculum for *Z. jamaicensis* originates from wild hosts in the orchard perimeter, where it is less subject to selection for resistance.

Resistance Management Tactics and Tools

Bacteria

The few products available to control bacteria limit RM strategies for bactericides. Tactics involve limiting the number of applications/season, optimizing the timing of applications for maximum benefit, and using host resistance or cultural management to minimize the size of populations that must be controlled with bactericides. Streptomycin-resistant strains of the fire blight pathogen, *Erwinia amylovora*, have been detected in most regions where streptomycin has been applied more than 2 to 3 times/season (Burr et al. 1993). Several disease models constructed over the past 15 yr allow apple and pear growers to optimize timing of their streptomycin sprays and thereby to increase the probability of satisfactory control of fire blight. At the same time, growers are advised to plant less susceptible varieties and to use various cultural practices to decrease host susceptibility and inoculum levels in their orchards.

Fungi

Fungal-resistance management strategies for fungicides include minimizing repeated exposure of fungal populations to the same fungicide (or fungicide class), using fungicides in prophylactic rather than in curative treatment timings, and using fungicides at rates sufficient to control the most resistant individuals in the wild-type population (Koeller 1996). The latter strategy is important because marginally resistant pathogens surviving low-rate applications can dominate fungal populations gradually and cause field failures of the fungicide. Curative treatments should be avoided because such treatments are applied only after a pathogen already is established in a crop, and often the fungal population is in the logarithmic phase of epidemic development. Selection pressure therefore is increased. Effectiveness of treatments, even in the absence of fungicide resistance, is improved when fungicides are timed to prevent initial infections.

Two strategies have evolved for avoiding repeated exposure of fungal populations to the same fungicide. According to one strategy, two or more fungicides with different modes of action are applied in a mixture. This strategy was used successfully to delay development of resistance to benzimidazole and de-methylation inhibitor (DMI) fungicides. According to the second strategy, applications of different fungicides are alternated. Both strategies are based on the idea that fungal populations will be slow to develop resistance to multiple fungicides applied at the same time or in close succession. Nevertheless, development of fungicidal as well as bactericidal resistance has resulted in significant control problems on numerous crops.

Genetically Modified Organisms

Genetically modified organisms contain genes from the same or from unrelated species introduced through biotechnology or genetic engineering. These organisms sometimes are called *transgenic*, a descriptor reflecting their heterogenetic origin. Arrays of GMOs have been or may be released into agriculture, forestry, and other settings. Those that suppress pests may select for resistant pest populations in the same way that chemical pesticides have selected for them. In 2001, more than 75 million a. of transgenic crops were planted, most of which were herbicide-tolerant crops such as corn, cotton, and soybean. Concerns about development of resistance to transgenic crops have focused on the deployment of cotton, corn, and potatoes containing *Bt* toxin (Gould 1998; McGaughey and Whalon 1992).

Insect-active *Bt*-toxin genes are derived from the soil bacterium *Bacillus thuringiensis*. Formulated into many different sprayable insecticides, the *Bt* microorganism represents one of the principal pesticides in organic and alternative crop production. Many consumers believe that *Bt* transgenic crops are a threat to the long-term viability of *Bt* because transgenic plants may select for insect resistance to *Bt* more quickly than would occur if the *Bt* organism were used only in spray applications.

Since the inception of *Bt* crops, the EPA has required or recommended insect resistance management (IRM) strategies for these systems. The EPA has determined the protection of the susceptibility of *Bt* is in the "public good" (EPA 1998, 2001; EPA and USDA 1999). The current set of IRM requirements is found in the EPA's 2001 reassessment of the risks and benefits of *Bt* crops document (EPA 2001). These plans require the planting of non-*Bt* refugia that will produce sufficient numbers of susceptible individuals to mate with any putative resistant individuals that emerge in the *Bt* fields, to dilute the effect of resistance should it arise in transgenic *Bt* crops. There are several important assumptions with the high dose/structure refugia strategy: (1) resistance to *Bt* will be conferred by a single recessive gene with two alleles resulting in three insect genotypes, (2) low initial frequency of resistance alleles (i.e., resistance will be rare), and (3) random mating between susceptible and resistant adults. These plans include the following elements:

1. a refugia/high dose strategy,
2. planting of non-*Bt* refugia greater than a certain minimum size to be determined by the effective dose and population genetics,
3. specified parameters of proximity of the refuge to the transgenic crop determined by the population biology and ecology (in particular, movement, mating, and ovipositional behavior),
4. an annual resistance monitoring program,
5. a plan for education and communication between user and seller,
6. a user compliance assurance program,
7. a remedial action plan to mitigate/eradicate resistance if it arises,
8. additional research,
9. annual reports to the EPA, and
10. additional modification of the RM plan, if necessary.

Although field resistance to *Bt* transgenic plants has not been a problem yet, insects certainly can develop resistance to *Bt* and to *Bt* crops (Tabashnik 1994). These RM plans therefore are warranted to protect a very valuable source of pest population suppression.

Insects and Mites

Deployment of synthetic organic insecticides and miticides in the 1940s was accompanied by development of broadscale resistance. In the 1970s, resistance was recognized as a major global challenge of modern pest management systems (Georghiou 1986). Today, few new insecticide or miticide products are introduced into the marketplace without label recommendations for RM.

Rate of resistance development and degree of RM difficulty are related to various features (Figure 2.9). Kennedy and Whalon (1995) have outlined many of these in terms of incentives and barriers to implementing RM programs in the field. Certain biological and ecological factors are beyond human control yet may have significant interactions with the costs and the operational aspects of RM programs or the environment.

Nematodes

Although populations of many nematode species have been reported as resistant to all nonfumigant nematicides (Viglierchio 1991), this is not a widespread problem. The very limited mobility of these soilborne pathogens restricts the dispersal of any population variants with nematicide resistance. A more frequent problem with fumigant and nonfumigant nematicides is accelerated microbial decomposition of the chemical in soils (Johnson 1998).

Weeds

More than 200 herbicide-resistant biotypes of weeds have appeared worldwide during the last four decades (Figure 2.10). Many herbicide-resistant weeds can be controlled today by means of transgenic systems in major crops such as corn, cotton, and soybean (Sherrick and Head 2000). For example, weed problems such as imidazolinone-resistant *Xanthium strumatium* and triazine-resistant *Chenopodium album* and *Amaranthus* species can be managed with HT crop varieties.

Increased selection pressure for resistant weeds when a given herbicide is applied for multiple years is an important consideration in the deployment of genetically engineered HT crops (Hess and Duke 2000). Beginning in 1997, glyphosate resistance has evolved in five weed species in seven locations worldwide: (1) goosegrass (*Eleusine indica*), Malaysia; (2) horseweed (*Conyza canadensis*), United States); (3) Italian ryegrass (*Lolium multiforum*), Chile; (4) rigid ryegrass (*Lolium rigidum*), Australia, United States, and South Africa; and (5) annual ryegrass (*Lolium multiflorum*) (Carpenter et al. 2002). Although the development of resistant weeds depends on the frequency of resistance in the general population, the potential for selection of resistant weeds is limited by the practice of mixing or rotating herbicides with different modes of action over time (Hess and Duke 2000). A key concern with HT crops is potential gene flow from resistant crop to susceptible weeds. Where crops and weeds are interfertile, risks for outcrossing usually are minimized by targeting the HT trait to the chloroplast (thus, HT becomes a maternally inherited trait and is not moved with pollen). Another approach for preventing movement of HT to related weeds is limiting the use of the trait to self-pollinating crops (Hess and Duke 2000).

In conclusion, any population suppression strategy, tactic, or tool may select for resistant individuals in the population. This is true even for relatively benign insecticides such as *Bt* and other relatively new, environmentally friendly, naturally derived, or even organic pesticides. Most newer pesticides control a limited number of pests by carefully targeting inhibition of specific biochemical pathways in the pest. But these target site inhibitors often are more susceptible to resistance problems than older, broader spectrum pesticides are.

Integrated pest management should continue as one of the principal strategies for preventing resistance development because IPM programs use diverse tactics to suppress populations. Pests have difficulty mounting an evolutionary response to multiple IPM tactics used concurrently or in close succession. Development of multiple resistance and cross-resistance in pest populations is a threat for nearly all pesticides. Thus, RM will continue to influence the way in which pesticides are used. Currently, there is much evidence for the success of IRM programs for *Bt* crops. The success of most RM programs for conventional pesticides, however, is anecdotal, but they will expand in importance as society and regulators recognize the threat posed by pest resistance to pesticides. Concerns about RM ultimately may generate a requirement that RM plans be submitted with other materials required for pesticide registration in a process similar to that already required by the EPA for registration of *Bt*-containing GMOs. As noted earlier in this section, the United States (PR Notice 2001-5) and Canada (DIR 99-06) have developed voluntary resistance management labeling guidelines based on rotation of mode of action. As implementation of this voluntary initiative occurs, more and more pesticide labels will include statements focusing on resistance management. Several new labels in both countries, especially for those undergoing a joint review process, are beginning to include statements focusing on RM, such as seasonal and temporal application sequences.

Certification and Regulation

IPM Certification

A certified IPM product has been awarded a label providing consumers information about the process by which that product was produced (Consumers Union 2002; IPMI 2002b). A certification in the form of a label is a process label, not an outcome label. The label "IPM" on a product tells the consumer that it was grown using IPM practices, whereas the label "organic" is used for products grown using few or no synthetic chemical pesticides. Both labels imply fewer and less damaging environmental effects than are associated with conventionally produced goods. Although this may be true, many consumers generally perceive certification such as *organic* and *IPM* to indicate that the product is safer to consume. By definition, this is not necessarily true with process labels. Goods produced with IPM tactics and organically grown foods can be just as harmful or harmless as conventional products because in IPM systems, chemical pesticides still are used in the production process and certain organically approved pesticides include potentially hazardous substances. Labels can be powerful marketing tools, however, and certain labels provide IPM growers with a premium over conventional growers. A few labels attempt to strengthen the local agricultural economy by focusing on local producers whereas other labels may convey the message that certain presumably more environmentally friendly production processes were used. Some examples follow.

Rainforest Alliance's ECO-O.K. Label

The ECO-O.K. agricultural certification program is managed jointly by the Rainforest Alliance and a network of Latin American partner organizations, under the umbrella of the Conservation Agricultural Network (Rainforest Alliance 2002). Each partner is a local conservation group committed to community-based conservation initiatives and research. The group has two primary labeling initiatives and is in the process of developing several more.

The Better Banana Project

The banana industry is an essential part of many Latin American and Caribbean economies. The Conservation Agricultural Network initiated the Better Banana Project in 1991 in an effort to diminish the negative environmental impacts of banana cultivation and to improve working conditions through certification of well-managed banana farms. Farms that comply with a comprehensive set of environmental and social guidelines are awarded the Better Banana seal of approval. Farms enrolled in the Better Banana Project are pioneers in the industry. These farms do not contribute to deforestation, protect the safety of workers, have minimized the use of agrochemicals, and are working to decrease soil erosion.

Since 1991, the Conservation Agricultural Network has succeeded in significantly altering the nature of banana farming in Costa Rica and is expanding its reach to other crops and countries rapidly. Enrollment in the program has decreased pollution of surface and groundwater significantly. More than 20% of banana production in Costa Rica and 50% in Panama have been awarded certification. Farm evaluations are under way in Ecuador and Colombia, and banana growers in Honduras and Guatemala are making necessary improvements to achieve certification in the near future.

The Eco-O.K. Conservation Coffee Project

Coffee is the most important export crop in the northern region of South America, both in terms of area cultivated and economic value. Coffee is of special interest to conservationists because it covers much of the region's richest ecological zones. The Conservation Agriculture Network (CAN) developed the Conservation Coffee Project to provide market incentives for the conservation of forested coffee farms. Farms meeting CAN'S comprehensive agricultural standards receive the ECO-O.K. seal of approval to distinguish their products in the marketplace. This program provides farmers with incentives to maintain their shade-grown coffee farms, thus preserving an important habitat and encouraging coffee farming that protects worker health and safety. (Additional certification guidelines from the Rainforest Alliance exist for orange juice and cocoa.)

Mothers and Others for a Livable Planet Core Value

The process of creating a meaningful certification system for CORE Values Northeast (CVN) began in 1995 (IPMI 2002a). The CORE Values Northeast Grower Committee decided that the central element of the certification program should be a farm plan. This plan would be a dynamic document intended to reflect growers' "effective" knowledge, that is, their abilities to make environmentally sound decisions given the conditions present at any time on their farms. The purposes of the Farm Plan are as follows:

- to determine if a grower is using an IPM approach to his/her farm management.
- to serve as a basis for information exchange among members of the CVN grower community and between the CVN growers and the research community, and
- to function as a tool in problem identification and to provide an innovative approach to problem solving.

This model ecolabeling program brings together Northeast farmers and consumers as partners in defining *ecologically responsible agriculture* through the accreditation of regional farmers using biointensive IPM production methods. Growers submit an annual farm plan and have an annual on-site inspection. The primary goal is to build public awareness of and demand for local, ecologically grown apples through consumer-oriented media and market-based educational strategies. This ecolabel is a point-of-purchase education tool providing information to consumers about the agricultural and environmental improvements being implemented by growers in the region. In 2001, Core Values became independent of Mothers and Others and is now led by an executive committee, with certification services provided by the IPM Institute (IPMI 2002b).

Partners with Nature

Partners with Nature was launched in 1993 and is an IPM certification program recognizing Massachusetts growers who have taken special steps in practicing IPM (University of Massachusetts 2000). It is a collaborative initiative between the Massachusetts Department of Food and Agriculture, the University of Massachusetts Extension Program, and the USDA Consolidated Farm Service Agency (CFSA). In 1997, Partners with Nature certified sweet corn, strawberries, potatoes, tomatoes, peppers, winter squash, and cole crops. In 1996, 50 farms across Massachusetts were IPM certified through the Partners with Nature program (University of Massachusetts 2000).

Stemilt Responsible Choice

In Europe, greater than 40% of apples and pears are produced under formal Integrated Fruit Production (IFP) systems (Stemilt Management 2002). Production standards are very detailed, consumers recognize the IFP logos, IFP fruit is placed preferentially by supermarkets, and growers may get a premium price. In the northwestern United States, the Responsible Choice program of Stemilt Growers, Inc. and the Hood River Grower/Shipper IFP Program are modeled after the European approach. Using IFP, Stemilt addresses all aspects of growing, harvesting, storing, and packaging fruit. With each stage of production, Stemilt has developed a program based on "Responsible Choice" and provides only fruit grown and packaged in an environmentally sound program.

The Food Alliance

The Food Alliance is a coalition of farmers, consumers, scientists, grocers, processors, distributors, farm worker representatives, and environmentalists working together to ensure that future generations have "Good Food for a Healthy Future" (Food Alliance 2002). The Food Alliance is an independent third party endorsing farms that meet strict requirements. A primary goal is to improve communication and understanding between consumers and producers, especially by means of market-based incentives.

The Cornell/Wegmans IPM Label

Wegmans Food Markets first approached Cornell University in 1994, seeking the means to offer its consumers IPM-grown sweet corn (Haining Cowles 1999). Customers were positive about the opportunity to choose IPM-grown, fresh-market sweet corn, and Wegmans decided to expand to IPM- grown canned and frozen vegetables in 1996. Comstock Michigan Fruit, Wegmans' supplier of processed fruits and vegetables, found that ten of its growers had already adopted most of the available IPM methods; thus, a partnership was founded. The Wegmans model incorporates a set of required IPM practices, developed by land-grant IPM experts with input from growers, private IPM experts, and others. Retailers identify eligible products with logos or brands licensed from the university and motivate consumer preference for IPM-identified products using educational/promotional materials provided by or developed with the land-grant university. Grower compliance is verified by third-party certifiers contracting with producers or retailers. Wegmans has an incentive for participating and investing in IPM education: its private-label products, labeled with the IPM logo, are value-added products that compete with enhanced effectiveness against national brands (Haining Cowles 1999). Wegmans increased consumer recognition of IPM from less than 11% to more than 19% in its first season, offering sweet corn identified as "IPM-produced" in stores, point-of-purchase brochures and displays, radio ads, and newspaper inserts (Green, T. 1999. Personal communication).

In conclusion, it has been argued that labeling that has involved the industry has contributed not only to consumer choices but, through associated environmentally friendly products, has served as a mechanism strengthening a particular market. (See Benbrook [2002] for an overview of activities creating IPM awareness among consumers, and VanKirk and Garling [2002] on the benefits and challenges to land-grant involvement in IPM labeling.) Critics of such practices have pointed out that companies make their own choices regarding what to include on labels without outside oversight (exceptions in the United States include the tobacco and alcohol industries), and that often "green" (or IPM) claims are exaggerated or even unfounded (Bruno 1992).[1] Ecolabeling through third-party oversight, on the other hand, generally is granted by an organization to applicants based on producers' voluntary compliance with standards established by groups outside the production process, such as environmental and consumer groups. This type of labeling, however, can give consumers the perception that something is "wrong" or at least "not right" with nonlabeled products.

Ecolabels in an International Context

The IPM label is an ecolabel much like the organic label or any other process label. Ecolabels recently have become a topic of discussion among the international organizations concerned with trade (Barham 1997). At the international level, ecolabels are associated most often with life-cycle assessments (LCA) and processing and production methods (PPM), which are evaluation criteria by which the label is granted and are generally the basis for government regulations. Such regulation for environmental, safety, and health purposes is permitted under multilateral trade

[1]For example, in 1996, 15 nongovernmental organizations formed the "Campaign to Defend EcoLabeling," which was led by Green Seal and the Institute for Agriculture and Trade Policy (IATP 1999), among others, in response to efforts by industry to put a stop to third-party labeling by working through the U.S. Trade Representative's Office. The campaign was led by the Grocery Manufacturers Association and Procter and Gamble Co., which sought to affirm industry control over labeling and asserted that third-party labeling constituted a violation of the principles of free trade. Although this issue has been settled, the two parties continue to organize (Barham 1997).

rules and typically does not cause problems for international trade when labels are specified for domestic use only. Thus, the last negotiations of the General Agreements on Tariffs and Trade in the Uruguay Round acknowledged the right of individual countries to set their standards within their own boundaries. What IPM is in the United States does not necessarily constitute IPM in other countries. Aside from the definitional difference, domestic PPM standards might cause problems for other countries should they be perceived as imposing trade restrictions on imports from other countries (e.g., the dolphin-tuna controversy). Countries with notably greater economic power could set the standards at home that less developed countries cannot meet, in effect creating nontariff barriers. How such schemes will affect international trade is being considered by several international governmental agencies.

Because individual PPM regulations have caused problems in international trade, specifically in terms of competitiveness, it has been suggested that the LCA (an alternative to ecolabeling, dealing with the entire life span of a product and the effects on the environment) should perhaps be broken into two phases, from the single "cradle to grave" approach to the two-pronged "cradle to export border"and "import border to grave" approach. This approach would allow both importing and exporting countries to set their own standards (Barham 1997).[2]

IPM Regulation

Because IPM has always been a voluntary endeavor, IPM regulation is a relatively new concept. In light of the federal government's increasing concern regarding pesticide residues on food and increased health awareness, and in light of the Food Quality Protection Act of 1996, IPM regulation has become a reality (See Chapter 14). Additionally, concerns about the environment encouraged the Clinton Administration to create a 75% goal for IPM participation. Thus, IPM regulation encompasses two important pieces of legislation.

The National Plan for the Development and Implementation of Integrated Pest Management Systems, developed by the USDA, land-grant universities, and other organizations, encompasses the goal to implement IPM methods on 75% of the nation's cropland by the year 2000 (GAO 2001; See Chapter 14). This goal, announced jointly by the USDA, the EPA, and the FDA in September 1993, represents a commitment by these government agencies to work with the states as well as with private sector partners to develop, and to help farmers implement, ecologically based pest management approaches that rely "less on chemical pest controls, are more sustainable, are equally efficient economically and still provide Americans with an economical, safe and plentiful food supply." The National Plan had four primary objectives: (1) to coordinate with land-grant universities and with public and private organizations at the state, regional, and national levels to achieve the 75% IPM implementation goal; (2) to develop a process for identifying grower needs and providing them support; (3) to develop methods and conduct programs to measure progress toward the 75% goal accurately and to assess the economic, environmental, public health, and societal effects of IPM implementation; and (4) to implement a communication and information exchange program that involves IPM stakeholders and enhances their understanding of the objectives of IPM and its related activities, as well as the progress, effects, and benefits of IPM (See Chapter 1).

There are actual performance goals for measuring the success of IPM in general and for producers to use, as well as for policymakers to consult when determining whether the goals of the legislation have been accomplished (See Chapter 14). Additionally, in order to qualify as an IPM producer, the grower must be using three or more of the performance goals outlined in what is called PAMS: Prevention, Avoidance, Monitoring and Suppression. Prevention, the practice of keeping pests from attacking a crop, includes tactics such as using pest-free seeds, irrigation scheduling to avoid disease development, and field sanitation procedures. Avoidance generally is practiced when pests are already present but potential damage to the crop can be avoided through cultural practices such as rotating crops, choosing cultivars with genetic resistance to the pest, and using trap crops and pheromone traps. Monitoring and appropriate identification of pests should be practices used as a basis for suppression activities. Examples of monitoring include scouting, soil testing, and weather monitoring. Suppression becomes necessary when pests have reached levels at which prevention and avoidance tactics have not been successful and economic loss needs to be avoided. Suppression practices are comprised of cultural, physical, and biological methods. Suppression tactics include cover crops and mulches, alternative tillage methods, cultivation and mowing for weed control, pheromone traps, and temperature

[2]As a methodology for granting labels, the LCA faces many obstacles in implementation, including high cost or unavailability of data and differing national methods of reporting data.

management. Biological control tactics such as mating disruption for insects should be considered before chemical alternatives are used. But where these options fail, carefully selected and targeted pesticide spraying may be applied only when necessary (See Chapter 1).

The other significant legislation affecting IPM is the Food Quality Protection Act. The USDA–Cooperative State Research, Education, and Extension Service (CSREES) Integrated Pest Management Programs were established as part of the FQPA implementation (See Chapter 14). This legislation consists of six components, most of which extend responsibility for implementation and success to state agencies. The following provides a short overview of the programs.

1. **Pest Management Alternatives Program**. The Pest Management Alternatives Program (PMAP) was established in 1996 to develop replacement tactics and technologies for pesticides under consideration for delisting by the EPA, for which effective alternatives are unavailable. This program is funded for short-term projects, with the objective of adaptive research and implementation of IPM practices. The focus is the replacement, on a single crop basis, not on a cropping system basis, of individual tactics in a pest management program.
2. **Extension IPM Implementation Program**. The Extension IPM Implementation Program (EIPM) supports applied research to disperse information about new technologies from researchers to producers as quickly as possible. A base program in each state assists in accelerating the transfer of proven IPM technologies for implementation as growers look for new approaches to managing pests, especially after key pesticides have been banned due to the FQPA or are lost due to resistance.
3. **Regional Integrated Pest Management Grants Program**. The Regional Integrated Pest Management Grants Program (RIPM) supports research of new pest management methods and the validation of these methods in the field, as well as the delivery of information to growers and other pest managers through education and training programs. Strong partnerships are key to the success of this program.
4. **National Agricultural Pesticide Impact Assessment Program (NAPIAP or PIAP)**. This program initially supported the assessment of pesticides and the effect of regulation of pest management options, encompassing education and research in cooperation with the USDA Office of Pest Management Policy, land-grant universities, producers, and the EPA. The provision of science-based data relating to the FQPA, pest management issues, and minor crop needs are the current focus of this program. The PIAP funds now are allocated to the four Regional IPM Centers (see Zurek 2002a, for information on these Centers: Western Region, North Central Region, Southern Region, Northeastern Region). Fulfilling the requirements of the FQPA is given a high priority.

The Pest Management Information Decision Support System

This program was developed to provide the foundation for integrated databases needed for the decision making process for questions regarding policy, research priorities and educational inquiries. The early prototype version could access the EPA Registration Data Base, the IR–4 Minor and Specialty Crop Data Base, the CRIS Database, and the National Center for Food and Agricultural Policy (NCFAP) database on pesticide use. This web-based program now is managed by the National Science Foundation (NSF) Pest Management Center (NSF 2002) and is interfaced with the four Regional Pest Management Centers (Zurek 2002a). The PMIDSS serves as the decision support system for the four regional pest management centers (and others) as they work to improved IPM Strategies and systems and to meet the FQPA requirements (Stinner 2002; Zurek 2002b).

Interregional Research Project Number 4 (IR–4)

The IR–4 project provides cooperation, grants, and scientific direction for field and laboratory research with the objective of developing data primarily on minor crops supporting registration packages to be submitted to the EPA.

Effects of Invasive Pests

Invasive pests are moving into all habitats—from bodies of water, to natural areas, forests, rangelands, pastures, and croplands (see Chapters 4 to 7). In the United States more than 2,000 nonnative plant species and a like number of nonnative insects and arachnids have been introduced from many countries (CNIE 1999a). These introductions of exotic plant pests, in addition to disrupting ongoing pest management programs, have resulted in devastating losses in crop productivity and altered natural landscape inhabitants. Invaders include weeds, a wide range of insects, viruses, bacteria, fungi, nematodes, reptiles, and mammals. For example, more than half of economically important arthropod pests in the United States have been introduced from other countries. A recent report indicates that an estimated 30,000 species of exotic organisms now exist in the United States (CNIE 1999a). Many fungi, bacteria, viruses, and nematodes have been introduced, but their origins often are unknown.

Annual costs due to and damages from nonnative pest species have been estimated at $123 billion/yr (CNIE 1999a). These heavy losses are due, in part, to the fact that natural antagonists and parasites of introduced crop pests often do not accompany pests to their new locations. This situation results in very rapid build-up and spread of newly introduced pests. Introduced pests are not limited to those organisms affecting various crop species. For example, the fire ants (*Solenopsis invicta*) now have become permanent residents of a wide range of states and counties in the southeastern United States. Estimated damage by the rapidly spreading fire ants to U.S. livestock, wildlife, and public health amounts to approximately $2 billion/yr (CNIE 1999b) (See Textbox 2.3).

Textbox 2.3. Fire ants and IPM (Courtesy of Ann Sorensen and Esther Day)

The imported fire ants (*Solenopsis invicta Buren* and *Solenopsis richteri* Forel) were accidentally introduced from South America into the United States in the 1930s. Since then, they have rapidly spread across the South and the West and established themselves in 14 states: Alabama, Arkansas, California, Florida, Georgia, Louisiana, Mississippi, New Mexico, North Carolina, Oklahoma, Puerto Rico, South Carolina, Tennessee, and Texas (Brenner 1999; Schmidt 1995; Weaver-Missick and Lee 1999). Fire ants have a great impact, especially on ornamental plants, sod, and landscaping industries because of problems associated with shipping infested plants into uninfested areas of the country. However, impacts on wildlife such as the northern Bobwhite quail in Texas (Allen, Lutz, and Demarais, 1995), other native ant species (Porter and Savignano 1990), and endangered species such as gopher tortoises and sea turtles (Weaver-Missick and Lee 1999) can also be significant. In addition, they can harm or even kill newborn calves and other vulnerable farm animals. Fire ants repeatedly sting potential prey, delivering a potent toxin that burns and leaves an annoying pustule on the skin. Although fire ants suffer from a limited gene pool, the absence of predators, their preference for disturbed areas, and their defensive and reproductive capabilities seem to have compensated for loss of genetic diversity (Schmidt 1995).

Fire ants have been a challenge to control through IPM, partly because they are so prolific and well-adapted and partly because they can sometimes be beneficial. Mirex initially was effective in controlling imported fire ants but also killed native ants; the material was banned in the early 1980s. They are highly visible in the South due to the number of mounds in yards and other public areas as well as invasions of hospitals, schools, and electrical boxes. This ant has no natural predators in the United States (Weaver-Missick and Lee 1999), and it is estimated that their population densities are about five times those in their native South America habitat (Brenner 1999). However, in certain situations, they may actually contribute to IPM programs. Fire ants feed primarily on other insects and related arthropods, which reduces the need for insecticides in commercial agriculture. For example, in cotton fields fire ants help control boll weevils, fleahoppers, cotton bollworms, pink bollworms, and tobacco budworms (Drees and Vinson 1998). In current IPM programs, control is achieved primarily through chemicals applied directy into the mound (drenches or insecticidal baits mixed with ant attractants). Nevertheless, the principle of establishing thresholds, or leaving some mounds to keep more ants from invading, prevails. In this way, IPM programs to control fire ants have raised public awareness of IPM.

Biological approaches to control the red fire ant are now being tested. In the past 2 years, USDA-ARS researchers have developed new tactics to combat the ants: a slow-acting disease and a decapitating fly. *Thelohania solenopsae*, a microorganism from South America, infects the ants and chronically weakens them. The pathogen ultimately infects all workers, eventually killing the colony. The other line of defense is the phorid fly (*Pseudacteon tricuspus*). These flies hover over the mound, then zoom in to puncture the ant's outer cuticle to deposit an egg underneath. The egg quickly hatches into a larva, which then moves into the ant's head. Once the larva is mature, it releases an enzyme causing the head to fall off. Because the fly can lay more than 100 eggs, it can make multiple attacks (Brenner 1999; Weaver-Missick and Lee 1999; Williams 1997). If these biological agents can be added to the existing IPM Tool Box, control efforts for this invasive pest will be enhanced.

Although damaging exotic pests occur throughout the United States, certain areas are more susceptible to invasion than others (CNIE 1999a). Specific factors favoring invasion and establishment of invasive pests include a mild climate, geographic isolation, high exposure or invasive rate, and natural-landscape disturbance. Thus, islands and long-isolated areas such as Hawaii and much of Florida have huge numbers of introduced species. Weeds, especially, have evolved to exploit these niches, often hitchhiking as contaminants in soil or plant materials and becoming established in freshly disturbed areas (CNIE 1999a).

Means of Spread of Invasive Plant Pests

The movement of most pests is passive, but they often can be distributed over hundreds or thousands of miles and become established in large numbers in brief periods. Introductions of nonnative pest species may be intentional or unintentional (CNIE 1999a). Kudzu, for instance, was brought to the United States to control soil erosion; other weeds were brought originally as ornamentals. In contrast, the gypsy moth was released accidentally in Massachusetts in the 1860s (Zimmerman 2000). Introduced crop and ornamental plants and seeds may contain viruses, bacteria, mycoplasmas, weed seeds, insects, and/or nematodes. Plant products such as animal feed or wood products also may serve as vehicles for dispersing invasive pests.

Viruses and mycoplasmas often are distributed over wide ranges by their insect vectors, in addition to being dispersed in infected plants and plant parts. This process may cause local epidemics as well as the long-distance movement of important plant pathogens. Long-distance dispersal and introduction of parasites by insects may involve movement of the latter by wind currents. Certain plant viruses also may be dispersed by windborne, virus-infected weed seeds such as those of dandelion.

Wind currents also effect extensive movement and facilitate establishment of numerous fungal pathogens, especially foliar pathogens, within regions as well as across countries. Somewhat surprisingly, air currents can move soilborne bacterial pathogens of potato such as the *Erwinia* species over hundreds of miles (Franc, Harrison, and Powelson 1985). A range of disease management models factor in the role of the wind and dispersal in short- to long-distance movement of pathogen spores. Invasive pathogens, facilitated by wind movement, include the potato late blight organism *Phytophthora infestans*, the tobacco blue mold fungus (*Peronspora tabacini*), coffee rust caused by *Hemileia vastatrix*, and many others (Main et al. 2001).

Examples of Current Invasive Plant Pest Problems

As suggested in a great number of publications and related web sites, invasive plant pests or weeds pose major threats to most U.S. agroecosystems and natural habitats (CAST 2000, 2002). The web site "Weeds gone wild: Alien plant invaders of natural areas" (Kwong 2000) provides a comprehensive listing of invasive species of natural areas in the United States. More than 500 species now are included in that list. The site also provides the names of management experts and other individuals and organizations that can provide assistance in the United States and worldwide. The federal weeds database focuses on weeds placed under the Federal Noxious Weed Act of 1975. Approximately 88 weed species have been added to this list since 1976. The magnitude of the threat of invasive weeds from various habitats is reflected in Chapters 4 (Forest IPM), 5, 6, and 7 in this report. Invasive weed pests are a serious problem in the United States, especially in western rangeland (CNIE 1999b), where leafy spurge (*Euphorbia esula*) often crowds out other vegetation in open areas of pasture or rangeland. When leafy spurge population levels reach 10 to 20%, cattle refuse to graze, resulting in major economic losses in the range of hundreds of millions of dollars. Introduced invasive plant species have been estimated to comprise as much as 47% of the total flora in most U.S. states, and certain publications indicate that 4,000 species of exotic plants now occur in this country (Kwong 2000). Estimated losses and costs associated with nonnative plants were as great as $20 billion/yr during the 1990s, a 300% increase over the previous four decades.

The witchweed parasite (*Striga asiatica*) of corn, sorghum, sugarcane, and other members of the grass family is an example of a major success story in the management of introduced weeds in the United States. This parasitic weed first was identified on North Carolina corn plants by plant pathology graduate student Akhtar Husain, a Ph.D. candidate at North Carolina State University in the mid-1950s. One witchweed plant may produce 500,000 seeds, which can remain viable in soil for as long as 15 yr (USDA–APHIS 1998). This weed was clearly a threat to the U.S. corn industry. Through an intensive eradication program initiated by APHIS in 1974, infested acreage has been decreased from 450,000 to less than

28,000 in 1995. Before most infested land was freed from this weed, farming activities such as soil sampling were restricted greatly by quarantine measures. North Carolina now manages the remaining eradication program, while APHIS participates in activities in South Carolina.

As has been indicated, greater than half of economically important arthropod pests in the United States have been introduced. An insect central to IPM in this country is the cotton boll weevil (*Anthonomus grandis*). This pest, introduced from Mexico in the late 1890s, exemplifies the importance that an introduced pest can have in IPM, including economic losses. Estimated amount of insecticides used to control this insect in 1973 amounted to approximately one-third of the total insecticide applied to agricultural crops in the country (USDA–APHIS 1998). Through an intensive long-term eradication program, weevil-free cotton acreage has increased from 35,000 in 1983 to 4,600,000 in 1998. The economic benefit/cost ratio for this program has been estimated at 12:1 on average for the United States and as great as 41:1 in certain regions of the country (USDA–APHIS 1998).

Although not always documented clearly, plant pathogens have affected and continue to affect IPM programs. Containment of the introduced potato golden nematode (*Globodera rostochiensis*) essentially to a few counties in New York State represents another success story for quarantine programs. But even though targeted by a federal quarantine program, the soybean cyst nematode (*Heterodera glycines*) invaded most soybean production areas in the United States. Currently, the research and extension efforts of nematologists and soybean breeders tend to focus on development of improved tactics and strategies for managing this genetically diverse pathogen.

Certain introduced pests overwhelm the capacities of humans to manage them. For instance, chestnut blight, caused by the fungus *Cryphonectria parasitica*, was discovered in the United States in 1904 (Schumann 1991; Vitousek, Loope, and D'Antonio 1994). By the 1930s, this disease had destroyed most of the chestnut population. Eventually the fungus eliminated the American chestnut, which previously had been the most abundant tree in the eastern United States (Vitousek, Loope, and D'Antonio 1994). Dutch-elm disease, induced by *Ophiostoma ulmni*, resulted in similar devastation of these stately trees in many U.S. cities (Schumann 1991).

In addition to problems resulting from the introduction of alien pests to given habitats or regions, major pest management problems are encountered as a result of the development or the introduction of pesticide-resistant and/or resistant variety attacking strains of pests. For example, aggressive strains of the fungus *Phytophthora infestans* have been dispersed over most of the potato- and tomato-producing regions of the world (Rubin, Baider, and Cohen 2001). Although new invasive pests may be the same species as those already present or even indigenous to an area, they possess new features such as resistance to pesticides, or even the capacity to transmit plant pathogens such as virus vectors or to attack a wider range of crops. The recent introduction of new biotypes or species of *Bemisia* white fly in the southwestern United States exemplifies the challenges that these altered plant pests pose to IPM programs. The white fly crisis in Arizona cotton, which involves the silverleaf white fly (*Bemisia argentifolii*), resulted in four to six times more insecticide applications than in years when insecticide efficiency against these pests was not negated by resistant strains (Dennehy 2000).

Continual use of given pest-resistant cultivars often results in the appearance of "resistance-breaking" populations of crop pests. This problem, which involves pests such as root-knot and cyst nematodes, can be averted or minimized by deployment of an appropriate cropping system that includes nonhost crops as well as susceptible and resistant varieties.

Reemergence of Old Plant Pests

Subtle shifts in production environments, including differences in crop varieties that minimize germplasm diversity or environmental factors, may bring about resurgence of specific plant pests. A classic example of this problem was the national corn leaf blight epidemic (caused by *Cochliobolus heterostrophus* [*Bipolaris maydis*]), which resulted from the widespread use of Texas male sterile cytoplasm in most corn varieties in 1970. Corn yield losses for that year were valued at more than $1 billion (Schumann 1991). Genetic resources were available, however, to solve that problem by the following year, but the amount of resistant seed was inadequate for all farmers. Recently, the U.S. wheat crop has encountered significant yield losses resulting from Fusarium head blight (caused by *Fusarium graminearum*). A relatively new disease of U.S. wheat, karnal bunt (*Tilletia indica*), also has received much attention.

New and widespread pest problems likely will be associated with the predicted global warming in the United States, with pests normally limited to temperate and to tropical regions invading the cooler regions of the country. Vitousek, Loope, and D'Antonio (1994) concluded that humans distribute so many organisms from their native ranges to places worldwide that this dispersal eventually will result in loss of the regional distinctiveness of the Earth's plants and animals. Decreased biodiversity can increase difficulties encountered in IPM by effecting less competition among given pests and their natural enemies (Altieri 1994).

Global trade, regrettably, may compound bioinvasion problems as pest species accompany shipping materials or actual products. Because of host pest interactions, however, climate change can have positive, negative, or neutral impacts on *pathosystems* (crop[s], associated pathogens, and their environment, which often includes humans) (Coakley, Scherm, and Chakraborty 1999). Clearly, the combination of new pest invasions, the recurrence of outbreaks caused by "old" pests, the appearance of pest strains resistant to available pesticides, and the development of pest populations overcoming host resistance will continue to pose challenges to IPM practitioners.

As reflected in a presidental executive order (Clinton 1999), invasive plants and animals in the United States are receiving increasing attention and research. This order includes definitions of key terms, related duties of federal agencies, a directive to establish an Invasive Species Council, an outline of the council's duties, and development and needed updates of an Invasive Species Management Plan. Additional financial resources were requested to support this endeavor as well as actually to control invasive plants and animals. Characterizations of invasive species in this report and elsewhere emphasize the magnitude and growing importance of these pests in the United States, as well as the challenges they represent for IPM. Many new IPM tools and innovative technologies (Henkens et al. 2000; Lawler and Harmey 1993; Louws, Rademaker, and deBruijn 1999; Stewart 2000) offer promise for keeping invasives under control.

Integrated Pest Management and Agricultural Bioterrorism

Bearing in mind the unprecedented "September 11th" tragedies in the United States, IPM specialists now must consider the tools and strategies needed to address potential agricultural bioterrorism. In addition to the necessity of having an abundant supply of high-quality, safe food, the value of the products generated by the agricultural industry ($1.5 trillion in 1998) and the number of people employed in agriculture (25 million) attest to the importance of this new issue (Cereijo 2001). With increasing concerns about threats to the U.S. food supply and the agricultural industry, "The Agricultural Bioterrorism Countermeasures Act of 2001" was introduced in the U. S. Senate in October 2001 (Hutchinson 2001). Examples of threats include mad cow disease and the foot and mouth disease, which already have been experienced in Europe. Approved U.S. legislation (CID 2002; Hawks 2002) outlined the purpose and approaches to addressing the bioterrorism threat. Many of the tools and strategies that may be essential as countermeasures to the threats of agricultural bioterrorism are yet to be developed.

Appendix A. Abbreviations and Acronyms

a.	acre
APHIS	Animal and Plant Health Inspection Service
Bt	*Bacillus thuringiensis*
bu	bushel
CIAT	International Center for Tropical Agriculture
CIMMYT	International Maize and Wheat Improvement Center
CIP	International Potato Center
CNIE	Committee for the National Institute for the Environment
CRIS	Current Research Information System
CSREES	Cooperative State Research, Education, and Extension Service
D-D	dichloropropane-dichloropropene mixture
DDT	dichloro-diphenyl-trichloroethane
DMI	de-methylation inhibitor
EIPM	Extension IPM Implementation Program
ELISA	enzyme-linked immunosorbent assay
EPA	U. S. Environmental Protection Agency
FAO	Food and Agriculture Organization
FDA	Food and Drug Administration
FQPA	Food Quality Protection Act
GDD	growing degree days
GIS	geographic information system
GMO	genetically modified organism
GPS	global positioning system

HT	herbicide-tolerant
ICARDA	International Center for Agricultural Research in Dryland Areas
IFP	integrated fruit production
IPM	integrated pest management
IPMI	IPM Institute
IR–4	Interregional Research Project Number 4
IRM	insect resistance management
IRRI	International Rice Research Institute
LCA	life-cycle assessments
NAPIAP	National Agricultural Pesticide Impact Assessment Program
NCFAP	National Center for Food and Agricultural Policy
NSF	National Science Foundation
PCR	polymerase chain reaction
P-days	physiological days
PIAP	Pesticide Impact Assessment Program
PMAP	Pest Management Alternatives Program
PPM	processing and production methods
RIPM	Regional Integrated Pest Management Grants Program
RM	resistance management
RTV	rice tungro virus
SAR	systemic acquired resistance
USDA	U.S. Department of Agriculture
USDA–APHIS	U.S. Department of Agriculture–Animal and Plant Health Inspection Service
ULV	ultra-low-volume
yr	year

Appendix B. Glossary

Allelopathy. The suppression of growth of one plant species by another due to the release of toxic substances.

Fallows. Bare soil or flooding of land to help eradicate certain pest populations.

Gene pool enrichment. Genetic evaluation and enhancement; the public research area underlying most improvements in host plant resistance.

Horizontal resistance. Genetic resistance governed by multiple genes that provide resistance to all pest races.

Kairomones. Plant-produced compounds that become incorporated into the body odor of herbivores.

Multilines. A practice in which breeders use mixtures of cultivars, each with different resistance genes.

Pathosystems. Crop(s), associated pathogens, and their environment, which often includes humans.

Pyramiding resistance genes. A technique in which breeders use several resistance genes in a single cultivar.

Semiochemicals. Compounds that are involved in communication among organisms.

Soil inhabitants. Pathogens that produce long-term survival structures or that have broad host ranges.

Soil invaders. Pathogens that attack members of a single or a few plant families and are unable to survive after infected host tissue has decayed.

Soil solarization. A process, widely used in tropical and subtropical regions, that uses clear polyethylene plastic mulches to warm soil temperatures.

Transgenic. Having chromosomes into which one or more genes from a different species have been incorporated either artificially or naturally.

Vertical resistance. Genetic resistance governed by single genes that provide resistance to specific races or biotypes of a pest.

Literature Cited

Agriculture and Agri-Food Canada, Cereal Research Centre. 1998. Wheat stem rust, <http://www.res2.agr.gc.ca/winnipeg/hd2a.htm> (18 December 2002)

Agrios, G. N. 1997. *Plant Pathology*. Academic Press, New York. 635 pp.

Allen, C. R., R. S. Lutz, and S. Demarais. 1995. Red imported fire ant impacts on northern bobwhite populations. *Ecol Appl* 3:632–638.

Allen, C. T. 1994. Pink bollworm management in Texas. Texas Agricultural Extension Service Publication ENTO, AGR 2–5. Texas A&M University System, College Station. 9 pp.

Alston, D. G., J. R. Bradley, Jr., D. P. Schmitt, and H. D. Coble. 1991. Response of *Helicoverpa zea* (Lepidoptera: Noctuidae) populations to canopy development in soybean as influenced by *Heterodera glycines* (Nematoda: Heteroderidae) and annual weed population densities. *J Econ Entomol* 84:267–276.

Altieri, M. A. 1994. *Biodiversity and Pest Management in Agroecosystems*. Food Products Press, An Imprint of the Haworth Press, New York.

Antonius, G. F. and J. C. Snyder, 1993. Trichome density and pesticide retention and half-life. *J Environ Sci Health Pestic Food Contam Agric Wastes B* 23:205–219.

Åstrand, B. 2000. Mechatronics in Agriculture: Row Guidance System for Agricultural Field Machinery. Licentiate thesis,

School of Electrical and Computer Engineering, Chalmers University of Technology, Göteborg, Sweden.

Ayers, A. D. 1948. Salt tolerance of birdsfoot trefoil. *J Am. Soc Agron* 40:331–334.

Baerveldt, A.-J. and B. Åstrand. 1997. Vision-guided mechanical weed control, <www.hh.se/staff/bjorn/mech-weed> (22 January 2003)

Baker, T. K. and T. F. Peeper. 1990. Differential tolerance of winter wheat (*Triticum aestivum*) to cyanazine and triazinone herbicides. *Weed Technol* 4:569–575.

Barbosa, P. and J. C. Schultz (eds.). 1987. *Insect Pest Outbreaks*. Academic Press, New York.

Barham, E. 1997. "What's in a Name?: Eco-Labelling in the Global Food System." Paper presented at the Joint Meetings of the Agriculture, Food, and Human Values Society and the Association for the Study of Food and Society, Madison, Wisconsin, June 5–8, 1997, <http://www.pmac.net/bbarham.htm> (18 December 2002)

Barker, K. R. and E. L. Davis. 1996. Assessing plant nematode infestations and infections. Pp. 103–136. In S. H. DeBoer (ed.). *Advances in Botanical Research* (incorporating *Advances in Plant Pathology*). Academic Press, London.

Barker, K. R. and S. R. Koenning. 1998. Developing sustainable systems for nematode management. *Annu Rev Phytopathol* 76:165–205.

Benbrook, C. 1998. IPM and the Marketplace, <http://www.pmac.net/ipm_mark.htm> (18 December 2002)

Berti, A., C. Dunan, M. Sattin, G. Zanin, and P. Westra. 1996. A new approach to determine when to control weeds. *Weed Sci* 44:496–503.

Blackburn, G. A. 1998. Quantifying chlorophylls and carotenoids at leaf and canopy scale: An evaluation of some hyperspectral approaches. *Remote Sens Environ* 66:273–285.

Boerboom, C. M. 1999. Nonchemical options for delaying weed resistance to herbicides in Midwest cropping systems. *Weed Technol* 13:636–642.

Boland, G. S. and L. D. Kuykendall (eds.). 1998. *Plant-Microbe Interactions*. Marcel Dekker, New York.

Boller, T. and F. Meins (eds.). 1992. *Genes Involved in Plant Defense*. Springer-Verlag, New York.

Bradley, E. and M. Duffy. 1982. *The Value of Plant Resistance to Soybean Cyst Nematodes: A Case Study of Forrest Soybeans*. Staff Report. Economic Research Service, U.S. Department of Agriculture, Washington, D.C. 42 pp.

Brenner, R. 1999. Hope for fire ant control. *Agric Res* 47:9.

Bridge, J. 1996. Nematode management in sustainable and subsistence agriculture. *Annu Rev Phytopathol* 34:201–225.

Bridges, D. 2000. Implications of pest-resistant/herbicide-tolerant plants. Pp. 141–153. In G. G. Kennedy and T. B. Sutton (eds.). *Emerging Technologies for Integrated Pest Management: Concepts, Research, and Implementation*. APS Press, Phytopathological Society, St. Paul, Minnesota.

Brooks, H. J. and G. Vest. 1985. Public programs on genetics and breeding of horticultural crops in the United States. *HortScience* 20:826–830.

Bruno, K. 1992. *The Greenpeace Book of Greenwash*. Greenpeace, Amsterdam, The Netherlands.

Buntin, G. D. and P. L. Raymer. 1989. Hessian fly (Diptera: Cecidomyiiadae) damage and forage production of winter wheat. *J Econ Entomol* 82:301–306.

Burns, J. R. and D. M. Benson. 2000. Biocontrol of damping-off of *Catharanthus roseus* caused by *Pythium ultimum* with *Trichoderma virens* and binucleate Rhizoctonia fungi. *Plant Dis* 84:644–648.

Burr, T. J., J. L. Norelli, C. L. Reid, L. K. Capron, L. S. Nelson, H. S. Akdwinckle, and W. F. Wilcox. 1993. Streptomycin-resistant bacteria associated with fire blight infections. *Plant Dis* 77:64–66.

Campbell, B. C. and S. S. Duffey. 1979. Tomatine and parasitic wasps: Potential incompatibility of plant antibiosis with biological control. *Science* 205:700–702.

Carlson, J. D., M. E. Whalon, D. A. Landis, and S. H. Gage. 1991. Evidence for long-range transport of potato leafhopper into Michigan. *Tenth Conference on Biotechnology and Aerobiology and the Special Session on Hydrometeorology*, Salt Lake City, Utah. Pp. 123–128.

Carpenter, J., A. Felsot, T. Goode, M. Hammig, D. Onstad, and S. Sankula. 2002. *Comparative Environmental Impacts of Biotechnology-derived and Traditional Soybean, Corn, and Cotton Crops*. Council for Agricultural Science and Technology, Ames, Iowa.

Center for Infectious Disease Research and Policy (CID). 2002. Congress approves wide-ranging bioterrorism preparedness bill, <http://www1.umn.edu/cidrap/content/fs/biosecurity/news/btbill.html> (23 January 2003)

Cereijo, M. 2001. *Threat: Animal and Agricultural Bioterrorism*, <http://www.amigospais-guaracabuya.org/oagmc107.html>

Charudattan, O. 2000. Current status of biological control of weeds. Pp. 269–288. In G. G. Kennedy and T. B. Sutton (eds.). *Emerging Technologies for Integrated Pest Management: Concepts, Research and Implementation*. APS Press, St. Paul, Minnesota.

Christensen, C. M. and R. A. Meronuck. 1986. *Quality Maintenance in Stored Grains and Seeds*. University of Minnesota, Minneapolis. 138 pp.

Clinton, W. J. 1999. Executive Order: Invasive Species. Washington, DC., <http://wric.ucdavis.edu/exotic/thewhitehouse.html>

Coaker, T. H. 1987. Cultural methods: The crops. Pp. 69–88. In A. J. Burn, T. H. Coaker, and P. C. Jepson (eds.). *Integrated Pest Management*. Academic Press, San Diego, California.

Coakley, S. M., H. Scherm, and S. Chakraborty. 1999. Climate change and plant disease management. *Annu Rev Phytopathol* 37:399–426.

Collins, W. and R. L. Phillips. 1991. Plant breeding training in public institutions in the United States. Report Number 591. Science and Education Resources Development, Cooperative State Research, Education, and Extension Service, U.S. Department of Agriculture, Washington, D.C.

Committee for the National Institute for the Environment (CNIE). 1999a. Harmful Non-native Species: Issues for Congress III (April 8), Washington, D.C., <http://www.cnie.org/nle/biodv-26b.html> (18 December 2002)

Committee for the National Institute for the Environment (CNIE). 1999b. Harmful Non-native Species: Issues for Congress VI (April 8). Washington, D.C., <http://www.cnie.org/nle/biodv-26e.html> (18 December 2002)

Consumers Union. 2002. Eco-labels: The Consumers Union Guide to Environmental Labels, <http://www.eco-labels.org/home.cfm> (18 December 2002)

Cook, R. J. 1987. Report on the Research Briefing Panel on Biological Control in Managed Ecosystems. National Academy Press, Washington, D.C.

Council for Agricultural Science and Technology (CAST). 2000. *Invasive Plant Species*. Issue Paper No. 13. CAST, Ames, Iowa.

Council for Agricultural Science and Technology (CAST). 2002. *Invasive Pest Species: Impacts on Agricultural Production,*

Natural Resources, and the Environment. Issue Paper No. 20. CAST, Ames, Iowa.

Davis, J. R., O. C. Huisman, D. T. Westermann, L. H. Sorensen, A. T. Schneider, and J. C. Stark. 1994. The influence of cover crops on the suppression of Verticillium wilt of potato. Pp. 332–343. In G. W. Zehnder, M. L. Powelson, R. K. Hannson, and K. V. Ramay (eds.). *Advances in Potato Pest Biology and Management.* The American Phytopathological Society, St. Paul, Minnesota.

Davis, J. R., O. C. Huisman, D. T. Westermann, S. L. Hafez, D. O. Everson, L. H. Sorensen, and A. T. Schneider. 1996. Effects of green manures on verticillium wilt of potato. *Phytopathology* 86:444–453.

Davis, R. F., A. W. Johnson, and R. D. Wauchope. 1993. Accelerated degradation of fenamiphos and its metabolites in soil previously treated with fenamiphos. *J Nematol* 25:679-685.

Denholm, I. and M. W. Rowland. 1992. Tactics for managing pesticide resistance in arthropods: Theory and practice. *Annu Rev of Entomol* 37:91–112.

Dennehy, T. J. 2000. Fulfilling the role of resistance management to preserve effectiveness of new insecticide technologies. Pp. 400–417. In G. G. Kennedy and T. B. Sutton (eds.). *Emerging Technologies for Integrated Pest Management: Concepts, Research, and Implementation.* APS Press, St. Paul, Minnesota.

Dennis, C. (ed.). 1983. *Post-harvest Pathology of Fruits and Vegetables.* Academic Press, New York.

DeSouza, K., J. Holt, and J. Colvin. 1995. Diapause, migration and pyrethroid-resistance dynamics in the cotton bollworm, *Helicoverpa armigera* (Lepidoptera: Noctuidae). *Ecol Entomol* 20:333–342.

Dinesen, I. G. and A. van Zaayen. 1996. Potential of pathogen detection technology for management of diseases in glasshouse ornamental crops. Pp. 23, 137, 170. In S. H. DeBoer (ed.). *Advances in Botanical Research* (incorporating *Advances in Plant Pathology*). Academic Press, London.

Drees, B. M. and S. B. Vinson. 1998. Fire ants and their management. Report Number B1536. Texas Agricultural Extension Service, the Texas A&M University System, College Station.

Dufour, R. and J. Bachmann. 1998. Integrated Pest Management. Appropriate Technology for Transfer to Rural Areas (ATTRA), <http://www.attra.org/attra-pub/ipm.html> (18 December 2002)

Dufour, R. and L. Greer. 1995. Farmscaping. Appropriate Technology Transfer for Rural Areas (ATTRA), <http://www.attra.org/attra-pub/farmscape.html> (18 December 2002)

Duvick, D. N. 1984. Genetic diversity in major farm crops on the farm and in reserve. *Econ Bot* 38:161–178.

Ellsbury, M. M., S. A. Clay, S. J. Fleischer, L. D. Chandler, and S. M. Schneider. 2000. Use of GIS/GPS systems in IPM: Progress and reality. Pp. 419–438. In G. G. Kennedy and T. B. Sutton (eds.). *Emerging Technologies for Integrated Pest Management: Concepts, Research and Implementation.* APS Press, St. Paul, Minnesota.

El-Zik, K. M. and R. E. Frisbie. 1991. Integrated crop management systems for pest control. Pp. 3–106. In D. Pimentel (ed.). *Handbook of Pest Management in Agriculture.* CRC Press, Boca Raton, Florida.

Farm Chemical Handbook (FCH). 2001. MeisterPro Information Resources, Salem, Massachusetts.

Fermin-Muñoz, G. A., B. Meng, K. Ko, S. Mazumdar-Leighton, A. Gubba, and J. E. Carroll. 2000. Biotechnology: A new era for plant pathology and plant protection. APSnet, <http://www.apsnet.org/online/feature/Bio-_Technology/Top.html> (23 January 2003)

Fleming, C. C. and T. O. Powers. 1998. Potato cyst nematode diagnostics: Morphology, differential hosts, and biochemical techniques. Pp. 91-113. In R. J. Marks and B. B. Brodie (eds.). *Potato Cyst Nematodes.* CAB International, Wallingford, United Kingdom.

Food Alliance. 2002. Homepage, <www.thefoodalliance.org> (18 December 2002)

Franc, G. D., M. D. Harrison, and M. L. Powelson. 1985. The dispersal of phytopathogenic bacteria. Pp. 37–49. In D. R. MacKenzie, C. S. Barfield, G. G. Kennedy, and R. D. Berger, with D. J. Taranto (eds.). *The Movement and Dispersal of Agriculturally Important Biotic Agents.* Claitor's Publishing Division, Baton Rouge, Louisiana.

Frey, J. K. 1996. National Plant Breeding Study. I: Human and financial resources devoted to plant breeding research and development in the United States in 1994. Report Number 98. Iowa Agriculture and Home Economics Experiment Station, Iowa State University, Ames.

Frey, J. K. 1997. National Plant Breeding Study. II: National plan for promoting breeding programs for Minor Crops in the U.S. Special Report Number 100. Iowa Agriculture and Home Economics Experiment Station, Iowa State University, Ames.

Frey, J. K. 1998. National Plant Breeding Study. III. National Plan for Genepool Enrichment of U.S. Crops. Report Number 101. Iowa Agriculture and Home Economics Experiment Station, Iowa State University, Ames.

Frick, B. and A. G. Thomas. 1992. Weed surveys in different tillage systems in southwestern Ontario field crops. *Can J Plant Sci* 72:1337–1347.

Garcia, J. 2000. 110 Years of Federal Biological Control. ARS/USDA News & Information. Washington, D.C., <http://www.ars.usda.gov/is/pr/2000/001124.htm>

Gavassoni, W. L., L. Tylka, and G. P. Munkvold. 2001. Relationships between tillage and spatial patterns of *Heterodora glycines. Phytopathology* 91:534–545.

Georghiou, G. P. (ed.). 1986. The magnitude of the resistance problem. Pp. 14–43. In *Pesticide Resistance: Strategies and Tactics for Management.* Committee on Strategies for the Management of Pesticide Resistant Pest Populations, National Academy Press, Washington, D.C.

Gianessi, L. and J. Carpenter. 1999. *Agricultural Biotechnology Insect Control Benefits.* National Center for Food and Agricultural Policy, Washington, D.C.

Godfrey, L. D., P. B. Goodell, E. Grafton-Cardwell, N. C. Toscano, and E. T. Natwick. 2001. Cotton bollworm. UC IPM Pest Management Guidelines: Cotton insects and mites. UC ANR Publications 3444. 3 pp.

Gould, F. 1998. Sustainability of transgenic insecticidal cultivars: Integrating pest genetics and ecology. *Annu Rev Entomol* 43:701–726.

Granett, J. and A. Retnakaran. 1977. Stadial susceptibility of eastern spruce budworm, *Choristoneura fumuferana* (Lepidoptera: Tortricidae) to insect growth regulator Dimilin. *Can Entomol* 109:893–894.

Green, S., S. M. Stewart-Wade, G. J. Boland, M. P. Teshler, and S. H. Liu. 1998. Formulating microorganisms for biological control of weeds. Pp. 249–281. In G. J. Boland and L. D. Kuykendall (eds.). *Plant-Microbe Interactions.* Marcel Dekker, New York.

Gutierrez, A. P., B. Wermelinger, F. Schulthess, J. U. Baumgartner, H. R. Herren, C. K. Ellis, and J. S. Yaninek. 1988a. Analysis of biological control of cassava pests in Africa. I. Simulation of carbon nitrogen and water dynamics in cassava. *J Appl*

Ecol 25:901–920.
Gutierrez, A. P., P. Neuenschwanter, F. Schulthess, J. U. Baumgartner, H. R. Herren, B. Wermelinger, B. Lohr, and C. K. Ellis. 1988b. Analysis of biological control of cassava pests in Africa. II. Cassava mealybug *Phenacoccus manigoti*. *J Appl Ecol* 25:921–940.
Haining Cowles, M. 1999. Cornell University IPM Program. An IPM Label on Supermarket Vegetables: A First for the Nation, <http://www.nysipm.cornell.edu/labeling/labels.html> (9 March 2002)
Hall, B. and G. Kuepper. 1997. Making the Transition to Sustainable Farming, <http://www.attra.org/attra-pub/trans.html>
Hall, F. Q. 2000. Delivering new crop protection agents within an IPM environment. Pp. 323–338. In G. G. Kennedy and T. B. Sutton (eds.). *Emerging Technologies for Integrated Pest Management: Concepts, Research and Implementation*. APS Press, St. Paul, Minnesota.
Hall, F. R. and J. J. Menn (eds.). 1999. *Biopesticides: Use and Delivery*. Humana Press, Totowa, New Jersey.
Hamm, J. J. and B. R. Wiseman. 1986. Plant resistance in a nuclear polyhedrosis virus for suppression of fall armyworm (Lepidoptera: Noctuidae). *Fla Entomol* 69:541–549.
Hanna, M., R. Hartzler, D. Erback, K. Paarlberg, and L. Miller. 1996. Cultivation: An effective weed management tool. University Extension PM-1623 (rev.). Iowa State University, Ames.
Hawks, B. 2002. Rules and regulations—Agricultural Bioterrorism Protection Act of 2002: Listing of biological agents and toxins and requirements and procedures for notification of possession. *Fed Regist* 67, No.155. USDA–APHIS, Washington, D.C.
Heald, C. M. 1987. Classical nematode management practices. Chapter 15. In J. A. Veech and D. W. Dickson (eds.). *Vistas on Nematology*. Society of Nematologists, Hyattsville, Maryland.
Heenan, D. P., A. C. Taylor, and A. R. Leys. 1990. The influence of tillage, stubble management and crop rotation on the persistence of great brome (*Bromus diandrus* Roth). *Aust J Exp Agric* 30:227–230.
Henkens, R., C. Bonaventura, V. Kazantseva, M. Moreno, J. O'Daly, R. Sundseth, S. Wegner, and M. Wojciechowski. 2000. Use of DNA technologies in diagnostics. Pp. 52–66. In G. G. Kennedy and T. B. Sutton (eds.). *Emerging Technologies for Integrated Pest Management: Concepts, Research and Implementation*. APS Press, St. Paul, Minnesota.
Hess, F. D., Sr. and S. O. Duke. 2000. Genetic engineering in IPM: A case study. Herbicide tolerance. Pp. 126–140. In G. G. Kennedy and T. B. Sutton (eds.). *Emerging Technologies for Integrated Pest Management: Concepts, Research, and Implementation*. APS Press, the American Phytopathological Society, St. Paul, Minnesota.
Hoffman-Hergarten, S., K. Hasky, M. Reitz, and R. A. Sikora. 1997. Induced systemic resistance by Rhozobacteria toward the cyst nematode, *Globodera pallida*, on potato. Pp. 292–295. In A. Ogoshi, K. Kobayashi, Y. Homma, F. Kodama, N. Kondo, and S. Akinow (eds.). *Plant Growth-Promoting Rhozobacteria: Present Status and Future Prospects. Proceedings of the Fourth Workshop on Plant Growth-Promoting Rhozobacteria: Present Status and Future Prospects*, OECD Joint Workshop, Sapporo, Japan.
Hollings, M. 1965. Disease control through virus-free stock. *Annu Rev Phytopathol* 3:376–396.
Horn, D. J. 1988. *Ecological Approaches to Pest Management*. Guilford Press, New York, NY. 285 pp.
Hoy, M. A. 2000. Current status of biological control of insects. Pp. 210–225. In G. G. Kennedy and T. B. Sutton (eds.). *Emerging Technologies for Integrated Pest Management: Concepts, Research and Implementation*. APS Press, St. Paul, Minnesota.
Hutchinson, C. M., M. E. McGiffen, Jr., H. D. Ohr, J. J. Sims, and J. O. Becker. 1999. Evaluation of methyl iodide as a soil fumigant for root-knot nematode control in carrot. *Plant Dis* 83:33–36.
Hutchinson, K. B. 2001. S. 1563, The Agricultural Bioterrorism Countermeasures Act of 2001. U. S. Congressional Record, October 17, 2001—Page S10796, <http://www.senate.gov/~Hutchison/speec212.htm>
Institute for Agriculture and Trade Policy (IATP). 1999. Eco-labels: Incomplete bibliography. In: *Adding Value through Environmental Marketing: Opportunities for Food Producers, Processors and Retailers*. Proceedings of a Conference, Madison, Wisconsin, December 6–7, 1999, <http://iatp.org/labels/envcommodities/eco-labels.html> (18 December 2002)
IPM Institute of North America (IPMI). 2002a. CORE Values Northeast; A Project of Mothers & Others, <http://www.corevalues.org/home.html> (18 December 2002)
IPM Institute of North America (IPMI). 2002b. IPM Standards, Certification, and Labeling for Ag and Community IPM, <http://www.ipminstitute.org/> (18 December 2002)
Jacobsen, B. J. and P. A. Backman. 1993. Biological and cultural plant disease controls: Alternatives and supplements to chemicals in IPM systems. *Plant Dis* 77:311–315.
Johansen, C. 1962. Principles of insect control. Pp. 138–158. In R. E. Pfadt (ed.). *Fundamentals of Applied Entomology*. MacMillan, New York.
Johnson, A. W. 1998. Degradation of fenamiphos in agricultural production soil. *J Nematol* 30:40–44.
Johnson, A. W., C. C. Dowler, N. C. Glaze, and A. M. Golden. 1992. Nematode numbers and crop yield in a fenamiphos-treated sweet corn-sweet potato-vetch cropping system. *J Nematol* 24:533–539.
Jones, R. L. and F. A. Norris. 1998. Factors affecting the degradation of aldicarb and ethoprop. *J Nematol* 30:45–55.
Julien, M. H. and M. W. Griffiths. 1998. Biological control of weeds: A world catalogue of agents and their target weeds. 4th ed. CABI Publishing, Wallingford, United Kingdom.
Kahn, R. P. 1991. Exclusion as a plant disease strategy. *Annu Rev Phytopathol* 29:219–246.
Katan, J. and J. E. DeVay. 1991. *Soil Solarization*. CRC Press, Boca Raton, Florida.
Kennedy, G. G. 2000. Perspectives on progress in IPM. Pp. 2–11. In G. G. Kennedy and T. B. Sutton (eds.). *Emerging Technologies for Integrated Pest Management: Concepts, Research, and Implementation*. APS Press, St. Paul, Minnesota.
Kennedy, G. G. and T. B. Sutton (eds.). 2000. *Conference on Emerging Technologies for IPM: Concepts, Research and Implementation*. APS Press, St. Paul, Minnesota.
Kennedy, G. G. and M. E. Whalon. 1995. Managing pest resistance to *Bacillus thuringiensis* endotoxins: Constraints and incentives to implementation. *J Econ Entomol* 88:454–460.
Kerry, B. R. and K. Evans. 1996. New strategies for the management of plant-parasitic nematodes. Pp. 134–152. In R. Hall (ed.). *Principles and Practices of Managing Soilborne Plant Pathogens*. APS Press, St. Paul, Minnesota.
Kloepper, J. W., R. Rodriguez-Kabana, J. A. McInroy, and R. W. Young. 1992. Rhizosphere bacteria antagonistic to soybean cyst (*Heterodera glycines*) and root-knot (*Meloidogyne incognita*) nematodes: Identification by fatty acid analysis and frequency of biological cotnrol activity. *Plant Soil* 139:75–84.
Koeller, W. 1996. Recent developments in DMI resistance. Pp.

301–311. In H. Lyr, P. E. Russel, and H. D. Sisler (eds.). *Modern Antifungal Compounds*. Intercept Limited, Andover, United Kingdom.

Koenning, S. R. and S. C. Anand. 1991. Effects of wheat and soybean planting date on *Heterodera glycines* population dynamics and soybean yield with conventional tillage. *Plant Dis* 75:301–304.

Koenning, S. R., D. P. Schmitt, and K. R. Barker. 1993. Effects of cropping systems on population density of *Heterodera glycines* and soybean yield. *Plant Dis* 77:780–786.

Koscelny, J. A., T. F. Peeper, J. B. Solie, and S. G. Solomon, Jr. 1991. Seeding date, seeding rate, and row spacing affect wheat (*Triticum aestivum*) and cheat (*Bromus secaninus*). *Weed Technol* 5:707–712.

Kuykendall, L. D. and F. M. Hashem. 1998. Enhancement of biological nitrogen fixation using strain-specific bacterial viruses for biological control of ineffective Bradyrhizobium or Rhizobium. Pp. 297–308. In G. S. Boland and L. D. Kuykendall (eds.). *Plant-Microbe Interactions*. Marcel Dekker, New York.

Kwong, O. 2000. *Weeds Gone Wild: Alien Plant Invaders of Natural Areas*. Plant Conservation Alliance, Washington, D.C., <http://www.nsp.gov/plants/alien/> (27 February 2003)

Lawler, C. and M. A. Harmey. 1993. Immunological detection of nematode antigens on the surface of a wood section. *Fund Appl Nematol* 16:521–523.

Levine, E. and H. Oloumi-Sadeghi. 1991. Management of diabrocticite rootworms in corn. *Annu Rev Entomol* 36:229–255.

Lewis, R. D. and J. H. Hunter. 1940. The nitrogen, organic carbon, and pH of some southeastern coastal plain soils as influenced by green-manure crops. *Agron J* 32:586–601.

Lewis, W. J., R. L. Jones, and A. N. Sparks. 1972. A host-seeking stimulation for the egg parasite *Trichogramma evanescens*: Its source and demonstration of its laboratory and field activity. *Annu Entomol Soc Am* 65:1087–1089.

Liebman, M., C. L. Mohler, and C. P. Staver. 2001. *Ecological Management of Agricultural Weeds*. Cambridge University Press, United Kingdom.

Locascio, S. J., J. P. Gilreath, D. W. Dickson, T. A. Kucharek, J. P. Jones, and J. W. Noling. 1997. Fumigant alternatives to methyl bromide for polyethylene-mulched tomato. *HortScience* 32:1208–1211.

Loper, J. E. and V. O. Stockwell. 2000. Current status of biological control of plant diseases. Pp. 240–256. *Emerging Technologies for Integrated Pest Management: Concepts, Research and Implementation*. APS Press, St. Paul, Minnesota.

Louws, F. J., J. L. W. Rademaker, and F. J. deBruijn. 1999. The three Ds of PCR-based genomic analyses of phytobacteria: Diversity, detection, and disease diagnosis. *Annu Rev Phytopathol* 37:81–125.

Luther, G. C., H. R. Valenzuela, and J. DeFrank. 1996. Effects of cruciferous trap crops on lepidopteran pests of cabbage in Hawaii. *Environ Entomol* 25:39–47.

Lyon, D. J., A. J. Bussan, J. O. Evans, C. A. Mallory-Smith, and T. F. Peeper. 2002. Pest management implications of glyphosate-resistance wheat in the western United States. *Weed Technol* 16:680–690.

Magdoff, F. and H. van Es. 2000. *Building Soils for Better Crops* (2d ed.). Sustainable Agiculture Network, Burlington, Vermont.

Main, C. E., T. Keever, G. J. Holmes, and J. M. Davis. 2001. Forecasting long-range transport of mildew spores and spread of plant disease epidemics. APSnet, St. Paul, Minnesota, <http://www.apsnet.org/online/feature/forecast/top.htm> (27 February 2003)

Martin, R. J. and W. L. Felton. 1993. Effect of crop rotation, tillage practice, and herbicides on the population dynamics of wild oats in wheat. *Austr J Exp Agric* 33:159–165.

McFadyen, R. E. C. 1998. Biological control of weeds. *Annu Rev Entomol* 43:369–393.

McGaughey, W. H. and M. E. Whalon. 1992. Managing insect resistance to *Bacillus thuringiensis* toxins. *Science* 258:1451–1455.

Meneley, J. C. 2000. Challenges to the commercialization of biological control technologies for IPM. Pp. 289–304. In G. G. Kennedy and T. B. Sutton (eds.). *Emerging Technologies for Integrated Pest Management: Concepts, Research, and Implementation*. APS Press, The American Phytopathological Society, St. Paul, Minnesota.

Michigan State University (MSU). 2002. Fruit Spraying Calendar, MSU E154, <http://www.msue.msu.edu/pestpubs/E154/> (27 February 2003)

Midgarden, D., S. J. Fleischer, R. Weisz, and Z. Smilowitz. 1997. Site-specific integrated pest management impact on development of esfenvalerate resistance in Colorado potato beetle (Coleoptera: Chrysomelidae) and on densities of natural enemies. *J Econ Entomol* 90:855–867.

Monteith, L. G. 1960. Influence of plants other than the food plants of their host on host-finding by tachinid parasites. *Can Entomol* 92:641–652.

Mortensen, K. 1998. Biological control of weeds using microorganisms. Pp. 223–247. In G. J. Boland and L. D. KuyKendall (eds.). *Plant-Microbe Interactions*. Marcel Dekker, New York.

Muid, B. 1993. Host plant modification of insecticide resistance in *Myzus persicae*. Ph.D. diss., University of Reading, United Kingdom.

Musgrave, M. E. and N. Ding. 1998. Evaluating wheat cultivars for waterlogging tolerance. *Crop Sci* 38:90–97.

Natarajan, K. 1990. Effect of leaf pubescence in cotton, *Gossypium hirsutum*, on the parasitism of whitefly, *Menisia tabaci* (Gennadius). *J Biol Control* 4:57–58.

National Academy of Sciences (NAS). 1972. *Genetic Vulnerability of Major Crops*. National Academy of Sciences, Washington, D.C. 307 pp.

National Research Council (NRC). 1992. *NEEM: A Tree for Solving Global Problems*. National Academy Press, Washington, D.C.

National Research Council (NRC). 1996. *Ecologically Based Pest Management: New Solutions for a New Century*. National Academy Press, Washington, D.C. 144 pp.

National Research Council (NRC). 2000. *The Future Role of Pesticides in U.S. Agriculture*. National Academy Press, Washington, DC.

National Science Foundation (NSF), Center for Integrated Pest Management. 2002. Virtual Center, <http://cipm.ncsu.edu/index.html> (18 December 2002)

Nester, E., L. Thomashow, M. Metz, and M. Gordon. 2002. 100 Years of *Bacillus thuringiensis*: A critical scientific assessment. A report from the American Academy of Microbiology, Washington, D.C.

Nicol, D., S. D. Wratten, N. Eaton, and S. V. Copaja. 1993. Effects of DIMBOA levels in wheat on the susceptibility of the grain aphid *Sitobion avenae* to deltamethrin. *Ann Appl Biol* 122:427–433.

Nicot, P. C. and D. I. Rouse. 1987. Precision and bias of three quantitative soil assays for *Verticillium dahliae*. *Phytopathology* 77:875–881.

North American Free Trade Agreement (NAFTA)–Technical Working Group on Pesticides. 2000. Status of harmonization of data requirements and test protocols for pesticide registration, <http://www.hc-sc.gc.ca/pmra-arla/english/pubs/dire-e.html> (27 February 2003)

Norton, G. A. and G. R. Conway. 1977. The economic and social context of pest, disease, and weed problems. Pp. 205–226. In J. N. Cherrett and G. R. Sagar (eds.). *The Origins of Pest, Parasites, Disease, and Weed Problems*. Blackwell, Oxford, United Kingdom.

Onstad, D. W., M. G. Joselyn, S. A. Isard, E. Levine, J. L. Spencer, L. W. Bledsoe, C. R. Edwards, C. D. DiFonzo, and H. Willson. 1999. Modeling the spread of western corn rootworm (*Coleoptera: Chrysomelidae*) populations adapting to soybean-corn rotation. *Environ Entomol* 28:188–194.

Panse, A., J. H. C. Davis, and G. Fischbeck. 1997. Yield formation in mixtures of rust resistant and susceptible plants of common bean. *J Agron Crop Sci Berlin* 178:111–116.

Pedigo, L. 1989. *Entomology and Pest Management*. Macmillan Publishing Company, New York.

Pedigo, L. P. 1999. *Entomology and Pest Management*. 3d ed. Prentice Hall, Upper Saddle River, New Jersey.

Pieters, A. J. and R. J. McKee. 1929. Green manuring and its application to agricultural practices. *Agron J* 21:985–993.

Pieters, A. J. and R. J. McKee. 1938. The use of cover and green-manure crops. Pp. 431–444. In *USDA Yearbook of Agriculture*. U.S. Department of Agriculture, Washington, D.C.

Pimentel, D. 1961. An evaluation of insect resistance in brocolli, brussels sprouts, cabbage, collards and kale. *J Econ Entomol* 54:156–158.

Porter, S. D. and D. A. Savignano. 1990. Invasion of polygyne fire ants decimate native ants and disrupts arthropod community. *Ecology* 6:2095–2106.

Price, P. W. 1986. Ecological aspects of host plant resistance and biological control: Interactions among three trophic levels. Pp. 11–30. In D. J. Boethel and R. D. Eikenbary (eds.). *Interactions of Plant Resistance and Parasitoids and Predators of Insects*. Halsted Press, New York.

Price, P. W. 1997. *Insect Ecology*. 3d ed. 1997. John Wiley & Sons, New York.

Price, P. W., C. E. Bouton, P. Gross, B. A. McPherson, J. N. Thompson, and A. J. Weis. 1980. Interactions among three trophic levels: Influence of plant on interactions between insect herbivores and natural enemies. *Annu Rev Ecol Syst* 11:41–65.

Quakenbush, L. S. and R. N. Andersen. 1984. Effect of soybean (*Glycine max*) interference on eastern black nightshade (*Solanum ptycanthum*). *Weed Sci* 32:638–645.

Rainforest Alliance. 2002. Rainforest Marketplace. Products from Farms Certified by the Sustainable Agriculture Network. New York, < http://www.rainforest-alliance.org/marketplace/eco-ok/index.html> (19 December 2002)

Read, D. P., P. P. Feeny, and R. B. Root. 1970. Habitat selection by the aphid parasite *Diaeretiella rapae* (Hymenoptera: Braconidae) and hyperparasite *Charips brassicae* (Hymenoptera: Cynipidae). *Can Entomol* 102:1567–1578.

Reigart, J. R. and J. R. Roberts. 1999. Recognition and management of pesticide poisonings. 735-R-98-003. Office of Prevention, Pesticides and Toxic Substances, U.S. Environmental Protection Agency, Washington, D.C.

Reiners, S., C. Petzoldt, and M. Hoffmann (eds.). 1999. *Integrated Crop and Pest Management Recommendations for Commercial Vegetable Production*. Cornell University, Ithaca, New York.

Risch, S. J., D. Andrew, and M. A. Altieri. 1983. Agroecosystem diversity and pest control: Data, tentative conclusions and new research directions. *Environ Entomol* 12:625–629.

Rodgers, T. H. and J. E. Giddens. 1957. Green manure and cover crops. Pp. 252–257. In *USDA Yearbook of Agriculture*. U.S. Department of Agriculture, Washington, D.C.

Roe, R. M., W. D. Bailey, F. Gould, C. E. Sorenson, G. G. Kennedy, J. S. Bacheler, R. L. Rose, E. Hodgson, and C. L. Sutula. 2000. Detection of resistant insects and IPM. Pp. 67–84. In G. G. Kennedy and T. B. Sutton (eds.). *Emerging Technologies for Integrated Pest Management: Concepts, Research, and Implementation*. APS Press, The American Phytopathological Society, St. Paul, Minnesota.

Roush, R. T. 1989. Designing resistance management programs: How can you choose? *Pestic Sci* 26:423–441.

Rubin, E., A. Baider, and Y. Cohen. 2001. *Phytophthora infestens* produces oospores in fruits and seeds of tomato. *Phytopathology* 91:1074–1080.

Ruesink, W. G. 1980. Economics of integrated pest management: discussion—An entomologist's view of IPM research needs. *Am J Agric Econ* 62:1014–1015.

Russell, G. E. 1978. *Plant Breeding for Pest and Disease Resistance*. Butterworths, London.

Russo, J. M. 2000. Weather forecasting for IPM. Pp. 453–473. In G. G. Kennedy and T. B. Sutton (eds.). *Emerging Technologies for Integrated Pest Management: Concepts, Research, and Implementation*. APS Press, The American Phytopathological Society. St. Paul, Minnesota.

Saxena, R. C., H. D. Justo, Jr., and E. L. Palanginan. 1988. Trap crop for *Nephotettix virescens* (Homoptera: Cicadellidae) and tungro management in rice. *J Econ Entomol* 81:1485–1488.

Schalk, J. M. and R. Ratcliffe. 1976. Evaluation of ARS [Agricultural Research Service] program on alternative methods of insect control: Host plant resistance to insects. *Bull Entomol Soc Am* 22:7–10.

Schmidt, K. 1995. A new ant on the block (fire ants). *New Scientist* 148:28–31.

Schubert, T. S., S. A. Rizvi, X. Sun, T. R.. Gottward, J. H. Graham, and W. N. Dixon. 2001. Meeting the challenge of eradicating citrus canker in Florida again. *Plant Dis* 85:340–356.

Schumann, G. L. 1991. *Plant Diseases: Their Biology and Social Impact*. APS Press, American Phytopathological Society, St. Paul, Minnesota.

Schuster, M. F., P. D. Calvin, and W. C. Langston. 1983. Interaction of high tannin with bollworm control by Pydin and Dipel. Pp. 72–73. *Proceedings of the 1983 Beltwide Cotton Production Research Conference*. National Cotton Council of America, Memphis, Tennessee.

Schweizer, E. E., D. W. Lybecker, L. W. Wiles, and P. Westra. 1993. Bioeconomic weed management models in crop production. Pp. 103–107. In D. R. Buxton, R. Sibles, R. A. Forsberg, B. L. Blad, K. H. Asay, G. M. Paulsen, and R. F. Wilson (eds.). *International Crop Science I*. Crop Science Society of America, Madison, Wisconsin.

Schweizer, E. E., P. Westra, and D. W. Lybecker. 1998. Seedbank and emerged annual weed populations in corn (*Zea mays*) fields in Colorado. *Weed Technol* 12:243–247.

Sherrick, S. and G. Head. 2000. General concepts, status, and potential transgenic plants in IPM. Pp. 96–199. In G. G. Kennedy and T. B. Sutton (eds.). *Emerging Technologies for Integrated Pest Management: Concepts, Research, and Implementation*. APS Press, American Phytopathological Society, St. Paul, Minnesota.

Smith, S. 1992. Farming activities and family farms: Getting the concepts right. *Joint Economic Committee Symposium on*

Agricultural Industrialization and Family Farms: The Role of Federal Policy, United States Congress, 21 October, 1992. 18 pp.

Southwood, T. R. E. 1978a. *Ecological Methods: With Particular Reference to the Study of Insect Populations*. Chapman and Hall, London.

Southwood, T. R. E. 1978b. Escape in space and time—Concluding remarks. Pp. 277–279. In H. Dingle (ed.). *Evolution of Insect Migration and Diapause*. Springer-Verlag, New York.

Southwood, T. R. E. and M. J. Way. 1970. Ecological background to pest management. Pp. 6–29. In R. L. Rabb and F. E. Guthrie (eds). *Concepts in Pest Management*. North Carolina State University, Raleigh.

Sparks, T. C., G. D. Thompson, H. A. Kirst, M. B. Hertlein, J. S. Mynderse, J. R. Turner, and T. V. Worden. 1999. Fermentation-derived insect control agents. P. 172. In F. R. Hall and J. J. Menn (eds.). *Biopesticides: Use and Delivery*. Humana Press, Totowa, New Jersey.

Stalker, H. T. and J. P. Murphy (eds.). 1991. *Plant Breeding in the 1990s—Proceedings of the Symposium on Plant Breeding in the 1990s*. CAB International, Wallingford, United Kingdom.

Stanton, G., D. L. Clement, and E. Dutky. 1999. *Pests and Diseases of Herbaceous Perennials: The Biological Approach*. Ball Publishing, Batavia, Illinois.

Stemilt Management. 2002. 'Responsible Choice'—What Does It Mean? <http://www.stemilt.com/index.php> (19 December 2002)

Stewart, S. J. 2000. Pathogen detection and pesticide uses. Pp. 85–93. In G. G. Kennedy and T. B. Sutton (eds.). *Conference on Emerging Technologies for IPM: Concepts, Research and Implementation*. APS Press, St. Paul, Minnesota.

Sticher, L., B. MauchMani, and J. P. Metraux. 1997. Systemic acquired resistance. *Annu Rev Phytopathol* 35:235–270.

Stinner, R. 2002. *Index of Data Sources for National Site (USDA Regional Pest Management Centers)*, <http://pmcenters.org/Pesticide> (27 February 2003)

Stirling, G. R. 1991. *Biological Control of Plant Parasitic Nematodes*. CAB International, Wallingford, United Kingdom.

Sullivan, P. 1998. *An Overview of Cover Crops and Green Manures: Fundamentals of Sustainable Agriculture*, <http://www.attra.org/attra-pub/covercrop.html> (27 February 2003)

Summy, K. R. and E. G. King. 1992. Cultural control of cotton insect pests in the United States. *Crop Protect* 112:307–319.

Tabashnik, B. E. 1994. Evolution of resistance to *Bacillus thuringiensis*. *Annu Rev Entomol* 39:47–49.

Takahashi, F. 1964. Reproduction curve with two equilibrium points: A consideration in fluctuations of insect populations. *Res Pop Ecol* 6:28–38.

Tamo, M. 1991. The interactions between cowpea (*Vigna unguicula* Walp.) and the bean flower thrips (*Megalurothrips sjostedti* Trybom) in the Republic of Benin. Ph.D. diss., Swiss Federal Institute of Technology, Zurich.

Tedford, E. and R. A. Brown. 2000. Integrating new fungicide technologies in IPM. Pp. 355–342. In G. G. Kennedy and T. B. Sutton (eds.). *Emerging Technologies for Integrated Pest Management: Concepts, Research and Implementation*. APS Press, St. Paul, Minnesota.

Thorpe, K. W. and P. Barbosa. 1986. Effects on consumption of high and low nicotine tobacco by *Manduca sexta* (Lepidoptera: Sphinigidae) on survival of gregarious endoparasitoid *Cotesia congragata* (Hymenoptera: Braconidae). *J Chem Ecol* 12:1329–1337.

Thurston, H. D. 1990. Plant disease management practices of traditional farmers. *Plant Dis* 74:96–102.

Turlings, T. C. J., J. H. Tumlinson, and W. J. Lewis. 1990. Exploitation of herbivore-induced plant odors by host-seeking parasitic wasps. *Science* 250:1251–1253.

Ullstrup, A. J. 1972. The impacts of the southern corn leaf blight epidemics of 1970–1972. [*Helminthosporium turcicum*] *Annu Rev Phytopathol* 10:37–50.

University of Massachusetts. 2000. Integrated Pest Management: Community Education, Partners with Nature, <http://www.umass.edu/umext/ipm/ipm_projects/education/partners_with_nature.html> (18 December 2002)

U.S. Department of Agriculture (USDA). 1999. Current Research Information System, Science and Education Resources Development, Cooperative State Research, Education and Extension Service, <http://cris.csrees.usda.gov/Welcome.html> (18 December 2002)

U.S. Department of Agriculture–Animal and Plant Health Inspection Service (USDA–APHIS). 1998. *Agriculture Fact Book 98*. Marketing and Regulatory Programs, Animal and Plant Health Inspection Service, United States Department of Agriculture, Washington, D.C., <http://www.usda.gov/news/pubs/fbook98/ch12b.htm > (19 December 2002)

U.S. Department of Agriculture–Animal and Plant Health Inspection Service (USDA–APHIS). 2003. Recently published APHIS rules and notices, <http://www.aphis.usda.gov/ppd/rad/web.report.html> (23 January 2003)

U.S. Department of Agriculture–National Agricultural Statistics Service (USDA–NASS). 1999. Trends in U.S. Agriculture, <http://www.usda.gov/nass/pubs/trends/cornhybridzation.htm> (23 January 2003)

U.S. Environmental Protection Agency (EPA). 1998. The Environmental Protection Agency's White Paper on *Bt* Plant-Pesticide Resistance Management. EPA Publication 739-S-98-001. EPA, Biopesticides and Pollution Prevention Division (7511C), Washington, D.C.

U.S. Environmental Protection Agency (EPA). 1999. Pesticide industry sales and usage. 733–R–98–001. Office of Prevention, Pesticides and Toxic Substances, U.S. Environmental Protection Agency, Washington, D.C.

U.S. Environmental Protection Agency (EPA). 2000. Harmonizing the Clean Air Act & Montreal Protocol Methyl Bromide Phaseouts, <http://www.epa.gov/docs/ozone/mbr/harmoniz.html> (19 December 2002)

U.S. Environmental Protection Agency (EPA). 2001. Biopesticides Registration Action Document *Bacillus thuringiensis* Plant-Incorporated Protectants (10/16/2001), <http://www.epa.gov/pesticides/biopesticides/pips/bt_brad.htm> (23 January 2003)

U.S. Environmental Protection Agency (EPA) and U.S. Department of Agriculture (USDA). 1999. EPA/USDA Position Paper on Insect Resistance Management for *Bt* Crops. June 18, 1999 (with minor revisions July 12, 1999), <http://www.epa.gov/pesticides/biopesticides> (6 March 2003)

U.S. Environmental Protection Agency (EPA) and U.S. Department of Agriculture (USDA). 2001. Pesticide Registration (PR) Notice to Manufacturers, Formulators, Producers, and Registrants of Pesticide Products, <http://www.epa.gov/oppmsd1/PR_Notice/pr2001-5.pdf> (23 January 2003)

U.S. Environmental Protection Agency (EPA) and U.S. Department of Agriculture (USDA). 2003. Pesticides: Biopesticides, <http://www.epa.gov/pesticides/biopesticides> (23 January 2003)

U.S. General Accounting Office (GAO). 2001. Agricultural pesticides: Management improvements needed to further promote integrated pest management. Report GAO–01–815, U.S. General Accounting Office, Washington, D.C.

U.S. Office of Technology Assessment (OTA). 1993. Harmful non-indigenous species in the United States. Assessment FB565. United States Office of Technology, Washington, D.C.

Valenzuela, H. R. 2000. Ecologically-based practices for vegetable crops production in the tropics. *Hortic Rev* 24:139–228.

Van Gessel, M. J. 2001. Glyphosate-resistant horseweed from Delaware. *Weed Sci* 49:703–705.

VanKirk, J. and L. Garling. [n.d.] IPM Labeling: A Discussion of Benefits and Challenges to Land Grant Involvement in IPM Labeling and Related Initiatives. Penn. State IPM Program, <www.nysaes.cornell.edu/ipmnet/indyintro.html> (9 March 2002)

Viglierchio, D. R. 1991. *The World of Nematodes*. D. R. Viglierchio, Davis, California.

Vitousek, P., L. Loope, and C. D'Antonio. 1994. Biological invasion as a global change, <http://www.gcrio,org/ASPEN/science/eoc94/eoc3/eoc3-summary.html> (23 January 2003)

Warrior, P., L. A. Rehberger, M. Beach, P. A. Grau, G. W. Kirfman, and J. M. Conley. 1999. Commercial development and introduction of DiTera, a new nematicide. *Pestic Sci* 55:376–379.

Waterworth, H. E. 1993. Processing foreign plant germplasm at the National Plant Germplasm Center. *Plant Dis* 77:854–860.

Weaver-Missick, R. and J. Lee. 1999. Ouch! The fire ant saga continues. *Agric Res* 9:48.

Wei, G., J. W. Kloepper, and S. Tuzun. 1996. Induced systemic resistance to cucumber diseases and increased growth by plant-growth promoting Rhizobacteria under field conditions. *Phytopathology* 86:221–224.

Wiles, L. J. and E. E. Schweizer. 1999. The cost of counting and identifying weed seeds and seedlings. *Weed Sci* 47:667–673.

Willey, R. W. 1979. Intercropping: Its importance and research need. *Field Crop Abstr* 32:1–10.

Williams, D. 1997. *ARS Quarterly Report*. Center for Medical, Agricultural and Veterinary Entomology, Gainesville, Florida.

Wilson, M. 1997. Biocontrol of aerial plant diseases in agriculture and horticulture: Current approaches and future prospects. *J Indust Microb Biotech* 19:188–191.

Wolfe, M. S. 1990. Intra-crop diversification, disease, yield, and quality. British Crop Protection Council. Farnham, United Kingdom. 45 pp.

Workneh, F., G. L. Tylka, X. B. Yang, J. Faghigi, and J. M. Ferris. 1999. Regional assessment of soybean brown stem rot, *Phytophthora sojae*, and *Heterodera glycines* using area-frame sampling: Prevalence and effects of tillage. *Phytopathology* 89:204–211.

Worsham, A. D. 1989. Current and potential techniques using allelopathy as an aid in weed management. Pp. 275–291. In C. H. Chou and G. R. Waller (eds.). *Phytochemical Ecology: Allelochemicals, Mycotoxins and Insect Pheromones and Allomones*. Monograph Series Number Nine. Academia Sinica, Taipei, Taiwan.

Young, L. D. 1998. Influence of soybean cropping sequences on seed yield and female index of the soybean cyst nematode. *Plant Dis* 82:615–619.

Ypema, H. L. and R. E. Gold. 1999. Kresoxim-methyl: Modification of a naturally occurring compound to produce a new fungicide. *Plant Dis* 83:4–17.

Zimmerman, K. 2000. Gypsy Moth in North America, <http://www.fsl.wvu.edu/gmoth/home.html>.

Zurek, E. 2002a. *About Pest Management Centers*. (USDA Regional Pest Management Centers). <http://www.pmcenters.org/about.html>

Zurek, E. 2002b. *Welcome to the National Site for the USDA Regional Pest Management Centers Information System,* <http://www.pmcenters.org/>

Related Web Sites

http://www.ams.usda.gov/nop/
http://schoolipm.ifas.ufl.edu
http://www.sarep.ucdavis.edu/concept.htm
http://www.attra.org/attra-pub/ipm.html
http://www.efn.org/~ipmpa
http://www.invasivespecies.org
http://ipmwww.ncsu.edu
http://www.ipm.ucdavis.edu/
http://www.sare.org/htdocs/docs/search.html [Sustainable Agriculture Network]

3 New and Emerging Technologies

New technologies based on computers, cell phones, biotechnology products, the World Wide Web, and so-called "smart" machines are affecting at an ever-increasing pace many disciplines, including agriculture, medicine, communications, aviation, navigation, education, and commerce. These new technologies are being applied to agriculture by private companies or through evolving public and/or private research (Figure 3.1). Likewise, *bioinformatics*, or the merging of genetic and information technologies, is affecting pest management as a component of integrated crop production. Three new emerging technologies likely to affect integrated pest management (IPM) adoption are (1) precision farming tools to deliver site-specific IPM, (2) genetically modified organisms (GMOs), and (3) integrated systems.

Site-Specific Integrated Pest Management

Precision Farming

Many food and fiber producers know that crop production and pest pressures can differ spatially across production landscapes (Figure 3.2). From the combine, farmers harvesting crops can see weedy patches or dry-knoll crop effects, for instance. Differences in crop yield potentials can be due to differences in soil type and quality, soil water-holding capacity, and/or soil fertility. Although crop yield rates differ within a field, growers traditionally apply fertilizers, pesticides, water, and seeds uniformly, without regard to site-specific needs or to the production potentials of different field zones.

Precision farming, or site-specific agricultural management, is being promoted actively by several agribusiness sectors. The goal of precision farming, which can make use of computers, global positioning system (GPS) receivers, advanced computer software, geographical information systems (GIS), management maps, decision support system (DSS) models, and variable-rate application equipment to modify farming input levels while monitoring final crop-yield, is to apply an appropriate amount of inputs at an appropriate time on an appropriate area. By using crop harvesters equipped with GPS and yield monitors, farmers can generate yield maps identifying spatial yield variations. Such yield data have helped farm-

Figure 3.1. To make variable-rate applications, modern farm equipment is equipped increasingly with computers, monitors, global positioning system receivers, and variable-rate controllers. Photo courtesy of P. Westra, Colorado State University.

Figure 3.2. As a result of many variable factors affecting crop production, crop growth and yield differ across fields and regions. Photo courtesy of P. Westra, Colorado State University.

Figure 3.3. Use of a crop harvester equipped with a yield monitor allows generation of a yield map showing yield variability within a field. Photo courtesy of P. Westra, Colorado State University.

ers appreciate a wide variety of variable aspects within their fields, as well as the need to implement site-specific management of fields (Figure 3.3).

Site-Specific Input Management

Fertilizer and Pesticide Application

Although practiced in most areas, high application rates of fertilizers and pesticides on all soils cannot be justified. Matching fertility and pesticide inputs to both soil type and potential crop-yield should increase profits for growers using these IPM strategies (Figure 3.4). After discounting for the cost of accurate sampling, growers can realize significant cost savings in certain instances by engaging in site-specific input management, especially when sampling cost can be spread over several years (Figure 3.5). Such prescription applications may lower input costs and protect environmental quality. More and more agriculture input dealers are offering variable applications of fertilizers and chemicals by means of specialized equipment (Fleming, Westfall, and Heermann 1999). Much pioneering site-specific research has been conducted with variable nutrient applications in fields with soils that differ in texture, organic matter, content, pH, and yield potentials.

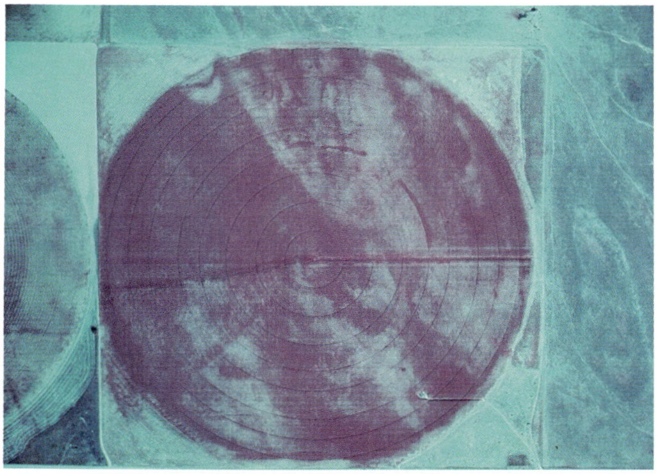

Figure 3.4. Knowledge of variation in soil type obtained from this kind of aerial photography is used to map different management zones across a field. Photo courtesy of P. Westra, Colorado State University.

Water Application

Irrigation is a very important factor in U.S. food production: the 15% of U.S. cropped acreage that is irrigated produces 40% of U.S. crop value. Approximately one-third of irrigated land (more than 17 mil-

Figure 3.5. Sampling for pest presence and density can improve IPM strategies to improve crop production. Photo courtesy of P. Westra, Colorado State University.

Figure 3.6. Precision irrigation and pesticide application will be possible with the new-generation irrigation systems under development. Photo courtesy of P. Westra, Colorado State University.

lion acres [a.]) is irrigated by self-propelled sprinkler systems, which can be automated readily. Computerized controls, now offered by all manufacturers of these systems, have tremendous potential to improve water control so that a crop environment conducive to pest management can be maintained. Computerized controls also have the potential to control chemical applications so that they are made only when needed (Duke et al. 1997) (Figure 3.6.).

Recently developed controls allow the irrigator to apply variable amounts of water to different portions of a field, perhaps to account for areas with more or less vegetation or to control spread of disease by limiting water application in the affected area. Networks of automated agricultural meteorological stations in many states now provide near-real-time data for prediction of pest outbreaks and for management of crop water applications (Doesken et al. 1998) (Figure 3.7). By integrating a sprinkler system control with a computer program estimating crop water needs (Buchleiter, Heermann, and Duke 1995), an irrigator can minimize over-irrigation problems such as ponded water from excessive runoff or waterlogged soils. This approach also decreases the amounts of leached water-soluble pesticides and soil-applied nutrients such as nitrate.

Figure 3.7. Automated weather stations can provide real-time, site-specific weather data that can be used to seed crop-management decision support systems. Photo courtesy of P. Westra, Colorado State University.

Prescription Crops

Traditionally, plant breeders have focused on crop yields, agronomic traits, or pathogen and/or insect resistance. More recently, plant breeding methods have been combined with new biological techniques to produce crop cultivars with unique properties such as resistance to pests, pesticides, heat, drought, and salinity. Precision-farming tools designed to identify and to describe variability across a field or a landscape—with respect to pest populations, soil attributes, and environmental conditions—will lend themselves readily to the planting, according to a prescription map, of a patchwork of crop cultivars across an area. Companies are engineering specialized farm equipment capable of planting crops at variable rates or of planting, in a single pass, different cultivars in different areas of a field.

Automated Data Acquisition

The high cost of spatial data collection is triggering widespread research into automated data collection. Sensing technologies can be categorized broadly, based on distance from the crop pest locus, as either remote or proximate.

Remote Sensing

To monitor ecosystems or to allow for the making of time-sensitive management decisions, remote sensing generally involves data collection from satellite or airborne platforms (Figure 3.8). The first satellite designed specifically to collect data from the earth's surface was Landsat 1, which was launched in 1972. Landsat 1 multispectral data have been used in agriculture, forestry, geology, land-use planning, and other areas. Hatfield and Pinter (1993) provide an excellent review of the use of remote sensing in the assessment of crop conditions.

New and emerging applications of this technology are being developed rapidly. Operational application

Figure 3.8. **Multiband cameras in airplanes or on ground rigs can be used to detect nutrient, water, or disease stress in crop plants. Photo courtesy of P. Westra, Colorado State University.**

of remotely sensed crop coefficients for estimation of site-specific crop water use is under development by engineers (Bausch 1995), and various scientists are developing and refining techniques using remote sensing to assess plant nitrogen status. Advanced computer vision, size, and shape characteristics can be used to identify grass and broadleaf weed species for spot spraying. This system mounted on field spraying equipment requires additional developing, refining, and testing under large-scale field conditions (Meyer et al. 1998) (Figure 3.9).

Moran, Inoue, and Barnes (1997) reviewed the potential for image-based remote sensing to provide spatially and temporally distributed information for site-specific crop management or precision farming. Topics reviewed, for which information on variability would be useful to precision farming, included soil moisture content, crop phenology, crop growth, crop water evaporation rate, crop nutrient deficiency, crop disease, weed infestation, and insect infestation. Current limitations for image-based remote sensing applications are due mainly to restricted spectral range, coarse spatial resolution, slow turnaround time, and inadequate repeat coverage. Upcoming commercial Earth-observation satellites, however, may provide the resolution, timeliness, and quality required for the monitoring of seasonally variable crop/soil conditions and for the managing of crops in a time-critical manner. Private and public entities will need to cooperate to develop infrastructure for incorporating remote sensing technology into precision farming. Data collection often will be contracted on a fee-for-service basis, with a single company providing the final product to the farmer.

Figure 3.9. **In the future, sensors may be mounted on tillage, spray, or irrigation equipment to collect data for data maps while other operations are being carried out. Photo courtesy of P. Westra, Colorado State University.**

Information obtained through remote sensing may not be available immediately for pest management purposes. Commercial sales of remote sensing information have created a new support industry for agriculture.

The IKONOS satellite used by Space Imaging Corporation will deliver one-meter resolution images recollected every three days to deliver, for marketing purposes, time-sensitive information such as crop growth stage, crop stress, and canopy coverage. Such time-sensitive information may be crucial to the early detection of disease, insect, drought, or fertility stress, especially in valuable horticultural crops.

Certain commercial companies offer essential site-specific information and management interpretations daily to growers and agribusinesses. For example, VantagePoint Network, an Internet-based crop management and record keeping system, offers growers a low-cost, easy-to-use information system that includes tracking of crop production activities, grain storage, and sales, as well as creation of farm maps, all in an environment keeping important information private for exclusive use by the grower. Aided by remote sensing data and by crop managers specializing in the interpretation of such data (Figure 3.10), users can plan, predict, and manage productivity and profits better than ever before. Companies such as VantagePoint Network often are *information multipliers*, which interpret data before delivering results to clients. Once trained, growers may be able to perform their own image analyses.

An exciting area of new remote sensing research involves the use of radar technology to collect site-sensitive weather information. The resolution of such multitemporal data is improved when radar output

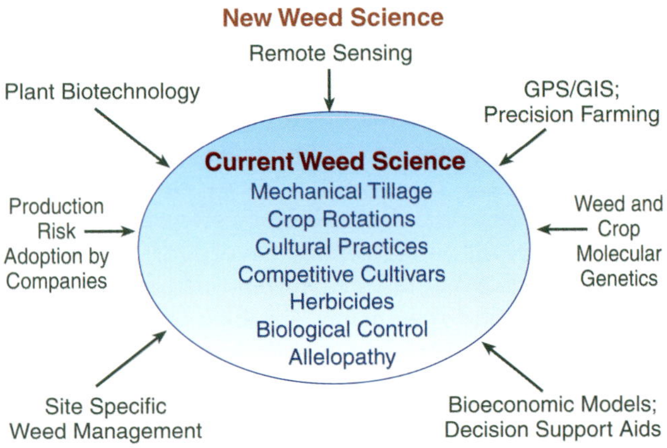

Figure 3.10. Many crop management disciplines will undergo radical change to adapt to new technologies' powerful effects on pest management and crop production. (Source: P. Westra, Colorado State University.)

is compared with both optical satellite images and GIS data.

In-Field Sensing

Proximate sensing typically is performed by means of tractor-mounted equipment. Such a system is best suited to stationary attributes, so its pest management applications have tended to focus on weed recognition. In-field weed recognition was first used to guide site-specific spraying of general-purpose herbicides such as glyphosate in unplanted fields. More recent research has used image processing based on leaf shape (Lee, Slaughter, and Giles 1999) and infrared reflectance. Because proximate sensors are close to their targets, they often control site-specific herbicide application equipment. In rare instances, weed control has been achieved by means of robots (Lee, Slaughter, and Giles 1999).

Figure 3.11. This image represents a 140-acre irrigated pivot with crop management zones developed from aerial photography and soils testing (See also Figure 3.4). Colors represent the following crop production potentials: blue, high; green, medium; purple, low; and brown, very low. (Source: P. Westra, Colorado State University.)

Field Management Zones

Technologies such as remote sensing, GPS/GIS, yield monitors, and computer management applications are changing the way growers look at crop production over a variable landscape (Fleming, Westfall, and Heermann 1999). Early site-specific crop management dealt with applications of fertilizer or lime, whose rates were varied according to the results of soil test samples taken on 2- to 4-a. grids in large fields (Fleming et al. 1998). Interpolation of results from grid samples allowed applications to be prescribed at distinct levels for different field areas.

Although grid sampling remains popular with certain farmers, the cost of intensive sampling is prohibitive under many conditions. To decrease costs, researchers are evaluating a field-zone management approach based on grower knowledge of production fields coupled with soil color assessment incorporating aerial photography of fields. This combination of information may help users identify different management zones requiring minimal ground truth to maximize crop yields (Figure 3.11).

Recent research has shown that production management zones often are associated with soil type or color (Fleming, Westfall, and Heermann 1999) and that black-and-white aerial photographs can be enhanced digitally to create contrasts based on soil color. Bare soil surface spectral properties were influenced largely by soil organic carbon and moisture. The grey-tone pattern in black-and-white aerial photographs might have been a reflection of these soil properties. Furthermore, soil color on photographic images might have correlated strongly with the soil's organic matter spatial distribution, as determined by grid soil-sampling. The semivariograms from the two methods of determining soil organic matter were similar; and, of greater significance, soil organic matter was correlated highly with winter wheat yields. Thus it would seem that soil organic matter measured by means of remote sensing can be used to identify different productivity zones within a field.

Spatial and Biological Strategies

Precision farming and management zones are important components of site-specific IPM, for pest populations can correlate with other field attributes. Integrating IPM with seed, fertility, and water management will allow farmers to practice integrated, sustainable agricultural production. New information-acquisition tools make it possible to gather and to analyze data regarding past, present, and future plant growing conditions at levels of detail and speeds never before contemplated. Two areas for innovative IPM are emerging as a result. First, site-specific IPM uses spatial information to focus pest management where needed. Site-specific IPM integrates data collection technologies (field sampling and sensing), reliable location data provided by satellite-based GPS, spatially referenced data management (GIS), and automated spatial application technologies. Second, developmental stage-specific IPM uses crop and pest biological information to predict when pest-control intervention is warranted (Swinton 1998). Typically, such information is embodied in computerized DSS models simulating pest-crop interactions. Certain models simulate individual plant-pest relations, whereas others predict infestation conditions. These models automate the threshold-based decision rules of classic IPM.

Pest scouting information typically is used to "seed" models. Because scouting likely will be more intensive with site-specific IPM than with uniform pesticide applications, scientists are studying pest distribution patterns to identify cost-effective sampling plans and are characterizing relations between pest populations and other field attributes to develop less-expensive directed sampling plans (Heerman et al. 1999). At the time of this writing, most pest management research remains at the stage of either pest spatial distribution studies or cost-effective spatial data collections. Although several simulation studies have indicated the feasibility of spatial pest management, few field experiments have been conducted of weed management (the most common application of site-specific pest management), and virtually none of insects, nematodes, or disease management (Swinton 2003).

Effective use of the models programmed with accurate pest and soil attribute data, together with site-specific IPM methods, permits growers to tailor pest controls. These technologies hold the promise of enhancing IPM adoption in a variety of cropping systems by integrating and automating certain complex decisions regarding pest management.

Agricultural Product Identity Preservation

Biotechnology and genetic engineering are being applied to crops to improve nutrient availability, pest and disease control, herbicide resistance, and tolerance to environmental stresses such as heat, drought, and salinity. Biotechnology also is expected to provide solutions for conserving finite natural resources such as fertilizers and petroleum-based inputs; improving food nutrition, taste, and texture; decreasing the incidence of food allergies; and producing pharmaceuticals and *nutraceuticals* (i.e., foods with specific health effects).

An emerging public and private concern is the identity preservation of agricultural products possessing novel plant traits or characteristics. These products likely will need to be identified and segregated, starting with planting, through harvesting, storing, processing, packaging, and distributing. This system already is in place for the producing and marketing of organic foods; expanding the system to include the major commodities corn, soybean, and wheat will require significant investment in infrastructure, however.

World Wide Web sources providing up-to-date details on new technologies, programs, and concepts in IPM are proliferating. Sites providing growers access to current information about weather, pest outbreaks, commodity prices, agricultural policy, and DSS models can only increase in importance. Examples of sites appear at the end of this chapter.

Transgenics and Resistance Management

Transgenic plant technology, which holds out the promise of precision genetics, exists to develop crop varieties that can be prescribed to mitigate specific agricultural pressures such as those involving disease, insects, soil moisture, soil quality, or weeds. Precision agriculture used in conjunction with custom-tailored varieties will provide growers with unprecedented opportunities to boost revenues by producing for niche markets.

Transgenic varieties of many economically important plants have been produced. The commercial success of transgenic varieties of row crops including soybean, corn, cotton, and canola has been demonstrated since their respective releases, beginning in 1996 (Figure 3.12). Other transgenic crops such as potato, squash, papaya, and tomato have been developed, but some, such as potato, are not being grown commercially. These crops have received agronomic traits such as virus resistance, insect resistance, and herbicide tolerance. Since their first demonstration in a laboratory in 1983, transgenic crop varieties have been adopted widely in the United States.

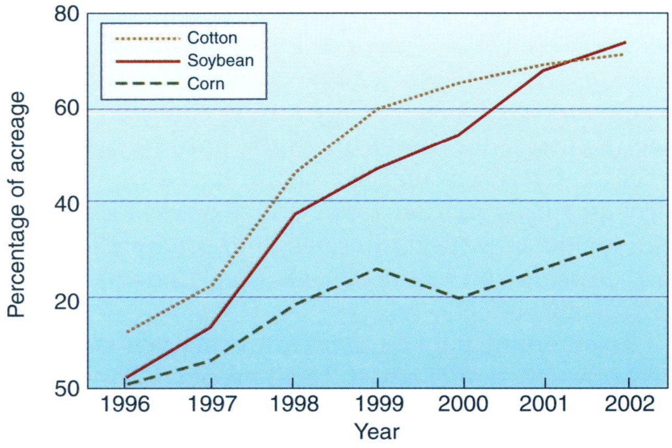

Figure 3.12. A dramatic increase in the extent of transgenic crop planting in the United States has given rise to new IPM technologies that decrease soil erosion and insecticide use. (Source: P. Westra, Colorado State University.)

Yet transgenic plant technology is in its infancy. Currently, only single-gene agronomic traits have been commercialized. The new traits, introduced by means of *Agrobacterium*-mediated gene transfer or particle bombardment-mediated gene transfer, have been expressed stably over multiple generations since the completion of a strenuous product development and breeder's selection process. Introduced traits behave as a single dominant gene and can be backcrossed easily into many elite varieties.

Herbicide-resistant crops have been developed in which single genes (such as those encoded by pat or bar genes) produce proteins that deactivate herbicides into nontoxic byproducts. Other modes of herbicide resistance are provided by single genes encoding for enzymes that normally are the targets of herbicides. For example, the CP4 5-enolpyruvyl-3-shikimate phosphate (EPSP) synthase gene encodes for a glyphosate-insensitive EPSP synthase, and the ALS gene encodes for a mutant acetolactate synthase gene insensitive to sulfonylurea herbicides (Duke 1996).

Herbicide-resistant crops can provide growers with significant weed control and crop safety advantages as well as less expensive and easier-to-use weed management practices (Carpenter and Gianessi 1999; Fulton and Keyowski 1999). When deciding whether to use an herbicide-resistant crop variety, growers often base decisions on the current cost of, and difficulty of, the weed control regime on their own farms. Another consideration is the fit of herbicide resistance with conservation tillage or no-till agriculture. For example, Roundup Ready soybean can make it easier for a farmer to adopt no-till management systems. Roundup Ready soybean, corn, and cotton have been commercialized in several countries and have been adopted rapidly by farmers because of the economic benefit and the time savings associated with the technology. Although released in the United States in the year 2000, Roundup Ready sugar beets have failed to gain commercial acceptance as a result of GMO market concerns. Widespread adoption of herbicide-resistant crops may affect the agricultural system, and wise management will be necessary to sustain their value (Hillyer 1999). The effects of the development of herbicide-resistant weeds or of the shifts to more-difficult-to-manage weeds can be minimized if management methods involving crop rotation and herbicide mixtures are used to minimize those risks associated with the technology.

Insect-tolerant crops have been developed through genetic engineering, and those that have been commercialized to date contain genes obtained from *Bacillus thuringiensis* (*Bt*). A wide variety of *Bt* proteins exists, and each protein has a specific spectrum of activity against specific classes of insects. Currently on the market are insect-tolerant cotton and corn. Insect-tolerant crop varieties have demonstrated both decreased need for pesticides and increased yields, especially when grown in areas with great insect pressure (Klotz-Ingram et al. 1999; Traxler and Falck-Zepeda 1999). As with herbicide-resistant crops, prescription of management practices such as crop refugia and crop rotation strategies must be followed to decrease the risks of insect resistance.

Crops that are herbicide and insect resistant have the potential both to decrease the total amount of chemicals required to control pests and to provide (1) alternative mechanisms for controlling weed and insect pests and (2) additional technologies that can be incorporated easily into IPM systems. New genetic technologies such as chemical induction of resistance traits and of genes for drought or stress tolerance will provide other means of tailoring variety to the agricultural management systems as well as to the land itself (Kendall et al. 1997). New traits will be produced not only by means of genetic engineering techniques, but also by means of discoveries generated by new techniques in plant genomics and bioinformatics. Genes for yield, disease resistance, and stress tolerance will be identified and moved by marker-assisted breeding into elite varieties from wild crop relatives faster and more precisely than possible with traditional plant-breeding techniques. New crop varieties, crop management practices, and precision farming techniques developed in tandem will help growers make the most of the productive potentials of land.

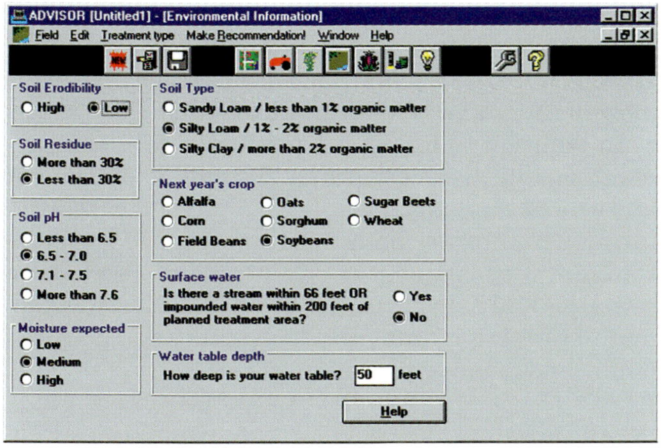

Figure 3.13. **Pest management decision support systems will be integrated into systems designed to help manage all aspects of food and fiber production. (Source: P. Westra, Colorado State University.)**

Integrated Systems

An integrated, whole-farm, systems approach to agricultural research and management is needed greatly. Sustainable agriculture has become a complex system demanding the consideration of many interrelated factors, processes, and institutions. Producers must be able to respond to fluctuations in weather and commodity prices, to trends in both federal and state legislation, and to perceptions entertained by the urban public. The ability to modify farm and ranch management practices wisely and rapidly—to take advantage of the global economy, new cropping, pest management, and tillage systems, and new legislation while protecting soil, air, and water resources—will determine whether an agricultural enterprise flourishes or perishes. Geospatial systems will provide a potential bridge among growers; crop consultants; seed and chemical salespeople; U.S. Department of Agriculture, U.S. Environmental Protection Agency, and university personnel; and private research groups. Such bridging benefits will affect education, usability, accessibility, and affordability.

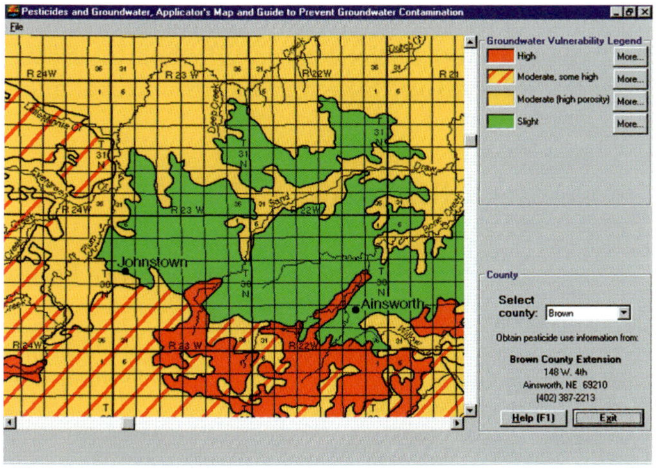

Figure 3.14. **Integrated systems can contain components that help growers manage fertility and pesticide inputs at levels protecting groundwater and the environment generally. (Source: P. Westra, Colorado State University.)**

Decision support systems are computer-based models incorporating data and user knowledge inputs to assist crop production management decisions (Figure 3.13). Over the past 20 yr, DSS models typically have focused on one aspect of crop production over one or more growing seasons. Examples of these models include WeedSOFT, developed by researchers at the University of Nebraska to help producers make weed management decisions while alerting them to environmental impact issues (Figure 3.14); and INTERCOM, an ecophysiological model analyzing crop-weed interference and competition. More recently, DSSs have evolved to incorporate several subcomponent models to help growers predict diversified crop or livestock production over several seasons.

GPFARM, a computer-based DSS providing fast and efficient crop and livestock management support at the whole-farm and ranch level, emphasizes the management of water, nutrients, and pesticides across a range of cropping, grazing, and integrated crop-livestock systems. GPFARM is targeted for use by agricultural consultants, computer-oriented producers, Extension personnel, and the National Resources Conservation Service (Figure 3.15). GPFARM also has strong links to economic and environmental analysis, site database generation, and site-specific management from which alternative sustainable management strategies can be developed and tested for each farm or ranch. An information system with direct ties to the Internet provides help to users as they develop and test alternative systems. GPFARM can simulate crop growth; soil processes; management changes; fertilizer and pesticide application effects (including those of amendments such as manure and sewage sludge); soil productivity; wind and water erosion; nutrient and pesticide runoff; nitrate and pesticide leaching; and extensive cropping, tillage, and livestock systems. Farm-enterprise economic budgeting procedures are used to determine farm profitability in terms of net costs and returns. Variable costs for each enterprise are calculated from the required production inputs, and a standard budget shows returns versus expenses.

Decision Support System for Agrotechnology

Transfer (DSSAT) is a computer-based software program combining crop, soil, and weather databases (and programs to manage them) with crop models and application programs, for the purpose of simulating multiyear outcomes of crop management strategies. Integration of these multiple input variables allows users to ask "what if" questions and to simulate results. Currently, the DSSAT shell is able to incorporate models of 15 different crops, including several cereal grains, grain legumes, and root crops. PLANETOR is a comprehensive environmental and economic farm-planning software program. To evaluate the effects of decreasing or otherwise changing pesticide, nitrogen, phosphorus, and/or manure applications, tillage systems, and/or crop rotations, PLANETOR combines site-specific environmental models with individual-farm financial planning data. Users enter detailed data regarding crop production practices and rotations, production costs, levels, and variability, and farm financial transactions and status. Then the software calculates the potential consequences of various production practices, evaluates alternative management plans, and compares the effects on soil erosion, nitrate leaching, phosphorus runoff, pesticide movement, and whole-farm profitability.

In addition to improving prediction and management capabilities, integrated systems management and the accompanying software packages require farmers to consider both the economic and the environmental consequences of alternative production scenarios and help farmers to acquire a more thorough understanding of integrated systems and their economic and environmental effects.

Figure 3.15. Pest numbers generated from well-designed scouting activities often are used to "seed" decision support systems. Increasingly, such pest population dynamic values are being collected remotely. (Source: P. Westra, Colorado State University.)

Appendix A. Abbreviations and Acronyms

a.	acre
Bt	*Bacillus thuringiensis*
DSS	decision support system
DSSAT	decision support system for agrotechnology transfer
GIS	geographical information system
GMO	genetically modified organism
GPS	global positioning system
IPM	integrated pest management

Appendix B. Glossary

Bioinformatics. Combining genetic and information technologies.

Information multiplier. Companies that interpret data before delivering results to clients.

Nutraceuticals. Foods with specific health effects.

Precision farming. Site-specific agricultural management.

Literature Cited

Bausch, W. C. 1995. Remote sensing of crop coefficients for improving the irrigation scheduling of corn. *Ag Water Manage* 27:55–68.

Buchleiter, G. W., D. F. Heermann, and H. R. Duke. 1995. Irrigation scheduling with climactic data. In *Proceedings of the Workshop on Computer Applications in Water Management. CO WRRI Info Serv* 79:61–65.

Carpenter, J. and L. Gianessi. 1999. Herbicide tolerant soybeans: Why growers are adopting Roundup Ready varieties. *AgBioForum* 2(2).

Doesken, N. J., H. R. Duke, B. L. Hamblen, J. Kleist, T. B. McKee, M. S. McMillan, and H. F. Schwartz. 1998. The Colorado agricultural meteorological network (COAGMET): A unique collaborative system supporting Colorado agriculture. Pp. 240–242.

In *Proceedings of the Twenty-Third Conference on Agriculture and Forest Meteorology*. American Meteorological Society, Albuquerque, New Mexico.

Duke, S. (ed.). 1996. *Herbicide-Resistant Crops. Agricultural, Environmental, Regulatory, and Technical Aspects*. Lewis Publishers, New York.

Duke, H. R., G. W. Buchleiter, D. F. Heermann, and J. A. Chapman. 1997. Site specific management of water and chemicals with center pivot irrigation systems. Pp. 273–280. In J. V. Stafford (ed.). *Precision Agriculture '97*. BIOS, Oxford, United Kingdom.

Fleming, K. L., D. F. Heermann, D. G. Westfall, D. B. Bosley, F. B. Peairs, and P. Westra. 1998. Precision farming: From technology to decisions. A Case Study. Pp. 1777–1783. In P. C. Roberg, R. H. Rust, and W. E. Larson (eds.). *Proceedings of the Fourth International Conference on Precision Agriculture*. Precision Agriculture Center, University of Minnesota, St. Paul.

Fleming, K. L., D. G. Westfall, and D. F. Heermann. 1999. Farmer developed management zone maps for variable rate fertilizer application. Pp. 917–926. In J. V. Stafford (ed.). *Proceedings of the Second European Conference on Precision Agriculture*. Odense Congress Center, Denmark, July.

Fulton, J. and L. Keyowski. 1999. The production of herbicide-resistant canola. *AgBioForum* 2(2).

Hatfield, J. L. and P. J. Pinter, Jr. 1993. Remote sensing for crop protection. *Crop Protect* 12:403–411.

Heerman, D. F., J. Hoeting, H. R. Duke, D. G. Westfall, G. W. Buchleiter, P. Westra, F. B. Peairs, and K. L. Fleming. 1999. Interdisciplinary irrigated precision farming research. Pp. 121–130. In J. V. Stafford (ed.). *Proceedings of the Second European Conference on Precision Agriculture*. Odense Congress Center, Denmark, July 15.

Hillyer, G. 1999. Biotechnology offers U.S. farmers promises and problems. *AgBioForum* 2(2).

Kendall, H. W., R. Beachy, T. Eisner, F. Gould, R. Herdt, P. Raven, J. Schell, and M. Swaminathan. 1997. Bioengineering of crops. Panel on Transgenic Crops, The World Bank, Washington, D.C.

Klotz-Ingram, C., S. Jans, J. Fernandez-Cornejo, and W. McBride. 1999. Farm-level production effects related to the adoption of genetically modified cotton for pest management. *AgBioForum* 2(2).

Lee, W. S., D. C. Slaughter, and D. K. Giles. 1999. Robotic weed control system for tomatoes. *Precision Agric* 1:95–113.

Meyer, G. E., T. Mehta, M. K. Kocher, D. A. Mortensen, and A. Samal. 1998. Textural imaging and discriminant analysis for distinguishing weeds for spot spraying. *Trans ASAE* 41:1189–1197.

Moran, M. S., Y. Inoue, and E. M. Barnes. 1997. Opportunities and limitations for image-based remote sensing in precision crop management. *Remote Sens Environ* 61:319–346.

Swinton, S. M. 1998. Precision farming as green and competitive. In S. S. Batie, D. E. Ervin, and M. A. Schulz (eds.). *Business-Led Initiatives in Environmental Management: The Next Generation of Policy?* Workshop of the American Agricultural Economics Association/Association of Environmental and Resource Economists/International Food and Agribusiness Management Association, Toronto, July 26, 1997.

Swinton, S. M. 2003. Site-specific pest management. Pp. 155-168. In F. den Hond, P. Groenewegen, and N. van Straalen (eds.). *Pesticides: Problems, Improvements, and Alternatives*. Oxford, Blackwell.

Traxler, G. and J. Falck-Zepeda. 1999. The distribution of benefits from the introduction of transgenic cotton varieties. *AgBioForum* 2(2).

Related Web Sites

Integrated Pest Management

Appropriate Technology Transfer for Rural Areas. 2001. Biointensive integrated pest management (IPM): Part one of two. Fundamentals of sustainable agriculture, July, <http://attra.org/attra-pub/ipm.html> (30 May 2002)

Emerge, Incorporated. 2002. Homepage, <http://emergeweb.com> (30 May 2002)

Gempler's, Incorporated. 2000. IPM almanac, <http://www.ipmalmanac.com> (30 May 2002)

Jointed goatgrass initiative: Bioeconomic model for jointed goatgrass research and management. 2002. <http://www.homepage.montana.edu/~jgg> (30 May 2002)

National Science Foundation–Center for Integrated Pest Management. 2002. North Carolina pest management information, <http://ipm.ncsu.edu> (30 May 2002)

U.S. Department of Agriculture–Cooperative State Research, Education, and Extension Service. 2002. Plant and animal systems, January 24, <http://www.reeusda.gov/ipm> (30 May 2002)

Site-Specific Management

California Polytechnic State University. 2001. Precision agriculture, December 14, <http://www.precisionag.org> (30 May 2002)

Dana, P. H. 2000. Global positioning system overview, May 1, <http://www.colorado.edu/ geography/gcraft/notes/gps/gps.html> (30 May 2002)

John Deer, Incorporated. 2002. Crop Tracer homepage, <http://www.vantagepoint.com> (30 May 2002)

University of Georgia–National Environmentally Sound Production Agriculture Laboratory. 2002. Homepage, <http://nespal.cpes.peachnet.edu> (30 May 2002)

Biotechnology

AgBioForum. 2002. Homepage, <http://www.agbioforum.missouri.edu/main.htm> (30 May 2002)

Pioneer Hi-Bred International, Incorporated. 2002. Biotech issues update, <http://www.pioneer.com/ biotech/default.htm> (30 May 2002)

Sciencemag. 2002. Homepage, <http://www.scientificjournal.com/side_bg.html> (30 May 2002).

4 Crop Production Systems

Although systems research offers ongoing challenges (Merrigan 2000), noteworthy progress has been made in incorporating integrated pest management (IPM) into most crop production systems (Fernandez-Cornejo and Jans 1999). The concept of a cropping system for pest management has been expanded to encompass many of the nontechnical tools described in Chapter 2. Antagonistic plants, trap crops, refugia, and cover crops are being used to enhance the activity of soilborne and foliage-inhabiting beneficials. Shifts in the time of planting and/or harvesting and in the timing and patterns of necessary pesticide applications have proved effective in IPM programs. The planting of grass and other cover crops or fallows, as well as other practices, are being incorporated into cropping systems to increase soil organic matter, to suppress crop pests, to decrease tillage so as to preserve existing organic matter, and to minimize erosion (Barker and Koenning 1998; Magdoff and van Es 2000). Many cropping systems that augment the activity of beneficial soil microorganisms have been developed (Barker and Koenning 1998; El Titi and Ipach 1989; Taylor and Rodriguez-Kábana 1999a,b; Zwart et al. 1994). The evolving philosophy of both IPM and the more recently proposed ecologically based pest management (EBPM) addresses the fact that significant levels of predation and/or parasitism are desirable insofar as they promote diversity and sustainability of agroecoystems (Cardina et al. 1999; Kennedy and Sutton 2000; NRC 1996). Thus, cropping systems are beginning to address soil and crop health as well as specific pest management goals.

Rotating certain crops that include nonhosts for given crop pests is a practice that has been used for centuries in subsistence and traditional agricultural systems. For example, six- to eight-year rotations and fallow have long been used to avoid the low potato yields associated with the potato cyst nematode (Barker and Koenning 1998). Crop rotation supports diversity in time and space and often is the favored means of managing soil-inhabiting pests such as nematodes. Still, the need to rotate depends on primary crop, location, and key pests. Corn may be grown successfully in monoculture in certain regions, but elsewhere monoculture may cause significant root pathogen problems. Certain *Brassica* spp. such as rape and mustard release glucosinolates in the soil, and these compounds and related decomposition products suppress the activity of numerous soilborne plant pathogens, especially nematodes. Crop rotations may be combined with other pest management practices to develop economic-loss estimates and models. While working with a peanut-cotton and velvet-bean rotation (plus or minus pesticides), Taylor and Rodriguez-Kábana (1999b) demonstrated the high cost of crop pathogens relative to the economic return from beneficials. In contrast to the yield losses caused by the white mold disease and root-knot nematodes, numbers of the beneficial microbivorous nematodes were positively related to yield of peanut as well as cotton.

Decreased tillage is becoming a widely accepted practice, largely for the purpose of soil conservation. Other benefits from decreased tillage include enhanced soil organic matter, with improved surface residue, soil structure, and water infiltration. Although causing soil compaction and altered pest problems, limited tillage favors earthworms and other beneficial fauna such as bacterivorous and fungivorous nematodes (Barker and Koenning 1998). Comprehensive integrated management farming systems that incorporate limited tillage, fertilization, and pesticide use, add organic manure, and seed clover cover crops were shown to cause a sixfold increase in number of earthworms (El Titi and Ipach 1989). Predatory mites and other beneficial microflora greatly increased in integrated plots while cereal cyst and stem nematodes were suppressed. Until recently, weed management concerned primarily the control of one or more weeds by means of a single technology (Elmore 1996; Liebman, Mohler, and Staver 2001). To limit herbicide application costs and adverse environmental effects, multiple tactics, or "tools," such as crop rotation, mechanical control, and crop competition are recommended (Liebman and Gallandt 1997).

Extensive effort has been directed at characterizing and modeling the many interactions of crop and pest systems. Many vegetable and fruit crop models focus on one or on a few pests within a similar taxonomic group. Crop simulation models have been developed for certain crops, but few of these models are comprehensive. For example, the CROPGRO–soybean simulation model has been used in the evaluation of various risks, including those related to water use (Alagarswamy et al. 2000) and associated pests. The goal of fully integrating key soybean pests in this system has not been reached yet. Statistical models have promise for predicting corn and soybean yields in given soils in Illinois (Garcia-Paredes, Olson, and Lang 2000). The erosion productivity impact calculator model shows promise for exploring the influence of rotation type (shallow vs. deep-rooted crops) and irrigation type on nitrogen leaching (Cavero et al. 1999). Although challenges related to an overall systems approach (including crop pests) remain, an integrated assessment model (Bland 1999) also has promise. At a practical as well as at a general research level, the cropping systems approach has been used successfully for numerous crops (Veseth 1995). In considerations of variable costs and input usage, improved climate forecasts may have value in this regard (Mjelde and Hill 1999). For example, Main and colleagues (2001) have created climate-based models for predicting development and spread of crop foliage pathogens. Ongoing, long-term research programs being conducted at major centers for integrated crop and plant systems, such as the center at Michigan State University, also have much promise (see Center for Integrated Plant Systems 2002).

This chapter illustrates effective pest management systems by presenting in each section concepts and examples of production systems for selected major crops. Pest management strategies (the overall plan) and tactics (the specific tools) may differ for crops in different geographic regions of the United States. Fortunately, this information is available readily in published form and on the Internet (English-Loeb 1999; Gray 2001; Van Kirk 1998). National web sites also are available on IPM generally (e.g., Consortium for Integrated Crop Protection 2002; USDA–CSREES 2002), on entomology (e.g., Fan 1998), and on crops (e.g., Bajwa 2000) that provide invaluable information. Descriptions of pest groups and their management also are available by crop, for most states. For example, the American Phytopathological Society published a series of plant disease compendia for many field, fruit, vegetable, and ornamental crops. For soybean, cotton, and corn diseases, see Hartman, Sinclair, and Rupe (1999); Watkins (1981); and White (1999). Additionally, a number of books published specifically on crop health are very useful (e.g., *Wheat Health Management* [1991], by R. J. Cook and R. Veseth). A recent issue of the *Journal of Crop Production* was dedicated to "expanding the context of weed management" (Buhler 1999), as was the recent book *Ecological Management of Agricultural Weeds* (Liebman, Mohler, and Staver 2001). Many publications on pest and crop management framed in ecological and environmental terms are forthcoming in the next few years. Current general use patterns of biological, cultural, pesticide, and monitoring practices in field crops are summarized in Table 4.1.

Field Crops

Field, including forage, crops generally are low-value crops/unit production area. To earn a reasonable profit, commercial producers must grow corn, soybean, small grains, cotton, and forage crops on large acreages. Producers' main interest is annual profit margin (Posner, Casler, and Baldock 1995), but they also are interested both in the effects of production strategies on the environment and in the results of decreased-input systems.

A tendency has existed to develop management strategies necessary to achieve economic and social goals without linking these strategies to biological factors and without investigating their interactions (Ghersa et al. 1994). But the cropping system on an individual farm is complex and becomes even more so as additional regional effects on pests are considered (Cardina et al. 1999). In the design of an effective sustainable cropping system, all relevant agronomic, biologic, ecologic (ideally on a regional basis), and economic factors are addressed. Specific factors such as crop production options depend on the region. Some areas produce mainly small grains; others produce a more complex mix of crops, including corn and soybean rotations; others may be forage-based; and still others may produce a mix of alternative land use systems, from forestry to intensive cropping. A major potential constraint in any cropping system is pests, including weeds, arthropods, pathogens, and vertebrates. Pests associated with field crops have unique behavioral characteristics due in part to the large areas planted to these crops. Long-distance pest dispersal often is enhanced because the host is available readily. But to maximize agroecosystem health, crop and pest management efforts must extend beyond a focus on pests, despite their importance. A multitude of biotic factors are involved in achieving a level of biodiversity that will help fill the ecological niches so essential to minimizing pest and nutritional problems.

Table 4.1. Pest management practices, field crops, 1996 (Fernandez-Cornejo and Jans, 1999)

Item	Corn	Soybean	Cotton	Fall potato	Winter wheat	Spring wheat	Durum wheat
	Percentage of planted acres						
Biological techniques							
Considered beneficial insects in selected pesticides	8	5	52	29	10	4	12
Purchased and released beneficial insects	*	*	*	0	*	*	0
Used pheromone lures to control pests	na	*	7	2	*	1	0
Used *Bacillus thuringiensis (Bt)*[1]	2.4	1.6	4.1	*	*	0	0
Cultural techniques							
Adjusted planting or harvesting dates[2]	5	6	25	7	19	11	13
Used mechanical cultivation for weed control	51	29	89	86	na	na	na
Used a no-till system	19	33	na	na	3	4	7
Crop rotations[3]							
Continuous[4]	18	11	67	2	42[5]	14	10
Rotation with other row crops[6]	54[7]	63[8]	15	2	2	2	0
Other[9]	28	26	18	96[10]	56[11]	83[12]	90[13]
Pesticide efficiency							
Alternated pesticides to control pest resistance	31	28	41	69	13	38	32
Monitoring							
Used pheromone lures to monitor pests[14]	1	*	33	3	*	4	1
Used soil biological testing to detect pests such as insects, diseases, or nematodes	2	3	9	46	2	0	0

*Less than 0.5%.

na = not available or not applicable.

[1] Percentage of insecticide-treated acres for *Bt.*

[2] Adjust planting dates only for corn.

[3] Crop rotations include 3 years: 1994, 1995, and 1996. Column crop heading indicates the crop planted in 1996.

[4] The same crop was planted in 1994, 1995, and 1996.

[5] Continuous same crop for winter wheat occurred in 1995 and 1996, for winter wheat planted in fall 1994 and for winter wheat planted in fall 1995.

[6] A crop sequence, excluding continuous same crop, where only row crops (corn, soybean, sorghum, cotton, and peanut) were planted for three consecutive years.

[7] 49% of corn planted acres was in rotation with soybean.

[8] 56% of soybean planted acres was in rotation with corn.

[9] *Other* excludes continuous same crop and rotation with row crops, and includes fallow or idle.

[10] 26% of potato planted acres was fallow in 1994 and 1995, and 70% was in rotation with other crops or fallow in 1994 or 1995.

[11] 40% of winter-wheat planted acres was fallow in fall 1994 and had winter wheat planted in fall 1995.

[12] 23% of spring-wheat planted acres was fallow in 1994 and had spring wheat in 1995, and 60% was in rotation with other crops or fallow in 1994 or 1995.

[13] 24% of durum-wheat planted acres was fallow in 1994 and had durum wheat in 1995, and 66% was in rotation with other crops or fallow in 1994 or 1995

[14] For corn, pheromone lures were used to monitor black cutworm.

Because agroecosystems are a complex of nested subsystems (Swanton and Murphy 1996), the greatest potential for anticipating pest and pathogen problems may arise if the system is viewed as a series of tightly linked food-based interactions (Bawden 1991). This ideal has yet to be achieved for field or forage crops, partly because of the ease of using one tool for targeting single pests and partly because of the limited options for pest management in these crops. The practicality of IPM inputs also is limited by the relatively low value of crops and the large areas in which they are grown.

Large cropping systems are ideal contexts in which to monitor systems to predict pest occurrences at or above an established threshold. Monitoring involves nutritional assessments as well as pest assays. New technologies improving the efficiency and accuracy of these processes are being developed (see Chapter 3), and growers will continue to enjoy increasing access to the ever-expanding information base.

The information and technology available should enable growers to better integrate into a system all aspects of crop and pest management. The most important step in using resources efficiently is to obtain site-specific information about soil nutrients and pests (MacRae 1998). Digital tools such as global positioning systems (GPS) and geographic information systems (GIS) allow for precise mapping of fields (see Chapter 3). These maps, along with yield monitors, have been key components in precision agriculture, the basis of which is the application of pesticides (MacRae 1998) and other inputs on a site-specific basis, and only as needed.

The steps for achieving IPM in field crops are similar to those for achieving IPM in other cropping systems, steps that are outlined in numerous easily accessed sources, both in print and on the Internet (MacRae 1996). Key components of IPM are monitoring and identifying pests and pathogens. Because basic information is useful in the formulation of management strategies, it always is helpful to know the life cycle or life history of organisms, as well as the epidemiologies of diseases and the ecologies of key pests. Management strategies need to take into account related costs, economic returns, and all-system effects. Finally, detailed records of inputs, observations, and yields are needed to optimize crop management programs.

Many pests and pathogens occur on a specific crop within a cropping system, but usually only a few pests create substantial problems at any given time. The decision to manage pests occurring in a specific crop within a cropping system should be made only after an assessment to determine the costs/benefits of their management. Because field crops typically are produced on large acreages, it is prudent to assess overall damage potential before taking corrective action. Some key pests of selected major field crops (corn, cotton, soybean) are discussed in this chapter. As discussed in the introduction to this chapter, comprehensive lists of state and regional pests are available for certain crops.

Weeds

Because of the wide range of weeds commonly affecting numerous field crops, treatment of this pest group is limited here to a general discussion. The importance of weeds is reflected by the annual expenditure of $6 billion on herbicides in the United States, and estimated crop losses to these pests still amount to $4 billion/year (yr) (Liebman 2001). Specific information about often-changing weed pests (Webster and Coble 1997) and their management for given crops within a state or region, or even globally, is available readily in crop production guides (Edmisten 1999; Heiniger 1999; Madden 1992) and on the Internet (Bajwa 2000; Gray 2001; VanKirk 1998). Because they compete for space, light, water, and plant nutrients, weeds can have a significant economic effect (Bridges 1994; Renner, Swinton, and Kells 1999) on field and forage crops. Herbicides have been the dominant tool for managing them (Buhler 1999), but herbicide options differ among crops. The most basic level of weed management involves herbicide use and/or cultivation (Elmore 1996). This approach emphasizes killing weeds. The aesthetic value of weed-free fields has been an important issue for growers. Despite the availability of more than 100 herbicides and the use of herbicide for some 40 yr, weeds frequently are perceived as more of a problem now than they were when herbicides first became available (Webster and Coble 1997). Nevertheless, the advent of genetically engineered herbicide tolerance in soybean, cotton, and corn, as well as of precision agriculture, is changing the philosophies and approaches to weed management (see Chapters 2 and 3).

Although the single-tactic approach has yielded short-term benefits, long-term benefits are more likely with a multitactic approach. Integrating multiple biological, mechanical, and chemical agents enhances the decline in seedbanks and in weed populations (Cardina et al. 1999). Field mapping and site-specific technologies are relatively recent approaches to weed management. For example, the goals of precision herbicide applicators are to detect the weed, to select an appropriate control, and to apply the proper rate, on a site-specific basis (Sudduth, Hummel, and Birrell 1997; see also Chapter 3).

An increased emphasis on weed biology and weed ecology research over the past ten years has generated new insights into the genetic complexity of weeds, the practical effects of weed seed dormancy, the influence of altered production systems (e.g., decreased tillage) on weed population dynamics, and the competitive abilities of different weeds as they affect both dryland and irrigated crops (Dunan, Westra, and Moore 1998; Schweizer, Westra, and Lybecker 1998; VanGessel, Schroeder, and Westra 1998; VanGessel et al. 1995a,b,c; VanGessel et al. 1998). Such information is being incorporated into complex weed bioeconomic models that simulate weed effects and facilitate related recommendations for weed control (Dunan, Westra, and Moore 1994; Dunan et al. 1996; Lindquist et al. 1996; Schweizer, Westra, and Lybecker 1994, 1998; Schweizer et al. 1994; VanGessel et al. 1995c). Future remote weed-sensing data may feed directly into such decision support systems to provide growers with site-specific weed management options (Berti et al. 1996; Lindquist et al. 1996). Increasing the reliance on herbicide-resistant crops has raised the issue of whether such technology will result in a weed shift to populations more tolerant of certain herbicides.

As has been noted, agroecosystems are complex, with each component of the system often affecting all other components. Weed management should evolve and become integrated with the total crop production system, which has physical, chemical, and biological components. Weed populations emerging within fields manifest disturbances caused by production practices. As summarized in Table 4.1, cultural practices can be central to integrated weed management. Bioeconomic weed management models such as "WEEDSIM" are proving invaluable in evaluating weed-management practices (Renner, Swinton, and Kells 1999). As emphasized in a recent book by Liebman, Mohler, and Staver (2001), ecology-based weed management has great potential, but much basic information still must be gathered before this approach can be implemented widely.

Corn Insects and Pathogens

Corn is the major cash-grain crop grown in the United States; it is produced for silage, grain, seed, and fresh-market produce. Of these products, *field corn* (corn grown for feed grain) occupies most of the total acreage (more than 65 million acres [a.]); approximately 78% of field corn is grown in the northcentral United States (Illinois, Iowa, Indiana, Minnesota, Missouri, Nebraska, Ohio, South Dakota, and Wisconsin, a.k.a. the Corn Belt) (Pedigo 1999).

A large complex of insects, pathogens, weed species, and vertebrates affects corn yield. Wild turkey and deer consume significant quantities of corn and cause additional losses by knocking down plants that, in consequence, cannot be combine harvested.

All told, insects probably cause more damage to corn than any other pest group does. More than 30 species or species groups of insects cause economic damage. Some of the most significant problems occur in field corn grown continuously on the same land. In the Corn Belt, that practice gives rise to problems from a key beetle complex comprising the northern corn rootworm (*Diabrotica barberi*) and the western corn rootworm (*Diabrotica virgifera*). Another key pest is the European corn borer (*Ostrinia nubilalis*), which damages field corn grown continuously or in rotation with other host crops. Together, these species account for most of the insecticide use in field corn in the Corn Belt and contribute to corn's ranking as the U.S. crop receiving the most insecticidal treatment (Pedigo 1999).

Corn is injured as a result of being chewed by insect larvae and by their tunneling inside roots. Root-damaged plants lodge during rain and windstorms (Pedigo 1999). Lodged plants lose yield because of modified architecture (goose-necking), and subsequent losses occur as a result of harvest difficulties.

The European corn borer, a species introduced into the United States more than 70 yr ago, is probably the most important pest in the corn-growing regions. Different genetic strains, or *ecotypes*, produce different numbers of generations; for instance, corn borers in Canada and northern Minnesota produce one generation, and borers in Georgia produce four generations. Two or three are common in the Corn Belt. The tunneling of larvae through the vascular system causes primary damage. Second-generation larvae cause subsequent damage as they tunnel in stalks and ear shanks. Ear shank tunneling causes ears to drop (Pedigo 1999).

The key element of management design for corn rootworms and the European corn borer is prevention, which begins in most regions with a plan of crop rotation to eliminate losses from the corn-rootworm complex. This plan would include alternating rotations with soybean or other nonhost crops such as oats, annually or on a three-year plan. The three-year plan is necessary in areas where extended diapause of northern corn rootworms or of soybean-laying strains of western corn rootworm occurs (Pedigo 1999). Incorporation of a *Bacillus thuringiensis* (*Bt*)-toxin gene greatly enhances the control of these important insects, but the issue of long-term use of the technology must be resolved.

Eventually, such a preventive program would eliminate completely the need for insecticides in corn. But when outbreaks occur, occasional therapeutic treatments are required for second-generation corn borers and for pests such as the black cutworm (*Agrotis ipsilon*). The therapeutic program relies on early detection, a process refined especially for second-generation corn borers (Pedigo 1999). Although computer models are available to aid decision making by insect managers, comprehensive information about the life systems of given insect pests and about the related effects of cropping systems is necessary for effective IPM (Kennedy and Storer 2000). These systems may aid growers in making decisions based on cost-benefit analyses (see Chapter 12).

Corn is susceptible to a number of plant diseases that limit crop yield and quality (White 1999). Diseases generally are subtle and go unnoticed, however. Host resistance to certain major diseases has minimized their effects.

Aflatoxin is a problem in corn during years when there are droughts and when other weather conditions favor the causal fungus (CAST 2003; Iowa Grain Quality Initiative 2002). Although aflatoxin is a chronic problem in southeastern states, effects on the corn market are much more pronounced when midwestern corn is affected. In 1988, the nine Corn Belt states produced 3.8 billion bushels of corn, or 78% of total U.S. production (Iowa Crop and Livestock 2000). Aflatoxin in these areas obviously limits the availability of clean stocks for export or for sensitive domestic food uses. A major problem with aflatoxin is that corn is delivered to the point of sale rapidly whereas aflatoxin is detected slowly. Thus, large quantities of corn with undetected toxins may be marketed. Although extreme temperatures (daytime highs and nighttime lows) and lack of rain are necessary conditions for creation of the crop stress favoring aflatoxin development, average weather conditions for a state or geographic area are ineffective means of predicting aflatoxin levels. Meaningful predictions depend on knowledge of localized weather data and of compounding agronomic factors (CAST 2003).

Although aflatoxin contamination occurs fairly frequently in the production of southeastern U.S. corn (the Corn Belt corn is affected only in weather-stressed years), the southeastern states contribute only a small percentage of total U.S. corn production. Country elevators in high-volume grain production areas should use sampling and analysis techniques compatible with handling constraints. Related decisions must be based on risk probabilities. Operators' legal options for handling grain must be flexible enough to accommodate variabilities in sampling, testing, and lot-to-lot aflatoxin level. Currently, the accurate identification of all contaminated lots poses a major challenge.

Cotton Insects and Pathogens

Insects and pathogens severely limit cotton production in the United States. Yield losses probably exceed 20% annually from damage due to these two pest groups. Seedling diseases and nematodes usually are severe in all cotton production regions. Fusarium wilt occurs in many production areas, and boll rots also are important. From planting until harvest, cotton is attacked by insects, with nearly 100 species being recorded as pests. The pink boll worm (*Pectinophora gossybiella*) and plant bugs (*Lygus* sp.) are the key insect pests of cotton in the western United States, whereas the boll weevil (*Anthonomus grandis grandis*) and certain plant bugs (Hemiptera: Miridae) dominate in Texas and the midsouth. As a result of the areawide eradication program covering much of the Southeast and West, the boll weevil nearly has been eradicated (see Myers, Savoie, and van Randen 1998).

In most areas, repeated insecticide application has caused replacement of formerly economically important pests with other species. For example, the corn earworm (*Helicoverpa zea*) and tobacco budworm (*Heliothis virescens*) have come to assume major pest status in most areas where host crops are grown (Pedigo 1999). Declining profits, resulting from programs of heavy insecticide use, have contributed greatly to changes in cotton-pest technologies. Although high yields can be maintained in long-season cotton, greater production costs significantly decrease profits.

The insect management program for cotton production in southern Texas, a program exemplifying advanced IPM strategies, is based on preventive tactics aimed at boll weevils. The basic tactic is early planting of short-season cotton cultivars, moderate fertilizer use, and appropriately timed irrigation. Additionally, because plant thinning is delayed or avoided, vegetative growth is suppressed and early fruiting stimulated. Thus the season of production and the period during which cotton is vulnerable to attack are shortened. Other important preventive tactics of the program include early harvest and stalk destruction. In some operations, defoliant use, after which stalks are cut and shredded, facilitates early harvest. These late-season tactics prevent further weevil reproduction and weaken or starve weevils entering hibernation; as a consequence, hibernating populations suffer higher than normal mortalities.

Such preventive tactics are supplemented by *scouting*, or pest surveillance, and by therapeutic treatments of organophosphate insecticides against boll weevil and cotton fleahopper (*Pseudatomoscelis seriatus*). Additionally, pyrethroids, among other insecticides, are applied when cotton bollworm and cotton budworm outbreaks occur. During application, natural-enemy conservation is practiced by means of close reliance on economic thresholds, use of lowest possible application rates, and, when possible, treatment of "hot spots" in a field. Use of insecticides against the bollworm/budworm complex is expected to diminish, however, as transgenic cotton cultivars expressing *Bt* toxin are planted and grown. In addition to conventional means of decision making in cotton pest management, computer programs also are available. One such program is TEXCIM, a user-friendly software package designed to forecast costs of pests, benefits of management tactics, and value of crops.

In an international context, most cotton diseases are regional in importance (Watkins 1981); for example, in alkaline soils, *Phymatotrichum* is always a threat in the southwestern United States and Mexico. The Columbia lance nematode is important in the southeastern United States. In contrast, seedling disease complexes are omnipresent in cotton production fields. Thus, growers need to be prepared to manage seed and seedling disease and, in addition, at least one other type of regional disease. It is essential for growers and crop advisors to recognize which diseases are common and to prepare to manage them as part of the cropping system.

Soybean Insects and Pathogens

Over the past 40 to 50 yr, soybean production has been constrained by pests whose numbers have been increasing with the intensification of cropping. The soybean cyst nematode (*Heterodera glycines*) (SCN) has emerged as one of the most important pests of soybean. Other pests occur and can cause severe damage locally; many others are associated with soybean but cause little or no loss in most production regions.

The SCN is causing billions of dollars of loss annually wherever soybean is grown throughout the world (Wrather et al. 1995; numerous web sites, including Hammond 1996). But despite the importance of SCN, many growers still do not recognize it as a problem. In soils with a potential for high levels of production, the nematode can cause a 30 to 40% loss without causing significant foliar symptoms. Because this pest can predispose soybean to other pests, including weeds and insects, managing it in the context of a total cropping system is crucial (Alston et al. 1991, 1993). In contrast to SCN, which causes major losses annually, many of the more than 100 pathogens affecting soybean are problems in specific regions and cause various degrees of damage from one season to the next (Hartman, Sinclair, and Rupe 1999).

Insect, like pathogen, problems usually were unimportant for early soybean growers. Today, however, growers often experience significant losses from insect pests. Thus, insects, pathogens, and weeds must be included in crop management IPM programs. For the insect portion of an IPM program, plant growth stage is the most important criterion because the relation between insect injury and crop damage depends on the stage during which injury occurs. As for foliar pathogens, spray schedules are based on plant growth stage. Injury from foliar pathogens and insects is usually not as detrimental to the soybean plant during the vegetative as during the reproductive stage. Because soybean response to insects depends on plant growth stage, economic thresholds for insect management depend on it as well. Growers and IPM practitioners therefore must recognize the crop development stage.

Major insect fauna associated with soybean frequently are grouped according to injury type. Types include *defoliators, pod feeders, stem feeders*, and *girdlers*. Insect problems in the United States generally follow a south-to-north line of severity. Insect pressure is the greatest in the southern states, with less pressure in the middle regions of the country; pressures generally are much less severe in the north central states (Hammond et al. 1991). Although economically important insect problems were quite rare in the northern states during the 1970s, they have become more frequent since the 1980s. For example, Mexican bean beetle (*Epilachna varivestis* Mulsant) caused noteworthy damage in the eastern regions of the Midwest during the 1980s; and bean leaf beetles (*Cerotoma trifurcata* [Forster]) became a problem throughout the Midwest in the later 1980s and the early 1990s (Pedigo 1999).

For all pest groups, the trend is toward preventive tactics such as the use of cover crops, trap crops, and resistant cultivars. The IPM philosophy probably is more advanced in soybean than in other crops. Nevertheless, many recommendations for a given pest remain to be integrated with those for other pest types and production systems.

Concerns for the Future

Integrated pest management theory for field crops is well developed, but progress in application, and especially in integration into full cropping systems, has lagged (Merrigan 2000). Most IPM integration has been within given pest disciplines. Although much progress has been made toward reaching the goals of the national IPM initiative established by the Clinton Administration (Fernandez-Cornejo and Jans 1999), the goal of certifying as IPM 75% of U.S. acreage in agricultural production has not been achieved (Merrigan 2000).

Merely thinking in terms of IPM practices will not result in the important advances needed to sustain a production system that must provide food for the world's rapidly growing human population, which now exceeds 6 billion. Countless factors in an ecosystem interact, affecting diversity of plant species and associated populations of pests and beneficial organisms. Moreover, both disturbed and natural ecosystems are confronted with an ever-increasing number of invasive species, and the genetic basis of crops used today is very narrow and vulnerable to pest attack and environmental stress (see Chapters 5, 6, and 7). To overcome these difficulties, IPM actions should be based on systems science and on whole-system management rather than on individual pests or other single inputs (Merrigan 2000). The approach also must contribute to development of a cropping system destined to maximize the resource utilization by crops while minimizing the adverse effects of pests and pathogens.

New and emerging technologies in pest management and crop production should give rise to increasingly efficient and sustainable systems (see Chapters 3 and 12). The goal of IPM must be more comprehensive than simply eliminating or suppressing one or more target pests, however. The goals of protecting field crops from pests (including habitat management generally [Landis, Wratten, Gurr 2000]) and of augmenting biodiversity as a key IPM component should be revisited (Altieri 1994). The IPM decision making process, including the process of receiving inputs from industry (Carroll 2000), should be improved by increased use of interactive simulation modeling and by the development and use of predictive systems. An additional, ongoing issue is that of invasive pests. For example, the soybean aphid (*Aphis glycines*), native to Asia, was detected in the United States for the first time in July 2000. Its distribution continues to expand and now includes multiple states in the Northcentral region, southern Ontario, and even western New York state.

Vegetable Crops

Vegetable crops and potatoes represent a major portion of the diet of most people and serve as important sources of vitamins, minerals, and fiber. The more than 40 vegetable crops grown in the United States also are important components of the agricultural economy. During 1998, the 25 principal vegetables grown in the United States (on 1.9 million a.) yielded 423 million hundredweight (cwt) and were valued at \$8.1 billion. The 10 principal vegetables grown for processing (on 1.5 million a.) yielded 15.5 million ton (t) and were valued at \$1.3 billion. Additionally, potato growers produced (on 1.4 million a.) 477 million cwt of tubers, which were valued at \$2.5 billion. Although clearly having the potential to profit from vegetable production, growers also typically must make a significant investment and bear a great deal of production risk. Costs associated with seed, fertilizer, pesticides, farm equipment, and labor have been increasing at a steady pace while on-farm values of vegetable crops to growers have remained comparatively flat. To illustrate the magnitude of risk associated with production, the growing of long-season russet-skinned potatoes now costs \$1,800 to \$2,000/a., and the crop must be protected from pest problems requiring a wide range of management strategies.

Vegetables are grown throughout the United States in a range of soils and environments, by means of a broad spectrum of management systems. For vegetable growers to succeed in today's competitive marketplace, they must produce high yields of foods free of damage from insects, pathogens, weeds, and other crop pests, as well as pesticide residues. Pest management often requires intensive, season-long management using many different tools and techniques. Consumers have grown to expect safe, affordable, blemish-free, high-quality vegetables year-round. At times, artificially high expectations and grading standards for fresh and processed vegetables lead to increased pesticide usage.

Many approaches to pest management are available for vegetable crops. The spectrum includes strict *organic production*, whereby nonchemical inputs conform to organic standards and requirements; *biologically based approaches*, which occasionally use chem-

Textbox 4.1. New vegetable varieties resist diseases: Multiple disease resistance benefits growers and consumers (Source: M. Haining Cowles, Cornell University 1999)

Resistant squash. A summer squash called 'Whitaker' received much media attention last year, along with the man responsible for its successful breeding: Cornell University horticulturalist Richard Robinson. This work of 10 years' duration, accomplished with the assistance of plant pathologists R. Provvidenti and H. M. Munger and research support specialists Joe Shail, has been supported in part by an IPM grant.

Why is 'Whitaker' such big news? Multiple resistance is the answer. 'Whitaker' is resistant to four significant diseases—three viral and one fungal. No other squash can resist this many diseases. Multiple resistance means decreased use of pesticides, control of diseases never before adequately controlled, and improved quality, higher yield, and longer storage life of the plant. Resistance to a single disease is not nearly as significant; a squash that resists one disease can be lost to another pathogen.

Robinson continued refining 'Whitaker' this year by attempting to add resistance to one more disease and to the cucumber beetle. He was assisted in this effort by Mike Hoffman of Cornell University's Department of Entomology. Robinson also worked on transferring 'Whitaker's' resistance to other squashes.

Resistant broccoli and cabbage. An effort similar to Robinson's is under way in the laboratory of Cornell's Elizabeth Earle, who is using a cell-culture procedure called *protoplast fusion* to transfer disease resistance into crucifer vegetables from other species. Following the initial fusion experiments, she began working with Cornell horticulturalist Mike Dickson to produce resistant, marketable-quality broccoli and cabbage. The researchers now have broccoli lines resistant to either blackrot or Alternaria leaf spot, and certain broccoli/cabbage crosses resistant to both. Six of the eighteen broccoli lines tested in 1998 showed good resistance to blackrot. Earle stated that this was an increase in percentage of resistance over earlier generations and could mean that resistance is becoming uniform in these strains.

Blackrot is a bacterial disease causing leaves to become discolored and brittle. When weather conditions favor its development, blackrot causes stunting, wilting, and even plant death. Alternaria leaf spot is a fungal disease appearing as dark spots sometimes covered with a black mold. This disease can render whole heads of cabbage worthless.

ical inputs, when needed, to decrease pest populations below economic injury levels; and *management approaches* using one or more chemical inputs in response to changes in pest pressures and environmental conditions. Current IPM programs are helping growers limit losses due to pest pressure; to increase biological inputs; and, wherever possible, to decrease pesticide inputs. Because of differences in pest pressures from one state or region to another, and because of the different techniques available to growers for managing plant pests, an array of IPM programs continues to be developed to solve specific needs.

Certain IPM programs, such as the planting of pest-resistant vegetable varieties, are straightforward and require few components for successful implementation (see Textbox 4.1 on fungal and viral-resistant squash varieties, and broccoli and cabbage varieties resistant to blackrot and/or Alternaria leafspot; and

Figure 4.1. Field scouting of sweet corn for plant diseases and insects. Photo courtesy of J. Gibbons, New York State IPM Program, Cornell University.

Figure 4.2. Scouting for insects and plant diseases. Photo courtesy of J. A. Wyman, University of Wisconsin.

Textbox 4.2. Screening vegetable varieties for disease resistance (Source: W. Stevenson, University of Wisconsin–Madison 1999)

Potato and carrot are two crops requiring relatively high levels of management that are commonly sprayed with fungicide several times during the growing season to protect against the fungi causing foliar blights. Early blight (*Alternaria solani*) and late blight (*Phytophthora infestans*) can defoliate potato vines in the field, thus decreasing yield and yield quality. These fungi also can infect potato tubers, thus leading to loss of quality and storability. Carrots are susceptible to other plant pathogenic fungi (Alternaria and Cercospora leaf blights) that progressively defoliate plants and lead to harvesting difficulties and to decreased yield and root quality. Fungicides are the primary method employed by growers for management of such diseases.

In breeding trials, yield potential and quality performance criteria often take precedence over disease resistance. To allow them to express their full potentials, breeding lines often are routinely treated with fungicides. Because of the intensive management practices used in these trials, differences in disease susceptibility may be lost in the selection process.

Several years ago, field trials were developed in Wisconsin for screening cultivars and breeding lines of both potato and carrot for field resistance to foliar diseases. Each year, more than 100 potato- and 50 carrot-plot entries in unsprayed field trials were evaluated weekly for susceptibility to common foliar diseases. Information provided to breeders has proved helpful in their efforts to develop cultivars with high levels of resistance to these diseases. Carrot growers are beginning to use this same information to block plantings according to disease susceptibility and thereby shrink fungicide spray programs to maximize the value of this resistance. Although breeding potato cultivars with multiple disease will require more time, breeders are making important progress in developing commercially acceptable lines with disease resistance to both early blight and late blight. The goal is to develop cultivars requiring substantially less or no fungicide inputs for disease management. Breeders are getting closer to realizing this goal.

Textbox 4.2 on fungal-resistant potato and carrot varieties) or intensive field scouting (see Textbox 4.3 on scouting for carrot leaf blights; see also Figures 4.1 to 4.4). Computer-based information about diagnostics is valuable material for identifying important pest problems before they have caused economic losses (see Figure 4.5 on the University of California [UC]–IPM Program Pest Management Guidelines for Barnyardgrass). Combined with crop rotation, careful field selection, cover crops, and tillage, the planting of pest-resistant cultivars adapted to local conditions and the scouting of fields form the basis for many successful IPM programs (see Textbox 4.4 on cabbage weed management, and Textbox 4.5 on cover cropping on onion fields). Weather data play an important role in many IPM programs dedicated to the monitoring and the forecasting of insect and disease problems. The World Wide Web now is used to disseminate crucial environmental data for incorporation into existing models in end-user computers or for immediate use in models on the Web. An increasing number of useful models are available, and many are accessible

Figure 4.3. Insect traps often provide key information when monitoring pest development. Photo courtesy of J. A. Wyman, University of Wisconsin.

Figure 4.4. Scouting onion for insects and diseases. Photo courtesy of C. Petzoldt, New York State IPM Program, Cornell University.

Textbox 4.3. Monitoring scheme critical for carrot leaf blights: Fungicides cut in half with scouting, proper timing (Source: M. Haining Cowles, Cornell University 1999)

When a 1997 plot comparison revealed no differences in carrot yield between a field receiving three fungicide treatments and one receiving eight, Cornell plant pathologist George Abawi recognized an urgent need for more work on carrot leaf blights. "I saw a tremendous opportunity to better control these diseases and also to save on fungicides," says Abawi. "There is a great need to educate growers about scouting and about withholding treatment until a certain level of disease severity has been reached."

This year's work, led by Abawi, made good on the opportunity and the need. Five commercial carrot fields were split into "IPM plots" and "Grower-managed" plots. The first treatment for leaf blight in IPM plots was made only when sampling showed infection on 25% of leaves. Subsequent treatments were applied at intervals of 10 to 14 days if scouting reports and weather conditions suggested a high probability of leaf blight development. Grower-managed plots were treated according to growers' standard practice.

Results were dramatic: Four of the IPM plots received zero, two, three, and three fungicide applications, or a total of eight, while corresponding grower-managed plots received six, four, seven, and eight applications, or a total of twenty-five. Both IPM and grower plots received six sprays at the fifth site, which was planted to the high susceptibility variety 'Eagle.'

Despite the much lower number of fungicide sprays applied in the IPM plots, incidence and severity of leaf blight was no worse in those plots than in the other sections of the fields. Furthermore, according to Abawi, "There were no detectable differences in yield and marketability of carrots grown under the IPM scouting program and carrots grown under the regular spray schedule at the sites we harvested."

An added bonus came with the discovery that carrot varieties differed greatly in their tolerance of leaf blight. Some—particularly 'Fullback' and 'Carson'—were highly tolerant; 'Carson' required no treatment at one site. Others—such as 'Eagle'—were very susceptible to blight. Armed with this new information, growers can cut down on fungicide applications and increase profitability by choosing certain cultivars.

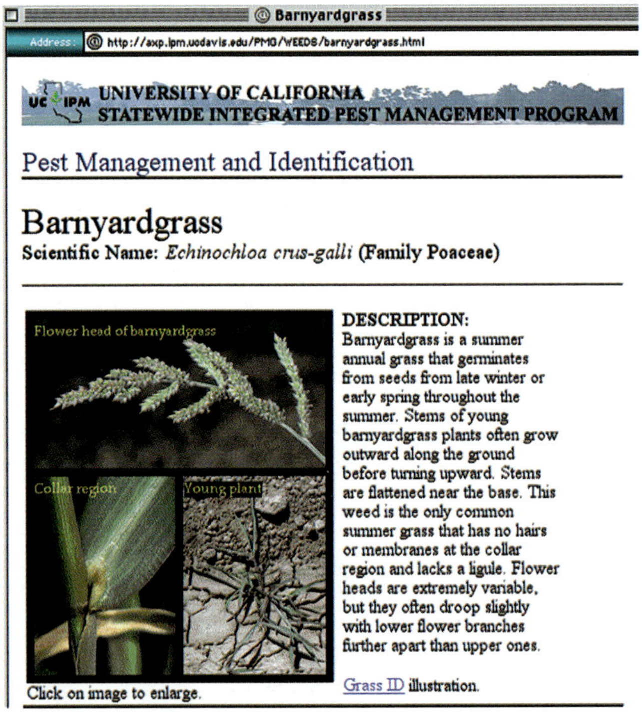

Figure 4.5. Description of Barnyardgrass on the University of California–IPM Program website (University of California 2003).

from the Web. The UC–IPM Program lists on its Web site several models for insects and mites beneficial to plants, for insects and mites harmful to plants, and for plant diseases, nematodes, and weeds. Many of these models provide insights into pest and crop development for vegetables and serve as bases on which to increase the intensity of field scouting or pest control activities (see Figure 4.6 on the UC–IPM Program's phenology model database for corn earworm and tomato fruitworm). The UC–IPM Web site also offers to approximately 350 weather stations current and historical weather data downloadable for use in pest and crop models. Many other states use the Internet to provide growers and consultants with weather information in addition to pest and crop information.

Each segment of the vegetable industry represents a continuum of adoption of IPM practices, ranging from systems considered chemically dependent (no IPM adoption) to those considered biologically dependent. With a goal of meeting the Clinton Administration's goal of 75% adoption of IPM practices, the National Potato Council initiated a survey of the U.S. potato industry in 2000 to determine where the industry was positioned on this adoption continuum (see Figure 4.7, "Calculating Grower IPM Adoption").

Textbox 4.4. Eliminating herbicides in New York state cabbage (Source: M. Haining Cowles, Cornell University 1999)

Good news on the IPM front came last year from a project on weed control in transplanted cabbage. Herbicide applications were eliminated; weeds were managed instead by a combination of cultivation and the planting of a cover crop between cabbage rows. Robin Bellinder, faculty member of the Fruit and Vegetable Science Department at Cornell University, found that as long as moisture conditions are adequate and cabbage receives enough nitrogen, yields in fields using these IPM methods are equivalent to those in fields treated with herbicides.

Variations on the theme included two cultivations versus three, plus either hairy vetch or spring oats as the cover crop; two cultivations versus three, with no cover crop; and one application of nitrogen fertilizer versus two. These treatments were compared with hand weeding, herbicide applications with no cultivation or cover crop interseeding, and no weed control at all (a check plot).

Following is a summary of findings:

1. The second nitrogen application increased cabbage yields for all treatments by an average of six tons per acre.
2. Three cultivations, either with or without interseeded cover crops, provided control equivalent to that of herbicides.
3. Two cultivations were insufficient as a weed management strategy, whether or not they were combined with a cover crop.
4. Cabbage interseeded with oats suffered the greatest yield reductions—about 30% less than yields in the herbicide-treated plots.

Bellinder is hopeful that this picture could look even brighter: "With further study focused on proper timing, I think we may see that two cultivations will be enough, meaning both cost and herbicide reductions."

Other approaches, including surveys, are being used to establish baselines of IPM adoption and to measure progress in adoption of IPM practices (see Textbox 4.6 on survey of sweet corn growers). Surveys such as these aid in determinations of adoption status and adoption progress but also help focus educational emphasis on IPM training. As growers continue to adopt new IPM methodologies, the industry is making an effort to provide economic incentives to adopters of what are often labor- and management-intensive programs (Lynch et al. 2000). Examples include programs for labeling foods "IPM-Grown" and for the use of eco-labels (see Textbox 4.7 on using eco-labels, and Textbox 4.8 on identifying potatoes grown under eco-label standards). Many IPM programs throughout the United States are focusing on moving the vegetable industry along the continuum toward increased adoption of a rich menu of IPM practices (Benbrook et al. 1996).

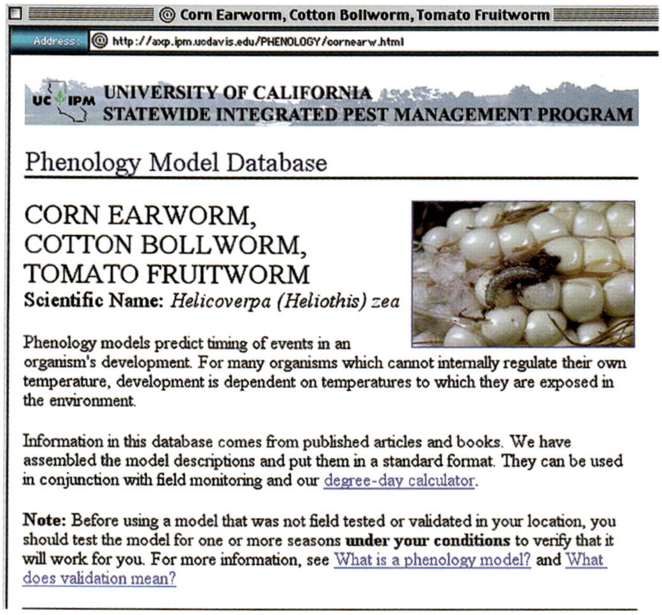

Figure 4.6. Access to phenology model database for corn earworm and tomato fruitworm from University of California–IPM Program website (University of California 2003).

Fruit Crops

Fruit crops pose special challenges for IPM because most are perennial, expensive to establish, and vulnerable to legions of damaging pests. Fruit crops

National Potato Council IPM Adoption Calculator

Potato IPM Points Available: 227 points
Advanced IPM Points Available: 109 points

No IPM	Emerging	Basic	Established	Advanced	Optimal
<50%	>50%	70%	80%	100%	100%
<114 points	<114 points	160 points	180 points	> 227 points	> 245 points
Chemically dependent, Low decision making	→ → → → → →			Biologically dependent, High decision making	

Figure 4.7. Beginning in 2000, the National Potato Council is using a pest management assessment survey to determine grower adoption of IPM practices (National Potato Council 2002).

Textbox 4.5. New York state muck onion growers see beneficial effects of Sudan grass cover cropping (Source: M. Haining Cowles, Cornell University 1999)

Sudan grass is a warm-season cover crop with extremely useful qualities: It builds and repairs soil damaged by compaction and depleted by overcropping, and evidently it suppresses harmful nematodes and root rot fungi. But New York's vegetable growers have until now lacked adequate information about how to get the most out of a rotation with Sudan grass. For instance, When should it be planted? When and how often should it be mowed?

Since 1993, Cornell Cooperative Extension educators have conducted demonstrations with Sudan grass used as a seasonlong rotational crop for onions in muck soil. In 1996 and 1997, on-farm harvest evaluations showed yield increases of 15 to 45% and stand count increases of 30 to 90% for rotated fields over nonrotated fields. Positive effects remain noticeable two years after rotation to Sudan grass, with average yield and stand count increases of 19 and 13%, respectively, over nonrotated fields. Onion size decreases slightly in rotated fields. A 10 to 20% decrease in planting rates after rotation should maintain adequate bulb size.

Previous studies have shown that Sudan grass decreases pest pressure by disrupting pest cycles and decreasing populations of soil-dwelling pests. No significant decreases in pest populations were seen in 1996, but analysis of rotated fields in 1998 did show a decrease in black mold incidence over that in nonrotated fields.

Although it is unclear at this point whether the improvements in onion yield and quality that follow rotation to Sudan grass are attributable to improved physical characteristics of soil, suppression of pests, or both, growers appreciate the improvements. They also appreciate the potential for decreased chemical pesticide use represented by such rotation. An average seasonlong pesticide load on 1 a. of onions in Orange County, New York, is 25 lb. An acre of Sudan grass, on the other hand, requires no pesticides. By planting 280 a. in Sudan grass in 1996, Orange County onion growers decreased countrywide pesticides usage by 7,000 lb (280 x 25).

One of the latest onion farms in Orange County has begun planting Sudan grass on 25% of its acreage each year. The grower will assure that each field has been rotated to Sudan grass within four years. Eight other growers also have incorporated Sudan grass into their production practices during the course of the IPM-funded Sudan grass project.

Vegetable growers and agribusinesses have received a fact sheet summarizing cultural practices for Sudan grass and data on yield and quality improvements attributable to it. The fact sheet includes the recommendations shown here in abbreviated form.

1. Sudan grass works best as a full-season crop, but it should not be sown before early June, when soils are warm and moist.
2. Seed should be sown at 40 to 50 lb/a. Drilling seeds seems to improve germination, especially under dry conditions. If seed is broadcast, it should be incorporated with a light disking.
3. Sudan grass will benefit from fertilization. Broadcasting 75 to 100 lb of nitrogen/a. should be adequate.
4. Sudan grass should be mowed at the height of 3 to 4 ft to encourage tillering and deeper root growth and to prevent stems from becoming too woody.
5. In the fall, before the first killing frost, Sudan grass, while still green, should be cut or chopped and disked thoroughly into the top 4 to 8 inches of soil. Fall disking allows ample time for decomposition, thus minimizing chances of nitrogen tie-up in spring.

destined for fresh consumption also must meet stringent marketing standards because even superficial blemishes on fruit are unacceptable to most consumers (Figure 4.8). Where pest-resistant varieties have been developed, acceptance of these varieties has been delayed at times, possibly as a result of the undesirable flavor or horticultural qualities. Unfamiliar varieties also may be comparatively difficult to market, especially in locations where consumers recognize and request known varieties (Merwin et al. 1994). Changing varieties or cultural practices is a slow process for fruit crops because plantings commonly have life expectancies of 12 to more than 30 yr.

Fruit crops usually must be protected with multiple applications of pesticides each season because a great number of major pests attack fruit and fruit trees. Over the past 60 yr, synthetic pesticides have been developed to control most insect, mite, and fungal pests on important fruit crops. Scientists pioneered the concept of IPM for tree fruits in the 1960s and the early 1970s, but most farmers then believed that pesticides had solved or soon would solve all major pest problems. From the 1940s through the early 1970s, many fruit growers applied pesticides on

Textbox 4.6. Measuring progress in IPM adoption: Survey of sweet corn growers provides data (Source: J. Tette, Cornell University 1999)

The "continuum of IPM" refers to the incorporation of IPM methods, starting with a chemically based calendar spray program and moving to an information-based program that relies on biologically intensive methods. By calculating their position along this continuum, growers in any cropping system can measure their progress in terms of IPM adoption.

In 1997, progress in IPM adoption was measured through a survey of fresh-market sweet corn growers, conducted as a collaborative effort with the New York State Agricultural Statistics Service. A total of 206 growers were asked fourteen questions about their IPM practices. Questions were related to IPM Elements for sweet corn, as defined by growers, processors, retail food distributors, and Cornell University researchers. The IPM Elements are the latest and best of IPM and integrated crop management practices for a particular crop. As can be seen in the Table, greater than 10% of 206 growers adopted 80% or more of all IPM elements. This group has progressed far along the IPM continuum. An additional 117 growers adopted 50 to 90% of the Elements, showing a high degree of adoption, but with some room for improvement. The remaining 30% may not have been able to adopt IPM practices more fully because their operations are not near sites where IPM methods have been demonstrated.

Data from the survey helped the New York State IPM Program by revealing that future educational outreach efforts need to emphasize weed mapping, nutrient testing before sidedressing additional fertilizer, and accurate record keeping.

Table: A measure of New York fresh-market sweet corn growers using IPM methods

Percentage of IPM Elements Adopted	Number of Growers Using IPM Elements (Total number of growers = 206)
≥90	9
80	17
70	26
60	52
50	39
40	34
30	10
≤20	19

a calendar basis regardless of pest populations. Farmer interest in IPM increased sharply in the mid and late 1970s as introductions of new pesticide chemistries slowed. Pesticide labels also became more restrictive as pest resistance to existing pesticides emerged as a significant problem and as adverse nontarget effects of pesticides became recognized more widely.

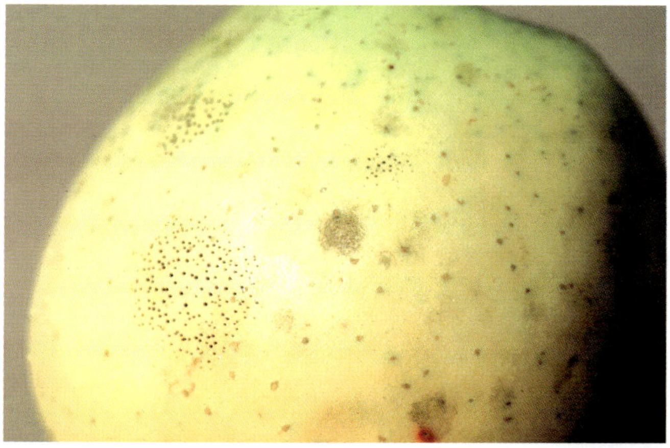

Figure 4.8. Flyspeck and sooty blotch diseases on apple fruit cause superficial defects unacceptable to consumers. These diseases are controlled by fungicides applied during summer. Photo courtesy of D. Rosenberger, New York State Agricultural Extension Service.

Since the 1970s, IPM strategies have focused on decreasing costs and minimizing dependence on broad-spectrum pesticides with adverse nontarget effects. Emergence of pesticide resistant pest populations has fueled farmer interest in preserving predators able to suppress these populations. Increasing costs for new pesticide chemistries have created an economic incentive for decreasing pesticide use, and farmers gradually have become more skilled at selecting pesticides and at timing applications so as to preserve parasites and predators. They also have become more adept at monitoring pest populations, delaying pesticide applications until populations reach economic thresholds, and timing applications to maximize pesticide efficacy. These developments have been facilitated by two successive national IPM programs: the National Science Foundation Huffaker Project and its successor, the U.S. Department of Agriculture (USDA)/Environmental Protection Agency (EPA) Adkisson Project (Whalon and Croft 1984).

After the Alar scare in 1989, many food processors (and especially baby-food processors) forced changes

Textbox 4.7. The growing market for IPM labels (Source: M. Haining Cowles, Cornell University 1999)

The labeling of food as "IPM-grown" began in 1995 with fresh IPM-grown sweet corn sold by a Wegmans grocery store near Rochester, New York. A follow-up survey completed by 300 Wegmans' customers showed substantial support for the new label. Four years later, the IPM labeling movement was burgeoning.

The Wegmans Initiative. Seeking the means to offer customers IPM-grown sweet corn, executives from Wegmans Food Markets, Inc., a Rochester, New York-based retail grocer, approached Cornell University in 1994. Cornell IPM extension educators offered training in IPM methods to growers who had been supplying fresh-market sweet corn to Wegmans. The training was funded and cofacilitated by PRO-TECH, a Cornell Cooperative Extension program whose mission was to enhance, through educational offerings, the sustainability and competitiveness of New York's fruit, vegetable, and ornamental horticulture industries.

When the initial experiment with IPM-grown sweet corn proved successful, Wegmans decided to sell it at additional stores and also to obtain IPM labels for canned and frozen vegetables. Comstock Michigan Fruit, the grower-owned company supplying Wegmans with processed fruits and vegetables, selected ten growers already practicing IPM to grow processing vegetables for marketing under IPM labels.

Once an agreement to market IPM vegetables was reached, interested growers, representatives from Wegmans and Comstock, and Cornell's IPM extension educators formulated "IPM Elements" for six processing crops: beets, cabbage for sauerkraut, carrots, peas, snap beans, and sweet corn. As of 1999, IPM Elements had been developed for 16 vegetable and fruit crops in New York.

The Elements are lists of agreed-upon IPM and crop management practices to be followed in producing crops to be sold under an IPM label. Growers are assigned points for adopting techniques in the Elements. Each grower must keep detailed records verifying use of the techniques, and a specific predetermined point total must be achieved to qualify a crop as "IPM grown." All farmers currently growing for IPM labeling have scored at or above the 80% level each year, or approximately 15% beyond the amount required to qualify for the IPM label.

Grower progress in adopting IPM Elements is documented by an independent third party and reported to Wegmans. The use of the New York State IPM Program logo on IPM labels ensures the integrity of participants in the labeling process. The logo is owned by the Cornell Research Foundation, which obtained a trademark on it and holds the licensing agreement for its use as a label, whether by Wegmans or by any other party.

Purposes of IPM labeling. Integrated Pest Management labeling encourages and rewards environmental stewardship on the part of growers and processors. This stewardship is expressed in part by decreased pesticide use. Data from several crops in New York state show that adoption of IPM Elements at the 80% level will result in pesticide use decreases of 30 to 50%. Data concerning the environmental impact of pesticides used on each crop also are being gathered. As shown in the Table, the data indicate decreasing environmental impact quotients (EIQs) over time. (Crops showing greater environmental effects in 1997 than in 1996 faced increased pest pressures in 1997.)

Table: Trends in decreasing the environmental impact of growing processed crops, measured by the Environmental Impact Quotient

Crop	EIQ Values 1996	EIQ Values 1997
Beets	72	66
Carrots	258	173
Kraut cabbage	45	74
Peas	23	27
Snap beans	114	110
Sweet corn	136	119

Integrated Pest Management labeling, New York and beyond. As of 1999, IPM labels could be found on bins of fresh corn, tomatoes, cherries, and asparagus and on cans of corn, peas, and beets in most of the 50+ stores owned by Wegmans Food Markets, Inc. in New York and Pennsylvania.

Eden Valley Growers, a central New York cooperative for sweet corn growers, is another licensed IPM logo user. Integrated Pest Management labels have been used at the farm stands of 50 strawberry growers who belong to the New York Berry Growers Association and on the canned and frozen vegetables processed by Agrilink Foods.

Certain growers and land-grant university scientists are developing IPM Elements for use in Wisconsin, Pennsylvania, and Hawaii. A nonprofit organization, the PM Institute of North America, was formed to coordinate labeling needs nationally. All these entities look to the New York program as their model.

Statistics about IPM labeling trends in New York State were gathered by an independent evaluator at the request of food processing companies. These statistics show increases in the numbers of growers and of acres producing crops for IPM labeling, and decreases in the environmental effects of growing these crops:

New York State producers growing for IPM labeling:
31 in 1996; 113 in 1997; 152 in 1998 (est.)

New York State acres growing IPM-labeled produce:
3,490 in 1996; 8,092 in 1997; 9,029 in 1998 (est.)

Textbox 4.8. "Healthy Grown" potatoes: A case example of collaboration (Source: D. Sexson, University of Wisconsin–Madison 2001)

The World Wildlife Fund (WWF) teamed up with the Wisconsin Potato and Vegetable Growers Association (WPVGA) and the University of Wisconsin (UW) to educate potato growers about using more biologically based pest management systems. The collaboration began as an effort to promote the development and adoption of biointensive IPM practices, to enhance habitat quality, to refine measurement systems for IPM adoption, to find marketplace incentives for ecologically produced potatoes, to identify policies and programs supporting IPM goals, and to maintain economically viable farming systems.

The collaboration has achieved significant progress toward decreasing the toxicity levels of pesticides used in potato production while increasing biointensive IPM adoption. Potato growers in Wisconsin achieved a 21% overall decrease in system toxicity from 1995 to 1999 (toxicity values for each pesticide are determined according to the relative environmental and human risks it poses). Of 11 specifically targeted high-risk pesticides, a 37% decrease in toxicity units was observed from the 1995 baseline to 1999 figures.

In the fall of 2000, ecological standards were written for potatoes. The ecostandards comprise (1) a biointensive IPM adoption section and (2) a toxicity score. In the first section are certain practices (such as rotation) that growers must adopt to be certified; without such adoption, growers are eliminated automatically. Other practices qualify growers for a bonus that increases their total number of points. The toxicity guidelines are written so that growers minimize the amount of high-risk pesticides they apply to a field in a given year. Toxicity unit totals are derived from a toxicity values scoring system developed by the WWF/WPVGA/UW collaboration. Growers must limit the total number of toxicity units during the year to be eligible for certification.

If growers meet the requirements of both sections, they qualify for the nonprofit ecolabel "Protected Harvest" (see www.protectedharvest.org). Potatoes grown in Wisconsin in 2001 that met the "Protected Harvest" standards were endorsed by the WWF and carried the panda logo on their bags. To maximize marketing efforts, it has been agreed that all potatoes grown under these standards will be sold under the brand name "Healthy Grown" (see www.healthygrown.com).

in fruit IPM by instituting pesticide use restrictions more stringent than federal pesticide label requirements. These processors purchased only crops treated according to their own guidelines and enforced their policies with extensive pesticide residue testing. Policies were developed primarily for marketing purposes and had little or no toxicological basis. Nevertheless, farmers producing fruit for such processors were obliged to adjust IPM strategies to meet the new pesticide restrictions. Similar adjustments have been made by farmers opting to grow fruit organically, for certified IPM programs, and for certain export markets in which standards are enforced by wholesale fruit buyers.

Numerous IPM strategies, economic thresholds, and predictive models have been developed over the past three decades and have been summarized in scientific and production-oriented publications (e.g., Agnello et al. 1999; Hogmire 1995; Ohlendorf and Clark 1999; Strand and Clark 1999; Sutton 1996). Evaluation of the acceptance of IPM strategies by commercial fruit producers has been more difficult to obtain. Prokopy et al. (1996) used relative degrees of integration to differentiate among levels of IPM. *First-level IPM* was defined as the use of multiple integrated tactics to manage a single class of pest. *Second-level IPM* was defined as the optimization of strategies for all classes of pests, and *third-level IPM* was defined as the integration of all aspects of crop production. Advancing to higher levels of IPM is exceedingly difficult with fruit crops because of the complexity involved in simultaneously optimizing control strategies for so many different pests.

Summarizing the details of pest management for any given fruit crop is difficult because of regional variations due to differences in pest complexes, varieties, climates, and marketing objectives. Pest management tools used against major pests for a few fruit crops are listed in Table 4.2.

The World Wide Web hosts many sites providing detailed information about IPM strategies for fruit crops. Most sites feature search functions allowing easy access to topics of interest, and useful links to related sites. Starting points for accessing fruit IPM on the Web include the following:

- West Virginia University's Kearneysville Tree Fruit Research and Education Center maintains a web site that contains extensive information and photography collections related to diseases, insects, and mites on tree fruit crops in the eastern United States. The site also contains useful links to other Web sites and to on-line newsletters covering fruit IPM (Biggs 2002).

Table 4.2. A checklist of availability, importance, and commercial acceptance of various IPM tools for managing major pests on several fruit crops

	Chemical pesticides	Biocontrols	Forecasting models to predict development emergence, or spray timing	Nonpheremone traps for population monitoring	Pheremone traps as monitoring tools	Pheremone disruption	Controls based on field scouting and economic action thresholds	Cultural controls or sanitation measures (including horticultural oils)	Resistant germplasm
Apple diseases									
Apple scab	A+ E+	R	A+					A–	A–
Powdery mildew	A+ E+	A–							A–
Fire blight	A+ E	A	A+ E					A+ E+	A–
Rust diseases	A+ E		A–					A–	A–
Black rot and white rot	A+ E		A–					A	
Bitter rot	A+ E							A	
Sooty blotch and flyspeck	A+ E		A				A–	A–	
Postharvest decays	A+ E	A–						A E+	
Apple insects and mites									
Plum curculio	A+ E		A	A–			A–	A–	
Apple maggot	A+ E		A	A			A+	A–	
Coddling moth	A E		A		A	A	A		
Oblique-banded leafroller	A+ E		A+		A	R	A		
Tufted apple bud moth	A+ E		A		A		A		
San Jose scale	A+ E		A	A			A–	A	
Leafhoppers	A+ E		A				A+	A–	
European red mite	A+ E	A+					A+	A+ E	
Peach and nectarine diseases									
Brown rot	A+ E+	R						A–	R
Peach leaf curl	A+ E+								A–
Powdery mildew	A+ E						A		
Bacterial spot	A E						A		A
Phomopsis (Constriction canker)	A							A	
Peach and nectarine insects									
Catfacing insects (Hemiptera)	A+ E+						A	A+	
Oriental fruit moth	A+ E+		A+		A+	A	A		
Peach tree borers	A+ E+		A+		A	A			
Cherry diseases									
Brown rot	A+ E+								
Powdery mildew	A+ E								
Cherry leaf spot	A+ E		A						
Bacterial canker	A E								
Cherry insects									
Cherry fruit fly	A+ E+		A	A			A		
Grape diseases									
Powdery mildew	A+ E+	A–	A					A– R	
Downy mildew	A+ E		A–					R	
Black rot	A+ E		A					A E–	
Bunch rot	A+ E+	A–	A–					A E–	
Phomopsis cane and leaf spot	A+ E+		R						
Grape insects and mites									
Grape berry moth	A+ E				A	A–	A		
Leafhoppers	A+ E+			A			A		
European red mite	A E–	A–				A			

Key

Commercial acceptance:
- A+ = available and widely used where pest is a threat
- A = available and used in some locations or situations
- A– = available, but rarely used as a management tool

Importance for managing pests:
- E+ = essential for most production regions and varieties
- E = essential for some production regions and varieties
- E– = an important tool, but it is often not essential for managing pests

R = research under way or completed

Source: Compiled by D. A. Rosenberger.

- Cornell University and the New York State Agricultural Experiment Station at Geneva maintain a web site that posts weekly and monthly newsletters for various crops during the growing season. Newsletter articles address issues as they emerge in the field. Articles provide life cycle information on pests and IPM options relevant to decision making at the grower level (Cornell Fruit Information Page 2002).
- The UC–Davis maintains a web site that contains extensive information on culture and pest control for nearly 50 different fruit and nut crops grown in California (Fruit and Nut Research 2000).
- Washington State University's Tree Fruit Research and Extension Center maintains a web site that covers research and educational efforts related to the Washington tree fruit industry, including reports on the areawide coddling moth management program (Tree Fruit Research 2002).

Figure 4.9. **Postharvest decays (in this instance caused by *Botrytis cinerea*) have been controlled with postharvest applications of fungicides, but improved sanitation and biocontrol products may provide alternatives in the future. Photo courtesy of D. Rosenberger, New York State Agricultural Extension Service.**

- David L. Green (2002) provides a wealth of information on bees and on the pollination of agricultural crops. His site also addresses concerns about pesticide poisoning of bees and how to avoid it.

The success of nonpesticide strategies for fruit depends on crop, pest complex, and geographic region. For example, the areawide coddling moth management program in the northwestern United States represents one of the most successful large-scale introductions of nonpesticide alternatives. In that program, pheromone disruption has controlled coddling moth on more than 7,700 a. of apples and pears, thereby decreasing the need for insecticide sprays and allowing improved survival of parasites and predators (Senft 1997). In some orchards, insecticides still are needed to control pests previously suppressed by coddling moth sprays. Pheromone disruption has proved less useful in eastern production regions where wild hosts and uncultivated apple trees sustain large populations of coddling moths and other pests that immigrate into orchards throughout the growing season (see Textbox 4.9 on IPM uses in apple orchards). Leafroller insects are more destructive than coddling moth in eastern orchards and have proved less susceptible to pheromone disruption.

Integrated pest management for postharvest storage and handling has received considerable attention from researchers, but nonpesticide solutions to postharvest problems are evolving only slowly. One strategy gaining widespread acceptance has been that of using a minimum 90-day period of low-oxygen storage to eliminate insect pests from apples designated for export to countries otherwise requiring fumigation with methyl bromide. Postharvest decay problems have been more intransigent (Figure 4.9). Sanitation measures that could limit exposure significantly to postharvest pathogens are often difficult to implement, so packinghouses have continued to rely mainly on postharvest fungicides. Numerous scientists and developers have attempted to develop biocontrols for postharvest diseases of citrus and pome fruits (Wilson and Wisniewski 1994), but no biocontrol product developed to date has generated broad commercial acceptance among packinghouse operators. Shelf life of the formulated biocontrols has been a limiting factor, and most have been unable to provide the consistent levels of decay control achieved with traditional fungicides. Postharvest IPM involving a combination of sanitation and biocontrol may gain wider acceptance in the future because many pathogens have developed resistance to available fungicides and because most companies have abandoned efforts to develop new postharvest fungicides.

One of the most successful IPM programs for fruit actually predates the emergence of IPM as a discipline. Certification programs were developed during the 1940s and 1950s to rid many fruit crops of viral problems that were causing extensive losses. Scientists recognized that most viruses were transmitted to new fruit plantings by way of virus-contaminated propagation materials. Pesticides have no direct effects on viruses, so farmers and scientists were forced to seek alternative control measures. Individual scientists and states designed programs to eliminate

Textbox 4.9. Integrated complexity: IPM for apple pests in Southeastern New York and Connecticut (Source: D. Rosenberger, Cornell University - Hudson Valley Lab 2001)

Nearly every orchard management decision has threads that ultimately impact pest management. Critical IPM decisions are made even before trees are planted. Orchard design affects the kinds and severity of pest problems that emerge later in the life of the orchard. Growers must select apple cultivars that will have consumer acceptance (i.e., be profitable) while at the same time avoiding cultivars for which pest damage or pest management costs will eliminate profits.

Drainage tile must be installed prior to planting in some sites to avoid Phytophthora root rot, a disease that kills tree roots in poorly drained soils; otherwise, trees in such soils may need annual applications of the fungicide metalaxyl to protect roots. Fencing may be required to keep out deer; otherwise, multiple applications of deer repellent sprays may be needed to prevent deer from eating buds and stunting trees. Choices concerning rootstocks, tree spacing, and training systems have long-term impacts on pest problems. Trees spaced too closely require excessive pruning; heavy pruning stimulates excessive vegetative growth; and lush growth favors development of many pests and inhibits effective spray coverage when pesticides are needed. Trees spaced too far apart will be unprofitable and require more units of pesticide per ton of apples produced.

Soon after planting, the orchard floor must be smoothed and seeded with a perennial grass cover. Uneven ground provides cover for vole populations that can later kill trees by eating the bark during winter. Poor establishment of the grass cover allows broad-leaf weeds such as dandelion to become established. Flowering weeds later compete with trees for pollinator insects and can also contribute to bee kills if pesticides are applied to apple trees when the weeds are in bloom.

A sod ground cover is established between rows to decrease erosion, but ground cover within the tree rows must be managed by mowing, tilling, or using herbicides. Selecting herbicides, rates, and timings so as to allow appropriate regrowth in late summer is just one more IPM decision.

Horticultural practices also are intertwined with IPM. Annual pruning is essential for removing diseased and broken limbs that would otherwise contribute inoculum for fire blight, black rot, white rot, and bitter rot. Pruning also opens the tree canopy and allows thorough pesticide coverage during the growing season.

To control diseases in apple orchards, growers must begin applying fungicides soon after bud break in the spring. However, the timing of the first application is critical. Starting too early may waste $20 to $40 per acre in unnecessary fungicide expenses whereas delaying just several days too long can result in major crop loss. Several monitoring and prediction tools are available to help growers optimize the timing of the first application.

Selecting the appropriate fungicides for disease control is a critical IPM decision. Some fungicides such as mancozeb are inexpensive but act only as protectants. Protectant fungicides must be applied before infections occur, whereas other fungicides have postinfection activity and can be applied up to four days after an infection period. The latter might appear to offer advantages for IPM programs because sprays can be applied only as needed based on weather conditions that have already occurred. Postinfection scheduling of fungicides, however, means that fungicide timing after rains becomes critical. Prolonged rains or windy conditions after rains can result in poor fungicide coverage, and the fungicide spray timing will seldom coincide with timings required for insecticide applications. Labor and equipment costs are therefore much greater for postinfection scab programs than when fungicides are applied as protectants and combined with insecticides at critical tree growth stages.

At the same time that apple growers are monitoring and spraying for apple scab, they must ensure simultaneous control of powdery mildew, cedar apple rust, quince rust, blossom-end rot, black rot, white rot, bitter rot, flyspeck, and sooty blotch. In addition, multiple applications of streptomycin may be needed during bloom for cultivars that are susceptible to fire blight, a bacterial disease that enters through blossoms and can quickly kill entire trees. Models are available for predicting the best spray timing for many of these diseases, but growers must integrate this wealth of information into cost-effective control programs. Each fungicide or fungicide combination has specific strengths and weaknesses depending on the diseases to be controlled, the apple variety, growth stage of the tree, and weather conditions past, present, and predicted.

Similar or greater complexity exists in managing apple arthropods. Degree-day models are available for most of the major insect pests on apples. Degree-day models predict the approximate time that pests will emerge. These predictions are often used as the basis for hanging pheromone or nonbaited sticky traps that can pinpoint exact dates of pest emergence. Control sprays are then suggested for a specific number of degree-days after the first emergence detected in orchard traps. In the Northeast, pheromone traps are commonly used for coddling moth, oblique-banded leafroller, and spotted tentiform leafminer. Traps baited with feeding attractants are used for apple maggot. Nonbaited traps are used for tarnished plant bug, mullein plant bug, European apple sawfly, and San Jose scale. Presence/absence sampling of leaves is useful to determining when mites, aphids, and leafhoppers have reached thresholds that require intervention with a pesticide. Careful selection of pesticides that preserve aphid predators means that aphicides are no longer needed in many orchards.

Apple growers can opt for biological control of mites by

Textbox 4.9. continued

introducing and fostering predaceous mites. Biological mite control can save growers $50 to $100 per acre annually by eliminating the need for miticide sprays. However, biological mite control can be maintained only in orchards where growers avoid pesticides such as synthetic pyrethroids that are toxic to predators.

All fruit growers in the Northeast must apply insecticide sprays to control plum curculio, a beetle that causes fruit scarring shortly after petal fall. Despite extensive research, no nonpesticidal alternatives exist for controlling curculio. Similarly, most growers depend on pesticides to control apple maggot flies during late summer. Methods for trapping apple maggot flies on baited spheres have been investigated for more than 20 years as an alternative to pesticide sprays. Maggot trap-out strategies are gradually becoming more feasible and effective, but no cost-effective alternative has yet emerged.

Over the past 20 years, apple growers have decreased the average number of pesticide applications per season and changed from calendar-based spraying to applications based on predictive models, orchard scouting, and pheromone traps. They have largely substituted newer "soft" pesticides for broad-spectrum pesticides that were equally lethal to both pests and beneficial insects. Integrated Pest Management is now an integral part of apple growing, but the apple farmer is the ultimate integrator.

viruses from propagating material, to maintain blocks of virus-free stock accessible to nursery operators, and to promote production and purchase of virus-free planting stock. These state-level programs led to the development of USDA Interregional Project 2 (IR–2) in 1955. The IR–2, which was reorganized as the National Research Support Project 5, maintained a repository of related virus-free apple, pear, and stone fruit propagating materials (WSU 2002).

In 1954, the NE–14 Regional Research Project "Virus Diseases of Woody, Deciduous Tree Fruits" was organized. For more than 35 yr, this project promoted cooperative research and information exchange among scientists throughout North America, and a very effective program has resulted. The recent detection in Pennsylvania of plum pox virus, a destructive stone fruit virus endemic in Europe, indicates that continued vigilance and tree fruit virus research are essential if the successes of the past 60 yr are to be maintained.

Although IPM programs for fruit crops have improved and expanded dramatically over the past 20 yr, continued progress and broader implementation can be expected over the next decade. The rate of progress will depend largely on two factors: the levels of federal and state funding designated for applied IPM-related research, and the number of private consultants working in fruit crops. Private consultants will play an increasingly important role in IPM implementation because fruit growers are recognizing that the expertise of consultants for managing complex IPM systems is just as essential to their operations as the expertise of a tax adviser for managing complex tax and estate planning issues is. Consultants are interdisciplinary integrators and bring with themselves a wealth of information gleaned from field observations over many years and many different farms. University-based research will provide the basis for improved fruit IPM in the twenty-first century, private consultants will be at the forefront of IPM implementation, and Cooperative Extension will provide a link between the two. Over the next decade and beyond, regional variations on this structure will be common, but the research-extension-consultant linkage will provide the most effective basis for broad implementation of IPM in fruit.

Integrated Forest Pest Management

Integrated pest management, integrated forest pest management (IFPM), forest health protection, forest health, and forest resource protection all are interrelated and have the common goal of protecting and sustaining forest resources. Ecosystem management for sustaining forest resources also has much to offer but sometimes fails to take pest problems into account (Boyce and Haney 1997). Although these approaches have similar goals, proponents of each approach espouse unique philosophies. With changes in both societal and individual perspectives, views shift of how forest resources are to be used appropriately. Some individuals may consider certain forest conditions threatening or unhealthy whereas other individuals, although recognizing the same conditions, deem the course of events healthy. Different convictions evolve out of ever-changing individual perceptions and/or organizational agendas (Allen 1994; Boyce and Haney 1997), and herein lies one of the greatest challenges to the protection of natural resources.

Little (1995) concluded that a new era has emerged in which trees are dying over large regions. Although most forests have exceptional "ecosystem resilience"

(Boyce and Haney 1997), major problems have developed in certain places; and exotic and endogenous pests (weeds, disease agents, and insects) and adverse environmental factors continue to pose threats to forests. For example, spruce budworm outbreaks have been documented in eastern Canada for approximately 300 yr (Allen 1994). Although insect and disease problems undoubtedly always have been part of the normally functioning spruce-fir ecosystem in this region, and although spruce budworms are a disturbance associated with a "healthy" forest, from an economic viewpoint these pests can be devastating. Seemingly as a result of a complex of pests (the primary being the balsam wooly aphid) and environmental problems (e.g., acid rain, which compounded damage from the 1980 to 1990s), Frazer fir and red spruce have exhibited a dramatic decline in the Appalachian Mountains, including North Carolina (Figure 4.10). Much natural regrowth has occurred in recent years, however (Figure 4.11).

Figure 4.10. Severe decline in Frazer fir in Mt. Mitchell, North Carolina in 1990. Photo courtesy of K. M. Hartman, Monsanto Agricultural Research.

Figure 4.11. Regrowth of Frazer fir in Mt. Mitchell, North Carolina in 1997. Photo courtesy of L. F. Grand, North Carolina State University.

Similarly, in presettlement forests, the southern pine beetle (SPB) was responsible for periodic perturbations that helped maintain unevenly aged forests and diverse plant species. These outbreaks were beneficial events in normally functioning southern pine ecosystems. But the SPB now is viewed as a pest because an economic value is placed on pine and because intensive management of pine forests has caused beetle populations to interfere with optimization of management objectives.

For purposes of this discussion, the term *IFPM* will be used with the understanding that overall forest health is of primary concern. Any pest causing change in management practices would be the focus of IFPM. Pest populations include weeds, plant pathogens, insects, and vertebrates with a significant effect on growth loss for pine (Figure 4.12). Branham and Hertel (1984) have presented IFPM recommendations for these agents.

Pest agents for change also are influenced by weather conditions such as prolonged drought, flooding, or lightning strikes, all of which make trees more susceptible to insects and pathogens. Integrated forest pest management is an aspect of EBPM in much the same sense that silviculture is an aspect of applied forest ecology (Hedden and Nebeker 1984). Based on this definition, IFPM takes into account two interacting factors: integrated control and integrated management of pest complexes.

Integrated control involves use of the appropriate combination of technologies to ensure economically efficient control while minimizing adverse effects on the forest ecosystem. Control techniques may include (1) natural or synthetic insecticides; (2) biological control agents such as parasites, predators, or pathogens; (3) behavioral chemicals for timing pest management activities or manipulating pest populations; (4) cultural techniques such as thinning to promote stand vigor, or sanitation logging to remove infested or infected material; (5) planting resistant pine species; (6) planting trees that have had genes inserted for resistance to destructive agents; and (7) related practices.

Integrated management of pest complexes is the simultaneous consideration of all potential pests during IFPM program development. This approach minimizes the total effect of pests on the forest. Examples of pest complexes include the development of annosus root rot after thinning in stands on sandy soils and the subsequent attack by bark beetles on the infected trees or the development of pitch canker in plantations with high populations of tip moths, where such populations are related to herbicide use (Hedden and Nebeker 1984). At times, a common management tactic might decrease the effects of more than one pest. Runion, Cade, and Bruck (1993) reported that the insecticide carbofuran decreased terminal shoot infection of the pitch canker fungus and terminal shoot damage by tip moths attacking loblolly pine in eastern North Carolina.

Integrated forest pest management is, however, only one component of an overall forest management program. Type and intensity of pest management practiced are determined, ultimately, by the landowner's objectives. For any successful management program to be developed, the management objective(s) must be determined. Lacking such a determination, management cannot perform a cost-benefit or any other type of analysis to determine whether proposed strategies are appropriate. It is accepted generally that the need for pest management increases with the intensity of forest management (Figure 4.13).

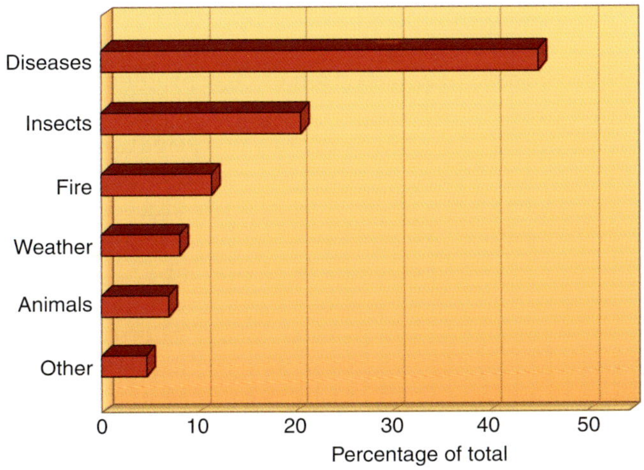

Figure 4.12. Agents of destruction and their associated proportion of growth loss in sawtimber (In Tainter and Baker [1996]; redrawn from Hepting and Jemison [1958]).

As the twenty-first century unfolds, the forested landscape will continue to change. More and more exotic forests are emerging. For instance, former agricultural farms are being converted to fiber farms, on which intensive silvicultural practices similar to traditional farming practices are being used. Many fiber farms are using former row-crop settings for the culture of hybrid poplar species such as eastern cottonwood (*Populus deltoides*), which are being grown on rotations briefer than 10 yr. Rotations such as these present special challenges to IFPM: new pest associations will be discovered, as evidenced by the case of root-feeding aphids appearing in a fiber farm in Missouri during the summer of 2000 and causing mortality to hybrid poplars during the first growing season (Nebeker, T. E. 2002. Personal communication). This case represents the first time such an association has been observed. Other challenges include the management of urban forests and their diverse pests as ever-increasing numbers of people move from cities into unpopulated, forested landscapes.

Prevention

Protecting forests from severe damage or mortality by harmful pests and other damaging agents by preventive methods is an ideal goal for forest managers. Six guiding principles should be consulted during the development of an IFPM plan. Recommended practices include the following:

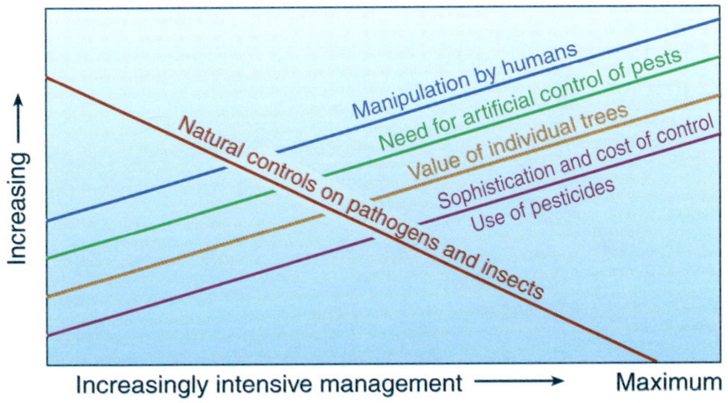

Figure 4.13. Characteristics of various forest management situations in regard to pest control practices (In Tainter and Baker [1996]; from National Academy of Sciences [1975]).

1. **Match the site with the appropriate tree species**. Trees planted on inappropriate sites seldom have sufficient vigor to deter or to withstand attack.
2. **Alter stand density**. If the stand's basal area exceeds the site index, the stand should be thinned to an appropriate level.
3. **Promptly salvage all lightning-struck, logging-damaged, diseased, or otherwise high-risk trees, and harvest overmature trees when pest activity is low**. Such trees constitute significant pest reservoirs and potential epicenters for future pest activity.
4. **Plant trees only in their natural range**.

Planting outside the range and off site causes stress, which increases susceptibility to attack.

5. **Minimize site and stand disturbances**. Exercise care in the use of heavy equipment, road layout, culvert location, and other construction projects. Changes in drainage cause tree stress.
6. **Harvest all mature trees at, or shortly after, rotation age**.

Prevention of losses from insects or pathogens follows implementation of forestry practices, with the attendant silvicultural practices, recommended for each tree species of interest. Preventive silvicultural practices include favoring the most resistant species, removing high-hazard trees, regulating stocking, and increasing tree species ecological diversity. An understanding of the interactions taking place among the pest, its host(s), and the environment is necessary (Figure 4.14).

Figure 4.14. Interaction of multiple components that lead to growth loss and/or mortality over a forested landscape (Tainter and Baker 1996).

Detection and Evaluation

Essential to any IFPM program are the processes of detection and evaluation. The oldest method of detection is that of nonexpert detection of a recent introduction of an exotic insect species. Haack and colleagues (1997) reported that on August 19, 1996, a Brooklyn resident notified the New York City Department of Parks and Recreation that all the Norway maple trees lining the street in front of his house were riddled with large holes. At first the resident thought the holes in the maples were the work of vandals with huge drills. He also reported seeing several large black-and-white beetles in the vicinity of the trees. The next day, a New York City forester sent one of the beetles to Cornell University for identification. Additional specimens soon were collected, and the insect was identified as the Asian long-horned beetle. Most early detection of pest problems occurs in this manner, with private citizens recognizing that something is abnormal and that investigation may be warranted. This scenario will continue to be replayed as additional exotic species are introduced into the United States through the increased importation of raw logs and others goods from abroad.

In addition to the happenstance recognition among lay people that "something is wrong," there are standard methods for detecting pest problems. Systematic ground surveys typically are associated with forest inventories and tree health monitoring. Powerful tools for examining forest attributes and health, aerial surveys allow large areas of forestland to be assessed with limited expenditures of time and labor. In the future, aerial detection of pest problems (defoliation or mortality) will incorporate remotely sensed data acquired from satellite platforms. The technologies now being used in precision agriculture, including GPS and GIS, have great potential for monitoring large- to small-scale forests for problem foci, as well as for data management and analysis (Boyce and Haney 1997). With the rapid advancement of technology within the realm of remote sensing, detection and recognition of present or potential problems may become more rapid and efficient. Like visual detection of defoliation and other forest pest indicators, acoustical detection may be incorporated into pest monitoring. Some pest organisms of economic concern spend much of their life cycle within their hosts and so cannot be detected visually. Using sonic and ultrasonic emissions to detect these hidden organisms will be part of future detection capabilities, especially as efforts are made to prevent the entry of exotic pests associated with raw log and wood imports.

Hazard Rating Systems

Hazard/risk rating systems also play an important role in IFPM. Conditions placing a forest at extreme, high, medium, or low risk of burning during the fire season are understood fairly well. The same is true of insects and disease. Hazard rating systems provide the ability to rate site and stand conditions in terms of potential for attack (very high to very low). Use of appropriate hazard rating systems, which have been developed for various insects and pathogens, can enhance management decisions and options. These decisions usually are derived from an understanding of site and stand conditions associated characteristically with specific insect and disease epidemics.

Numerous hazard-rating systems incorporating site and stand information (Billings and Bryant 1982; Hicks et al. 1980; Honea, Nebeker, and DeAngelis 1987; Ku, Sweeney, and Shelburne 1980; Kushmauel et al. 1979; Lorio and Sommers 1981; Mason and Bryant 1984; Sader and Miller 1976) have been developed for the SPB for various regions of the Southeast. Honea, Nebeker, and Straka (1987) conducted an economic evaluation of seven SPB hazard rating systems. These have been recommended since as decision criteria for harvest or thinning-age determinations regarding nonindustrial private forestlands. The use of such systems in harvest decision making can result in pine stands' being harvested or thinned before or after *economically optimal rotation age*, or before or after the age that maximizes net present value. Substantial opportunity costs can result. Recommendations to use any hazard rating system should be accompanied by guidelines reflecting the economic implications of nonoptimal harvest or thinning.

Kessler, Heller, and Hard (1981) demonstrated a hazard rating system indicating the susceptibility of stands to Douglas fir tussock moth defoliation, including the probability of defoliation. For fir and red spruce stands infested with the spruce budworm in northern Maine, New Hampshire, and Vermont, a potential hazard rating system incorporating cambial electrical resistance has been developed (Davis, Shortle, and Shigo 1980).

Hazard rating systems also have been developed for diseases. For annosus root rot, Baker et al. (1993) developed a classification and a regression tree analysis for assessing the hazard of pine mortality caused by *Heterobasidion annosum*. Alexander (1989) presented annosus root disease hazard rating, detection, and management strategies for the southeastern United States. Hazard rating systems also have been developed for dogwood anthracnose (Langdon et al. 1991), white pine blister rust (Rust 1988), and little leaf disease (Oak 1985), to mention a few.

In IFPM, the land manager or owner must know something about the vulnerability of his or her lands to a variety of potentially destructive agents. To this end, hazard ratings can be of great assistance, just as risk rating systems can be with respect to fire. Hazard ratings must be viewed as another piece of information in the decision making process concerning the management of valuable resources as a whole; management decisions should not be based on these ratings alone, for negative economic, ecological, and social consequences of decisions made with limited input—especially when based solely on hazard rating values—can result.

Actions and Consequences

Problems or potential problems must be defined in terms of forest management objective(s) stated and subsequent impact(s) quantified. Impacts must be definable and measurable in terms of the stated management objective or objectives. Impacts can be defined in many different ways but are, broadly speaking, ecologic, economic, or social. From an IFPM perspective, impact must be defined specifically so that appropriate actions can be taken and balanced against objectives.

An IFPM option that always must be considered is that of doing nothing, or of "letting nature run its course." What are the implications of doing nothing? From a historical point of view, periodic outbreaks can be expected as a result of population fluctuations. That rates of mortality and/or growth loss in forests will be in line with past trends and continue into the future also can be expected. With increases in acreage of host type, however, proportional increases in pest population activity and associated losses may result. Growers also must be aware of the potential introduction of exotics as forestry practices change and intensify.

Direct intervention, of which pesticide usage provides a helpful example, is another IFPM approach to consider. In many instances, use of pesticides, or biocides, may be preferred in the short term because they make possible the immediate control, through mortality, of unwanted organisms. Pesticides have numerous disadvantages, however, and long-term ecologic, economic, and social costs must be taken into account during considerations of pesticide use in forests.

Other direct management options include salvage removal. Land managers and owners usually prefer salvage removal because it allows for the removal and use of infested/infected trees, thus providing the landowner with a degree of economic return. For effective salvage, affected material must be removed as soon as possible after detection. Rapid reaction will increase greatly the probabilities that the causal organism will be removed from the site and that additional spread in the area will be prevented. This advantage is of special interest in urban forest settings, where the value of individual trees is quite high. Other direct options are cutting and leaving; cutting, piling, and burning; and using various *trap-tree* and *trap-out* techniques (i.e., baiting selected trees to attract insects and then destroying insects and infested trees; and setting general area, pheromone-baited insect traps).

Indirect intervention, yet another IFPM approach, manipulates the pest population by disrupting the normal colonization processes. Applicable methods include pheromone aggregation or antiaggregation tactics, to silvicultural treatments. Indirect treatments are preventive and have as objectives the maintenance of tree growth above, and of mortality below, a predetermined level (Figure 4.15). This level often is referred to as an *action threshold*, above which management activity takes place to prevent effects such as mortality and/or growth reduction. The threshold also might be aesthetic: for instance, the beholder may consider a certain amount of defoliation acceptable, but once it approaches the threshold level, action must be taken (Figure 4.15).

Insofar as they incorporate tactics of standard forestry operations, silvicultural treatments are preferred actions among forest managers. Such tactics include preparing sites, managing water, thinning, burning in a prescribed manner, and fertilizing.

Future Concerns

Of crucial importance to the forests of North America is the introduction of exotic pests/invasive plant and pest species. Recognizing that invasive plants and pests pose an increasing threat to U.S. forests, the Invasive Species Council released its first Management Plan in July 2000 (Campbell 2002).

In the future, soil also will need to be considered carefully. Soil is a nonrenewable resource often overlooked in IFPM programs. Insofar as it is the cornerstone of ecology for terrestrial ecosystems (Boyce and Haney 1997), soil inhabitants (including insects, parasitic fungi and nematodes, and beneficial microbes/fauna) also warrant closer study (Barker and Barker 1998; Görres et al. 1997). Geostatistical analyses and mapping should facilitate characterization of mineralization and nutrient cycling, including the effects of soil microflora/fauna and soil-related factors such as organic-matter content and soil moisture (Görres et al. 1997). Among the soil organisms used as indicators of soil and plant health, *entomopathogens*, or nematodes and fungi-parasitizing insects, may have potential as biological indicators of deforestation and land use (Barker and Barker 1998).

Doubtless, many technologies as yet undiscovered also will be used to assist in IFPM programs. The ways in which remote sensing can be used to monitor forest health have not been exhausted, and the manner of data acquisition also will change, although system biology must not be taken for granted. Species still will need to be identified if prescriptions for management are to be made. To avoid decreasing biodiversity, growers will need to pay attention to details of the living landscape.

The importance of information will continue to grow, and it will be obtained from an increasing number of sources. Internet web sites will serve as one type of source regarding IFPM; indeed, they already provide a great deal of information. Web sites that currently are providing useful information and are likely to be maintained during the next decade include Canadian Forest Service (2002), Center for Integrated Pest Management (2002), University of Massachusetts Bookstore (2002), and USDA (2000).

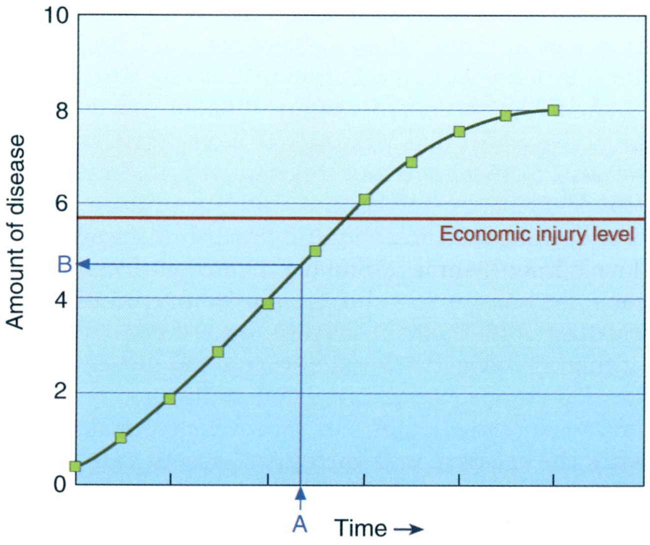

Figure 4.15. A disease-control threshold diagram showing (A) time and (B) amount of disease at which control must be applied to keep disease level below the economic injury level (Tainter and Baker 1996).

Organic Systems/Sustainable Agriculture

Although productivity for key crops in several countries has more than doubled during the last five decades, the need for food (based on projected population increases) likely will increase by more than 50% during the next 50 yr (Avery 1995; Gold 1999). Much enhanced productivity has been the result of the advent of synthetic fertilizers and pesticides enabling producers to augment yields and quality readily. But concerns about synthetic inputs are generating interest in organically grown foods.

Additional crop productivity likely will not come as the result of bringing new lands into cultivation, especially in the United States. Total land devoted to cropland in the contiguous 48 states increased only slightly for the period 1945 to 1993 (Anderson and Magleby 1997). In contrast, total land devoted to all agricultural uses decreased sharply while total land devoted to special uses increased sharply (Anderson and Magleby 1997). Since 1970, more than 30 million a. of productive farmland has been lost to urban development; some 155,000 U.S. farms were lost or consolidated during the decade between 1987 and 1997 alone (Gold 1999).

From 1945 to 1993 in the United States, agricultural production increased mainly as the result of improved conventional farming systems, and output increased more than twofold (Anderson and Magleby 1997; see Figure 4.16). It should be pointed out, however, that the almost-stable input figure for this near half century is somewhat misleading. Agricultural producers minimized cost increases by substituting durable equipment and capital for labor. Moreover, pesticide usage increased an average of 6.2%/yr; fertilizer usage, 1.7%/yr; and energy inputs, 0.8%/yr. These data offer clues as to how agriculture has, over the last five decades, been fulfilling rapidly growing demands for increased food. In light of declining available cropland and increasing populations—especially worldwide—a need for sustainable cropping systems is nondebatable. Nevertheless, the characteristics of farming systems that contribute to sustainability with minimally negative effects on the environment and human health have been and continue to be discussed vigorously by parties with divergent viewpoints. Although they have certain EBMPs in common, organic farming and sustainable agriculture are very different approaches to food and fiber production. Rather than focusing on such divergences, this section addresses recent progress made in both organic food production systems and sustainable agriculture systems and assesses, briefly, the outlook for each, including the role of IPM.

Organic Cropping Systems

Background and Terminology

Foods promoted as *organically grown* have been produced in the United States since the late 1940s (NOP 2002). Food has, of course, been "organically grown" for centuries. The term *organic farming* was used initially in England, by Lord Northbourne (Gold 1999). According to a 1980 USDA definition, the term *organic farming* refers to a production system that avoids or largely excludes synthetically produced fertilizers, pesticides, growth regulators, and livestock feed additives; or genetically modified organisms (GMOs), which are bred by means of genetic engineering techniques.

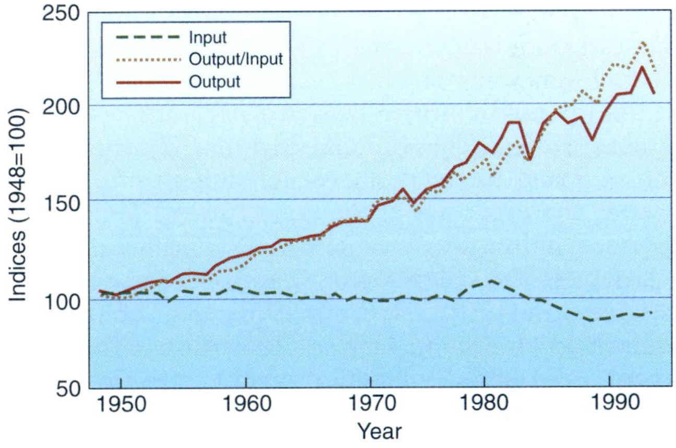

Figure 4.16. Productivity growth in U.S. agriculture, 1943 to 1993 (after Anderson and Magleby 1997).

To the greatest extent possible, organic farming systems rely on crop rotations, crop residues, animal manures, legumes, green manures, off-farm wastes, mechanical cultivation, mineral-bearing rocks, and biological pest control to maintain soil productivity and tilth, to supply plant nutrients, and to control insects, weeds, and other pests (Gold 1999). Having begun as garden plots on a few scattered farms, the organic food sector has expanded greatly, growing at a rate of 20%/yr. Sales now amount to some $5 billion to $6 billion/yr. Such a rate of growth cannot be sustained indefinitely.

A marketplace premium of 10 to 30%, and in some instances more than 200%, for organic foods often is an important reason for growers' choosing to produce by organic methods (Greer and Diver 2000; Liebman 2001). The total number of organic farmers in the United States is approximately 12,200, most of whom are small-scale producers and cannot benefit from the economies of scale associated with corporate-style farming of traditional crops. In fact, organic production often has variable returns for farmers and regions and is not a viable means of increasing profits for all growers (Liebman 2001). Still, the number of organic farmers is increasing at a rate of 12%/yr. With the current growth in the organic product industry, premiums may diminish greatly or disappear over time. Nevertheless, should public concerns over GMOs or biotech issues such as product labeling continue to grow, premiums could last indefinitely. Additional motivating factors for producers of organic products include the potential for decreasing input costs, minimizing negative environmental impacts, and improving safety and function or diversity of agroecosystems (Greer and Diver 2000).

An updated characterization of organic foods in the United States, "The National Organic Program (NOP) Final Rule," includes the establishment of an NOP under the direction of the Agricultural Marketing Service, a division of the USDA (NOP 2002, December 21, 2000 65 FR 80548). The NOP document includes an extensive list of definitions, including a definition for *organic matter*. According to the document, *organic* is "a labeling term that refers to an agricultural product produced in accordance with the act and regulations in this part" (p. 5). In 1995, a related definition of *organic farming* was developed jointly by the USDA National Organic Standards Board (NOSB) and the NOP: Organic production is "a production system that is managed in accordance with the Act and regulations in this part to respond to site-specific conditions by integrating cultural, biological, and mechanical practices that foster cycling of resources, promote ecological balance, and conserve biodiversity" (p. 5). *Organic* is a labeling term denoting products produced under the authority of the Organic Foods Production Act (OFPA) (Gold 1999; NOP 2002).

The primary purpose of the NOP and the final rule is, therefore, to provide regulations ensuring that organically labeled products meet consistent national standards. The Final NOP Rule also should prove helpful with unexpected pest epidemics, for it provides a crop insurance system. All farms, wild-crop harvesting operations, and handling operations selling agricultural products as organically produced will be affected. The only organic producers exempted from the plan are those selling products valued at < $5,000/yr. The OFPA and the NOP require that these organic foods originate from farms or handling operations certified by a state or a private agency accredited by the USDA (NOP 2002).

Development of the new NOP and the related final rule took many years. After establishing the 15-member NOSB and publishing the proposed rule in 1997, the USDA received 275,603 public comments suggesting how the rule should be reviewed. The current voluminous document is available on the Internet (NOP 2002).

The need for the final rule stemmed from the rapid growth of the organic farming sector. The NOP Final Rule is expected to facilitate European acceptance of NOP-certified organic products. Annual growth rates of 25 to 30% in organic farming also have occurred in Europe and in Japan over the last few years. By the year 2006, annual organic food markets in Europe are projected to reach approximately $58 billion, and in the United States, $47 billion (NOP 2002).

Organic Production Systems and Integrated Pest Management

The principal guidelines for organic production are to use materials and practices enhancing the ecological balance of natural systems and integrating the parts of the farming system into an ecologic whole. Although they cannot ensure that products are completely free of harmful residues, organic agriculture practices can be used to minimize pollution of air, soil, and water. Organic food handlers, producers, and retailers adhere to standards maintaining the integrity of organic agricultural products. The primary goal of organic agriculture is to optimize the health and productivity of interdependent communities of soil life, plants, animals, and people (Gold 1999; NOP 2002).

The presence of adequate soil organic matter has many benefits, including improved soil tilth, water-holding capacity, soil-nutrient cycling, activity of beneficial soil organisms, and general crop growth (Magdoff and van Es 2000).

The severely restricted use of chemical pesticides and the exclusion of genetically engineered crop varieties affect IPM programs in organic production, under which the pesticides permitted must generally be derived from organic materials. The Pesticide Data Progam (PDP) will be used to establish residue tolerances, which likely will be 5% of the EPA tolerances for major pesticides. Synthetic pesticides are prohibited unless specifically allowed on the national list, which is recommended by the NOSB and approved by the Secretary of Agriculture (see NOP 2002).

Greenhouse Systems

Most horticultural and greenhouse practices/technologies for organic and conventional production systems are quite similar, but the two major components that differ most substantively involve pest management and soil fertility (Greer and Diver 2000). In regard to both, IPM relies heavily on cultural practices and biological controls, but in regard to soil fertility IPM also relies heavily on animal wastes and green-manure crops and legumes.

Numerous parasitoids and predators of insects are becoming available for use in the management of greenhouse insects, mites, and other arthropods (Gill, Clement, and Dutky 1999; Greer and Diver 1999); strategies include commercial packages of parasitoids and predators (Greer and Diver 1999; see Chapter 2). As described in various publications addressing IPM for greenhouses, additional products are likely to be marketed in the future (Gill Clement, and Dutky 1999; Greer and Diver 1999). For ground-culture greenhouse systems, weed control in organic production can be a challenge. Current tactics available for nonherbicidal weed management within the greenhouse include a combination of mechanically cultivating in the traditional manner, hoeing, steam pasteurizing, solarizing, mulching with plastic/organic materials or another type of fiber, and poultry grazing. An example of poultry grazing involves the use of jungle fowl that pick off insects from vegetable plants without damaging fruits and graze on associated weeds (Greer and Diver 2000).

Management of soilborne pathogens, including nematodes, requires a number of carefully developed practices for greenhouse and general nursery production (Diver 1998). For organic production, primary management tactics include solarization, steam pasteurization, and use of disease-suppressive potted mixtures. Preparation of these mixtures typically involves incorporating specially prepared organic amendments or compost and inoculating the composts and/or potting media with selective microbes such as *Trichoderma, Gliocladium, Bacillus* spp., and *Pseudomonas* sp. (Diver 1998). Plant-growth and health-promoting microbes, such as certain *Pseudomonas* sp. and mycorrhizal fungi, also have potential in this regard. Composted bark is an example of a disease-suppressive component that may be added to soil mixes. Disease-suppressive bark mulches have been so successful in Ohio (Hoitink and Fahy 1986) that methyl bromide has not been used in the nursery industry for two decades in that state (Diver 1998). Plant root pathogens such as *Pythium* sp. usually are suppressed through general competition in these mixes whereas other fungi, including *Rhizoctonia* spp., may require specific microbial antagonists. Of interest are extracts from a pine bark mulch fortified with numerous microbes recently shown to induce, in cucumber and in *Arabidopsis*, systemic acquired resistance to foliar pathogens (*Colletotrichum orbiculare* and *Pseudomonas syringae* pv. *maculicola*, respectively [Zhang et al. 1998]).

Field Organic Systems

Integration of management practices resulting in increased agroecosystem biodiversity is crucial to successful organic production on a field scale. Enhancement factors or practices include rotations, cover crops, no-tillage, composts, green manures, other organic matter additions, windbreaks, intercrops, and agroforestry (Altieri 1994). Related functions and components that must be considered in efforts to promote biodiversity are summarized in Figure 4.17.

Pest control on highly susceptible crops continues to pose a challenge in all cropping systems. The use of one key practice, such as the addition of large amounts of organic matter, may fail to control given plant pests adequately. For example, the addition of 16 t (wet wt) of swine manure/a. did not control root-knot nematodes on tomato but did suppress *Fusarium* sp. and *Sclerotium rolfsii* (Bulluck 1999). Additionally, levels of *Trichoderma* sp., enteric bacteria, fluorescent pseudomonad bacteria, and culturable bacteria were greater with organic amendments than in conventional production systems. Thus, very susceptible crops are problematic in an organic production system, especially on a short-term basis (i.e., < 2 to 4 yr), and should be grown only with careful planning and management.

Polyculture is one of the prime management strategies of organic farming (Altieri 1994; see Chapter 2 for information on various types of polyculture). In addition to requiring fewer inputs than conventional monocultures do, polycultural production systems generally result in increased biodiversity in agroecosystems. In contrast, monoculture often results in the build-up of severe infestations of soil pathogens such as root fungi and/or plant parasitic nematodes.

Lists of botanicals and other pesticides for use in organic production are available, including those posted with the Final NOP Rule (Adam 2001; NOP 2002).

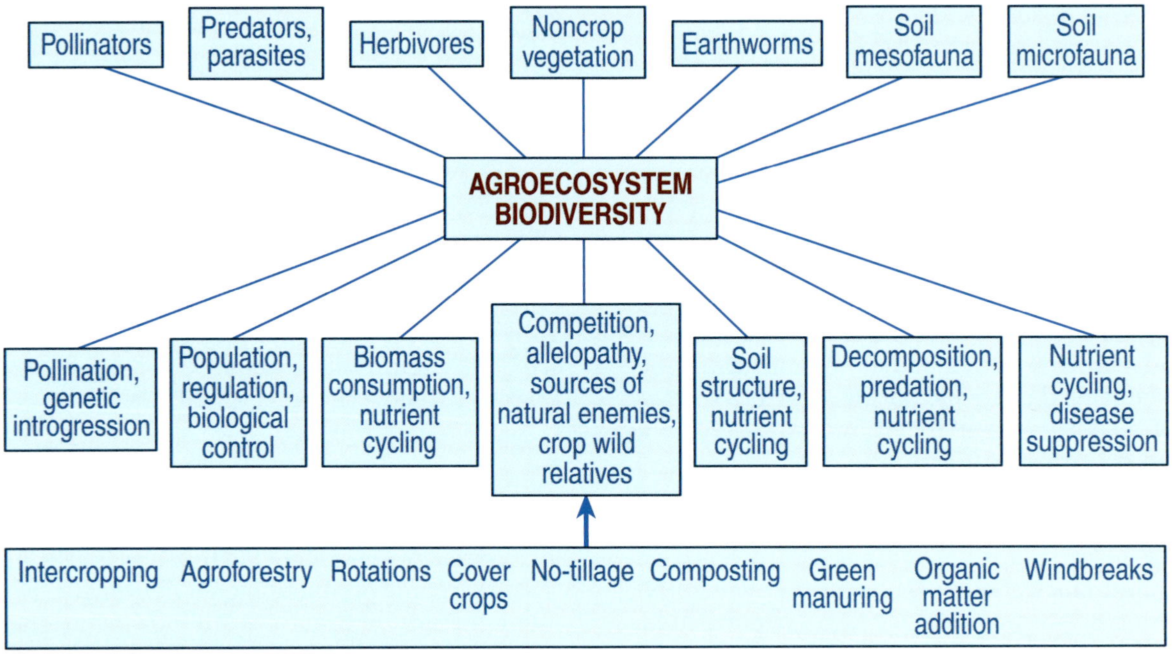

Figure 4.17. The components, functions, and enhancement strategies of biodiversity in agroecosystems (after Altieri 1994).

These products include insecticides such as "Neem" (NOP, Tables 2.5 to 2.9), as well as fungicides based on natural plant-product chemistry (see Chapter 2 for lists of current biopesticides and related product availabilities). In addition, the EPA (2003) has provided extensive information on labeling and registration of pesticide products under the NOP.

Sustainable Cropping Systems

Over the last five decades, the twofold increase in agricultural productivity on fewer acres in the United States (Figure 4.16) could be viewed as evidence that U.S. agriculture is, in fact, sustainable. At the same time, however, excessive use of chemical pesticides and fertilizers as well as problems such as soil erosion have many negative effects on the environment and often on agricultural productivity (Schaller 1991). Undoubtedly, over the centuries, farmers have been concerned about agricultural sustainability and have done much to achieve that goal. Yet one of the first official federal policies addressing agricultural sustainability came as a part of the Food Security Act of 1985, which authorized the Low-Input Sustainable Agriculture (LISA) Program for research and education. The LISA Program sometimes resulted in confusion, with certain parties suggesting that external inputs of pesticides and fertilizers eventually should be eliminated in agriculture production. Many other terms also have been used in reference to conserving natural resources, protecting the environment, and enhancing the health and safety of consumers. These quite divergent terms include *alternative agriculture, organic, regenerative, biological, ecological, biodynamic, low-input, natural, reduced input, regenerative agriculture, and sustainable agriculture* (Gold 1999; Schaller 1991).

A key development related to enhancing support and interest in agricultural sustainability has been the widespread recognition that environmentally sound farming practices must be profitable if they are to be adopted. Additionally, certain synthetic chemicals may be essential to ensuring sustainability of intensive farming systems (Schaller 1991). These two points clearly differentiate the concept *sustainable agriculture*, especially as it has developed over the last decade, from the more input-restrictive concept of *organic food production*. As discussed earlier in this report (Chapter 3), new technologies such as precision agriculture and GMOs offer much hope for the future of a sustainable agriculture. Nevertheless, biodiversity in reference to cropping systems, soil biology, and pest management must be given a high priority in conventional as well as in organic production systems (Altieri 1994; CAST 1999; Ingham et al. 1985). During the last 14 yr, the USDA's Sustainable Agriculture Research and Education Program, including its range of regional research and educational projects, has contributed much to the advancement of sustainable agriculture and to the understanding of soil health (Magdoff and van Es 2000).

In the United States, one of the first definitions of *sustainable agriculture* was published by the American Society of Agronomy in 1989: "A sustainable agriculture is one that, over the long term, enhances environmental quality and the resource base on which agriculture depends; provides for basic human food and fiber needs; is economically viable; and enhances the quality of life for farmers and society as a whole" (p. 3) (Norman et al. 2002). Benbrook (1991), in discussing basic concepts and operational definitions, concludes as follows: "Sustainable agriculture, which is a goal rather than a distinct set of practices, is a system of food and fiber production that

1. improves the underlying productivity of natural resources and cropping systems so that farmers can meet increasing levels of demand in concert with population and economic growth;
2. produces food that is safe, wholesome, and nutritious, and that promotes human well-being;
3. ensures an adequate net farm income to support an acceptable standard of living for farmers while also underwriting the annual investments needed to improve progressively the productivity of soil, water, and other resources; and
4. complies with community norms and meets social expectations." (p. 3)

Norman et al. (2002) concluded that the term *sustainable agriculture* refers to an integrated system of plant and animal production practices applied in a site-specific manner, with these long-term objectives: (1) to satisfy human food and fiber needs; (2) to enhance environmental quality and the natural resource base toward which the agricultural economy depends; (3) to make the most efficient use of nonrenewable resources and on-farm resources and to integrate, where appropriate, natural biological cycles and controls; (4) to sustain the economic viability of farm operations; and (5) to enhance the quality of life for farmers and society as a whole. These five components of the official definition of *sustainable agriculture*, which were authorized under the 1990 farm bill, emphasize productivity, quality of environment, efficient use of nonrenewable resources, viability of economy, and quality of life (Norman et al. 2002). Emphasis is on the long term for all components. Either exploitation of any nonrenewable resource such as groundwater or excessive use of fossil fuels would result in a farm's being considered nonsustainable in the long term. Gold (1999) has provided a compilation of various definitions and concepts related to sustainable agriculture.

Components of Sustainable Agriculture

The goals for sustainable agriculture can be categorized in terms of four conceptual themes. These include (1) farming and natural resources; (2) plant production practices; (3) animal production practices; (4) and economic, social, and political contexts (Feenstra, Ingels, and Campbell 1997). Stewardship of land and natural resources, which involves maintaining and enhancing these crucial resources for the long term, is essential for sustainable agriculture. A systems approach requires an understanding of sustainability and provides tools for exploring the many interconnections between farming and other environmental facets (Feenstra, Ingels, and Campbell 1997). The overall system thus encompasses individual farm and local ecosystems as well as communities affected locally and globally by farming. From the systems perspective, the consequences of farming operations and practices on the environment as well as on human communities must be considered. The systems approach also requires an interdisciplinary approach to research and education and solicits input from farm workers, consumers, policymakers, researchers, and teachers.

Sullivan (2001) concluded that sustainable agriculture has much to gain from holistic management, including a strategy for driving decisions that reflect natural functions and ensure that farming remains sustainable. Four key natural processes discussed by Sullivan include water cycle, mineral cycle, biodiversity, and energy flow. Like natural ecosystems, landscapes with adequate ground cover or plants give rise to minimal run-off and soil erosion. Similarly, minerals are wasted minimally when cycled through the biological system, as in the movements from microorganisms to soil, plants, animals, and back to soil (Barker and Koenning 1998; Ferris, Venette, and Lau 1996; Ingham et al. 1985). Sustainable farming operations should use this natural mineral cycle and minimize off-farm purchase of minerals. The third natural process, biodiversity, although comparatively complex, is crucial to sustainable agriculture. Crop rotation is a key step toward increasing biodiversity on any farm. In addition to stabilizing the system, enhanced biodiversity also limits problems related to plant pests (e.g., weeds, insects, and various pathogens) (Altieri 1994; CAST 1999). A fourth natural process involves the flow of energy from the sun through the biological system (Sullivan 2001). When soils are bare, no sunlight is converted into energy by plants; when two or more crops are grown per year, solar energy collection is maximized. Cropping systems including two or more types of plants increase leaf area and thereby capture more energy from the sun than systems including a single type of plant do.

Integrated Pest Management and Sustainable Agriculture

Integrated pest management, including pest management practices preventing damage to the environment, is an essential component of sustainable agriculture. The original IPM concepts were conceived with one of the goals' being the development of EBPM practices (Kennedy and Sutton 2000; Rabb and Guthrie 1970). Over the last 30 yr, IPM programs have resulted in a decline in pesticide use on cotton, but, until recently, overall pesticide use continued to increase on most crops. Because of the continuing reliance on crop pesticides, emphasis on deployment of systems relatively committed to EBMPs has increased greatly in recent years (Altieri 1994; Boland and Kuykendall 1998; Kennedy and Sutton 2000; NRC 1996). Research and experience in the United States and Europe have shown that IPM/integrated cropping systems can control pests effectively and can contribute to the build-up of beneficial organisms and to the suppression of soilborne pests. For example, populations of cyst and stem nematodes on cereals were suppressed more effectively by an integrated farming system with altered tillage and sowing techniques (with a clover cover crop), fertilization, organic manure, and decreased pesticide applications than by a conventional production system (El Titi and Ipach 1989). In a related study, earthworms constituted 17.6% of total biomass in an integrated cropping system but were completely absent in a conventional system (Zwart et al. 1994). Protozoa also were favored by the integrated system.

An aim of this section has been to emphasize the benefits of integrating practices that have been discussed in much greater detail in Chapters 2 and 3 and throughout Chapter 4 itself. Although much progress has been made in developing sustainable cropping systems, flexibility is essential in the initial choice of IPM tactics during the transition from conventional to sustainable systems (Hall and Kuepper 1997). When pest species fluctuate greatly and cause crop losses until equilibrium with natural antagonists has been effected, pesticide use often is needed to minimize economic losses. Crop insurance also can help limit economic and production risks.

Outlook

Emerging concepts and technologies related to IPM, organic farming, and sustainable agriculture will require new perspectives for research, teaching, outreach, and production agriculture. This outlook will include a "systems approach" to food, an approach that traditionally has not been considered. For example, the benefits of cropping systems' increasing levels of soil organic matter will extend beyond the build-up of microbes suppressing plant pathogens. Related soil carbon sequestration can help minimize projected global increases of atmospheric carbon dioxide (CAST 2000). Furthermore, deployment of ecology-based cropping and IPM practices can contribute to long-term restoration ecology by facilitating recovery of degraded lands (Dobson, Bradshaw, and Baker 1997). To achieve the ecologic goals of sustainable agriculture, practices for crop and pest management need to be directed at (1) suppressing incidence and intensity of a wide range of crop pests, (2) increasing soil organic matter, and (3) enhancing soil and crop health. Cropping systems that include rotations of primary crops with cover crops and green manure crops contribute much to the biodiversity of soil microflora and microfauna. Specific cropping systems as well as cover crops must mesh with the requirements of the primary crop and, of course, the farmer. Intensive long-term research projects focusing on the many facets of conventional and alternative cropping systems, such as research projects in California (Drinkwater et al. 1995; Ferris, Venette, and Lau 1996; Madden 1992), Michigan (Cavigelli et al. 2000), and other states (Fernandez-Cornejo and Jans 1999; Lipson 1997; NRC 1996) and countries (El-Titi and Ipach 1989; Zwart et al. 1994), will contribute much to the sustainability of food production in the United States. An extensive list of books and other resources on sustainable agriculture is available through the Sustainable Agriculture Network of Burlington, Vermont (see Magdoff and van Es 2000).

Appendix A. Abbreviations and Acronyms

a.	acre
Bt	*Bacillus thuringiensis*
EBPM	ecologically based pest management
EPA	U.S. Environmental Protection Agency

GIS	geographic information system
GMO	genetically modified organism
GPS	global positioning system
IFPM	integrated forest pest management
IPM	integrated pest management
IR–2	Interregional Project 2
LISA	low-input sustainable agriculture
NOP	National Organic Program
NOSB	National Organic Standards Board
OFPA	Organic Food Production Act
SCN	soybean cyst nematode
SPB	southern pine beetle
t	ton
UC	University of California
USDA	U.S. Department of Agriculture
wt	weight
yr	year

Appendix B. Glossary

Action threshold. The level above which management activity takes place to prevent effects such as mortality and/or growth reduction.

Biologically based approaches. Processes whereby chemical inputs occasionally are used, when needed, to decrease pest populations below economic injury levels.

Economically optimal rotation age. The age maximizing net present value.

Ecotypes. Different genetic strains.

Entomopathogens. Any parasite (bacteria, fungi, nematodes, viruses, etc.) that attacks and induces disease in insects.

Field corn. Corn grown for feed grain.

First-level IPM. The use of multiple integrated tactics to manage a single class of pest.

Integrated control. Use of the appropriate combination of technologies to ensure economically efficient control while minimizing adverse effects on the forest ecosystem.

Integrated management. The simultaneous consideration of all potential pests during IFPM program development.

Organic. A labeling term denoting products produced under the authority of the Organic Foods Production Act.

Organic farming. A production system that avoids or largely excludes synthetically produced fertilizers, pesticides, growth regulators, and livestock feed additives; or genetically modified organisms that are bred by means of genetic engineering techniques.

Organic production. Process whereby nonchemical inputs conform to organic standards and requirements.

Pest management approach. Process whereby one or more chemical inputs are used in response to changes in pest pressure and environmental conditions.

Scouting. Pest surveillance.

Second-level IPM. The optimization of strategies for all classes of pests.

Sustainable agriculture. An agricultural method is one that, over the long term, enhances environmental quality and resource base on which agriculture depends; provides the basic human-food and -fiber needs; is economically viable; and enhances the quality of life for farmers and society as a whole.

Third-level IPM. The integration of all aspects of crop production.

Trap-out techniques. Use of general area, pheromone-baited insect traps.

Trap-tree techniques. Use of bait in selected trees to attract insects, followed by subsequent destruction of insects and infested trees.

Literature Cited

Adam, K. 2001. Suppliers of organic and/or non-GE seeds and plants: Horticulture resource list, February, <http://www.attra.org/attra-pub/altseed.html> (13 June 2002)

Agnello, A., J. Kovach, J. Nyrop, H. Reissig, D. Rosenberger, and W. Wilcox. 1999. Apple IPM: A guide for sampling and managing major apple pests in New York State. Publication Number 207. New York State Integrated Pest Management Program, Geneva.

Alagarswamy G., P. Singh, G. Hoogenboom, S. P. Wani, P. Pathak, and S. M. Virmani. 2000. Evaluation and application of the CROPGRO-Soybean simulation model in a Vertic Inceptisol. *Agric Syst* 63:19–32.

Alexander, S. A. 1989. Annosus root disease hazard rating, detection, and management strategies in the southeastern United States. *USDA For Serv Gen Tech Rep* 116:111–116.

Allen, D. C. 1994. Healthy forestry—Healthy forests: A balancing act. *J For* 92:60.

Alston, D. G., J. R. Bradley, Jr., D. P. Schmitt, and H. D. Coble. 1991. Response of *Helicoverpa zea* (Lepidotera: Noctuidae) populations to the canopy development in soybean as influenced by *Heterodera glycines* (Nematoda: Heteroderidae) and annual weed population densities. *J Econ Entomol* 84:267–276.

Alston, D. G., D. P. Schmitt, J. R. Bradley, and H. D. Coble. 1993. Multiple interactions in soybean: Effects on *Heterodera glycines* egg populations. *J Nematol* 25:42–50.

Altieri, M. A. 1994. *Biodiversity and Pest Management in Agroec-*

osystems. Food Products Press, an Imprint of the Haworth Press, Inc., New York.

Anderson, M. and R. Magleby (eds.). 1997. Agricultural resources and environmental indicators, 1996–97. Agricultural Handbook Number 172. Agricultural Research Service, U.S. Department of Agriculture, Washington, D.C.

Avery, D. T. 1995. *Saving the Planet with Pesticides and Plastic: The Environmental Triumph of High-Yield Farming*. Hudson Institute, Indianapolis, Indiana.

Bajwa, W. I. 2000. Database of IPM Resources, August 7, <http://www.ippc.orst. edu/cicp/index.htm> (13 June 2002)

Baker, F. A., D. L. Verbyla, C. S. Hodges, Jr., and E. W. Ross. 1993. Classification and regression tree analysis for assessing hazard of pine mortality caused by *Heterobasidion annosum*. *Plant Dis* 77:136–139.

Barker, C. W. and G. M. Barker. 1998. Generalist entomopathogens as biological indicators of decomposition and agricultural land use impacts on Waikato soils. *New Z J Ecol* 22:189–196.

Barker, K. R. and S. R. Koenning. 1998. Developing sustainable systems for nematode management. *Annu Rev Phytopathol* 36:165–205.

Bawden, R. J. 1991. Systems approaches to agricultural development: The Hawksbury experience. *Agric Sys* 40:153–176.

Benbrook, C. M. 1991. Introduction. Pp. 110. In *Sustainable Agricultural Research and Education in the Field*. National Academy Press, National Resource Council, Washington, D.C.

Benbrook, C. M., E. Groth, III, J. M. Halloran, M. K. Hansen, and S. Marquardt. 1996. *Pest Management at the Crossroads*. Consumers Union, Yonkers, New York.

Berti, A., C. Dunan, M. Sattin, G. Zanin, and P. Westra. 1996. A new approach to determine when to control weeds. *Weed Sci* 44:496–503.

Biggs, A. R. 2002. Kearneysville Tree Fruit and Education Center. Homepage, <http://www.caf.wvu. edu/kearneysville/wvufarm1.html> (13 June 2002)

Billings, R. F. and C. M. Bryant. 1982. Southern pine beetle field guide for hazard rating, prevention and control. Circular 259. Texas Forest Service, College Station. 10 pp.

Bland, W. L. 1999. Toward integrated assessment in agriculture. *Agric Sys* 60:157–167.

Boland, G. J. and L. D. Kuykendall (eds.). 1998. *Plant-Microbe Interactions and Biological Control*. Marcel Dekker Incorporated, New York.

Boyce, M. S. and A. Haney (eds.). 1997. *Ecosystem Management: Applications for Sustainable Forest and Wildlife Resources*. Yale University Press, New Haven, Connecticut.

Branham, S. J. and G. D. Hertel (eds.). 1984. *Proceedings of the Integrated Forest Pest Management Symposium*. Continuing Education, University of Georgia, Athens. 281 pp.

Bridges, D. C. 1994. Impacts of weeds on human endeavors. *Weed Technol* 8:392–395.

Buhler, D. D. 1999. Expanding the context of weed management. *J Crop Prod* 2:1–7.

Bulluck, L. R., III. 1999. Effects of synthetic and organic soil amendments on soil biological communities, physical and chemical factors, and vegetable production. PhD dissertation. North Carolina State University, Raleigh.

Campbell, F. 2002. Invasive species, <http://www.americanlands.org/forestweb/invasive.htm> (13 June 2002)

Canadian Forest Service. 2002. Integrated Pest Management at the Canadian Forest Service, <http://www.glfc.cfs.nrcan.gc.ca/ipm> (13 June 2002)

Cardina, J., T. M. Webster, C. P. Herms, and E. E. Regnier. 1999. Development of weed IPM: Levels of integration for weed management. *J Crop Prod* 2:239–267.

Carroll, B. N. 2000. Development in transition: Crop protection and more. Pp. 339–354. In G. G. Kennedy and T. B. Sutton (eds.). *Emerging Technologies for Integrated Pest Management—Concepts, Research, and Implementation*. APS Press, American Phytopathological Society, St. Paul, Minnesota.

Cavero, J., R. E. Plant, C. Shennan, D. B. Friedman, J. R. Williams, J. R. Kiniry, and V. M. Benson. 1999. Modeling nitrogen cycling in tomato-safflower and tomato-wheat rotations. *Agric Sys* 60:123–135.

Cavigelli, M. A., S. R. Deming, L. K. Probyn, and D. R. Mutch (eds.). 2000. Michigan field crop pest and ecology management. Michigan State University Extension Bulletin E–2704. Extension, Michigan State University, East Lansing.

Center for Integrated Pest Management. 2002. Homepage, <http://ipmwww.ncsu.edu/cipm/center_ sites.html> (13 June 2002)

Center for Integrated Plant Systems. 2002. Homepage, <http://www.cips.msu.edu> (13 June 2002)

Consortium for Integrated Crop Protection. 2002. Homepage, <http://www.ipmnet.org/cicp.html> (13 June 2002)

Cook, R. J. and R. J. Veseth. 1991. *Wheat Health Management*. APS Press, St. Paul, Minnesota.

Cornell Fruit Information Page. 2002. Homepage, <www.fruit.cornell.edu> (15 January 2003)

Council for Agricultural Science and Technology (CAST). 1999. *Benefits of Biodiversity*. Task Force Report Number 133. Council for Agricultural Science and Technology, Ames, Iowa.

Council for Agricultural Science and Technology (CAST). 2000. *Storing Carbon in Agricultural Soils to Help Mitigate Global Warming*. Issue Paper Number 14. Council for Agricultural Science and Technology, Ames, Iowa.

Council for Agricultural Science and Technology (CAST). 2003. *Mycotoxins: Risks in Plant, Animal, and Human Systems*. Council for Agricultural Science and Technology, Ames, Iowa.

Davis, W., W. Shortle, and A. Shigo. 1980. Potential hazard rating system for fir, *Abies balsamea,* and red spruce, *Picea rubens*, infested with budworm *Charistoneura fumiferana* using cambial electrical resistance in northern Maine, New Hampshire, and Vermont. *Can J For Res* 10:541–544.

Diver, S. 1998. Disease suppressive potting mixes: Pest management technical note. Appropriate technology transfer to rural areas, June, <http://www.attra.org/attra-pub/dspotmix.html> (13 June 2002)

Dobson, A. P., A. D. Bradshaw, and A. J. M. Baker. 1997. Hopes for the future: Restoration ecology and conservation biology. *Science* 277:515–522.

Drinkwater, L. E., D. K. Letourneau, F. Workneh, A. H. C. van Bruggen, and C. Shennan. 1995. Fundamental differences between conventional and organic tomato agroecosystems in California. *Ecol Appl* 5:1098–1112.

Dunan, C. M., P. Westra, and F. D. Moore, III. 1994. A plant process-economic model for wild oats management in irrigated barley. *Agric Sys* 45:355–368.

Dunan, C. M., P. Westra, F. Moore, and P. Chapman. 1996. Modeling the effect of duration of weed competition, weed density and weed competitiveness on seeded, irrigated onion. *Weed Res* 36:259–269.

Dunan, C. M., P. Westra, and F. D. Moore, III. 1998. A plant process economic model for weed management decisions in irrigated onion. *J Am Soc Hort Sci* 124:99–105.

Edmisten, K. L. 1999. 1999 cotton information. AG–417 (Revised). Cooperative Extension Service, North Carolina State Univer-

sity, Raleigh.
Elmore, C. L. 1996. A reintroduction to integrated weed management. *Weed Sci* 44: 409–412.
El Titi, A. and U. Ipach. 1989. Soil fauna in sustainable agriculture: Results of an integrated farming system at Lautenback, F.R.G. *Agric Ecosyst Environ* 27:561–572.
English-Loeb, K. 1999. Elements of New York state IPM, <htpp://www.cornell.edu> (13 June 2002)
Fan, J. 1998. Field crop insect pests, <http://www.isis.vt.edu/~fanjun/text/Link_pest1. html> (13 June 2002)
Feenstra, G., C. Ingels, and D. Campbell. 1997. What is sustainable agriculture? December, <http://www.sarep.ucdavis. edu/concept.htm> (13 June 2002)
Fernandez-Cornejo, J. and S. Jans. 1999. Pest management in U.S. agriculture. Agricultural Handbook Number 717. Economic Research Service, U.S. Department of Agriculture, Washington, D.C.
Ferris, H., R. C. Venette, and S. S. Lau. 1996. Dynamics of nematode communities in tomatoes grown in conventional and organic farming systems, and their impact on soil fertility. *Appl Soil Ecol* 3:161–175.
Fruit and Nut Research and Information Center. 2000. Homepage, July 18, <http://fruitsandnuts.ucdavis. edu/crops.html> (13 June 2002)
Garcia-Paredes, J. D., K. R. Olson, and J. M. Lang. 2000. Predicting corn and soybean productivity for Illinois soils. *Agric Sys* 64:151–170.
Ghersa, C. M., M. L. Roush, S. R. Radosevich, and S. M. Cordray. 1994. Coevolution of agroecosystems and weed management. *BioScience* 44:85–94.
Gill, S., D. L. Clement, and E. Dutky. 1999. *Pests and Diseases of Herbaceous Perennials*. Ball Publishing, Batavia, Illinois.
Gold, M. V. 1999. Sustainable agriculture: Definitions and terms, September, <http://www.nal.usda. gov/afsic/AFSIC_pubs/srb/srb9902.htm> (13 June 2002)
Görres, J. F., M. J. Dichiaro, J. B. Lyons, and F. A. Amador. 1997. Spatial and temporal patterns of soil biological activity in a forest and an old field. *Soil Biol Biochem* 30:219–230.
Gray, M. 2001. Integrated pest management, <http://www.ipm.uiuc.edu> (13 June 2002)
Green, D. L. 2002. Pollination homepage, <http://www.pollinator.com/index.htm> (13 June 2002)
Greer, L. and S. Diver. 1999. Integrated pest management for greenhouse crops: Pest management systems guide, January, <http://www.attra.org/attra-pub/gh-ipm.html> (13 June 2002)
Greer, L. and S. Diver. 2000. Organic greenhouse vegetable production: Horticulture systems guide, January, <http://www.attra.org/attra-pub/ghveg.html> (13 June 2002)
Haack, R. A., K. R. Law, V. C. Mastro, H. S. Ossenbruggen, and B. J. Raimo. 1997. New York's battle with the Asian long-horned beetle. *J For* 95:11–15.
Hall, B. and G. Kuepper. 1997. Making the transition to sustainable farming: Fundamentals of sustainable agriculture, March, <http://www.attra.org/attra-pub/trans.html> (13 June 2002)
Hammond, R. B. 1996. Soybean insect IPM, January 25, <http://impworld.umn.edu/ chapters/hammond.htm> (13 June 2002)
Hammond, R. B., R. A. Higgins, T. P. Mack, L. P. Pedigo, and E. J. Bechinski. 1991. Soybean pest management. Pp. 341–472. In D. Pimental (ed.). *CRC Handbook of Pest Management in Agriculture*. 2nd ed. CRC Press, Inc., Boca Raton, Florida.
Hartman, G. L., J. B. Sinclair, and J. C. Rupe (eds.). 1999. *Compendium of Soybean Diseases*. 4th ed. APS Press, The American Phytopathological Society, St. Paul, Minnesota.
Hedden, R. L. and T. E. Nebeker. 1984. Integrated pest management in pine stands (0–5 years)—Insects. Pp. 39–53. In S. J. Branham and G. D. Hertel (eds.). *Proceedings of the Integrated Forest Pest Management Symposium*. Continuing Education, University of Georgia, Athens.
Heiniger, R. W. 1999. Corn production guide 1999. AG–594. Cooperative Extension Service, North Carolina State University, Raleigh.
Hepting, G. H. and G. M. Jemison. 1958. Forest production: Timber resources for America's future. Forest Research Report Number 14. Forest Service, U.S. Department of Agriculture, Washington, D.C.
Hicks, R. R., Jr., J. E. Howard, K. G. Watterston, and J. E. Coster. 1980. Rating forest stand susceptibility to southern pine beetle in east Texas using probability functions. *For Ecol Manag* 2:269–283.
Hogmire, H. W., Jr. (ed.). 1995. Mid-Atlantic orchard monitoring guide. Publication NRAES–75. Northeast Regional Agricultural Engineering Service, Ithaca, New York.
Hoitink, H. A. and P. C. Fahy. 1986. Basis for the control of soilborne plant pathogens with composts. *Annu Rev Phytopathol* 24:93–114.
Honea, C. R., T. E. Nebeker, and J. D. DeAngelis. 1987. Selection of southern pine beetle hazard rating systems for Mississippi. Pp. 259–266. In *Proceedings of the Fourth Biennial Southern Silvicultural Research Conference*. General Technical Report SE–42. Forest Service, U.S. Department of Agriculture, Washington, D.C.
Honea, C. R., T. E. Nebeker, and T. J. Straka. 1987. Economic evaluation of seven southern pine beetle hazard rating systems. *J Environ Manage* 24:259–266.
Ingham, R. E., J. A. Trofymow, E. R. Ingham, and D. C. Coleman. 1985. Interactions of bacteria, fungi, and their nematode grazers: Effects on nutrient cycling and plant growth. *Ecol Monogr* 55:119–140.
Iowa Crop and Livestock Reporting Service. 2000. Iowa agricultural statistics. Iowa Department of Agricultural Statistical Reporting Service, Des Moines.
Iowa Grain Quality Initiative. 2002. Homepage, <http://www.extension.iastate.edu/ pages/ grain> (13 June 2002)
Kennedy, G. G. and N. P. Storer. 2000. Life systems of polyphagous arthropod pests in temporally unstable cropping systems. *Annu Rev Entomol* 45:467–493.
Kennedy, G. G. and T. B. Sutton (eds.). 2000. *Emerging Technologies for Integrated Pest Management: Concepts, Research, and Implementation*. APS Press, American Phytopathological Society, St. Paul, Minnesota.
Kessler, B. L., R. C. Heller, and J. S. Hard. 1981. A demonstration for hazard rating susceptibility of stands to Douglas-fir tussock moth defoliation on the Paluse Ranger District: Probability of defoliation. Report 81–88. U.S. Department of Agriculture, Forest Service, Washington, D.C. 45 pp.
Ku, T. T., J. M. Sweeney, and V. B. Shelburne. 1980. Site and stand conditions associated with southern pine beetle *Dendroctonous frontalis* outbreaks in Arkansas—A hazard rating system. *South J Appl For* 4:103–106.
Kushmauel, R. J., M. D. Cain, C. E. Rowell, and R. L. Porterfield. 1979. Stand and site conditions related to southern pine beetle susceptibility. *For Sci* 25:656–664.
Landis, D. A., S. D. Wratten, and G. M. Gurr. 2000. Habitat management to conserve natural enemies of arthropod pests in agriculture. *Annu Rev Entomol* 45:175–201.
Langdon, K., C. Parker, M. Windham, S. Powell, and K. Johnson.

1991. A preliminary hazard rating for dogwood anthracnose in the Southern Appalachians. *USDA For Serv South Rep* R8–PR 20:34–38.

Liebman, M. 2001. Weed management: A need for ecological approaches. Pp. 1–39. In M. Liebman, C. L. Mohler, and C. P. Staver (eds.). *Ecological Management of Agricultural Weeds.* Cambridge University Press, Cambridge, United Kingdom.

Liebman, M. and E. R. Gallandt. 1997. Many little hammers: Ecological management of crop-weed interactions. Pp. 291–343. In L. E. Jackson. *Ecology in Agriculture*. Academic Press, New-York.

Liebman, M., C. L. Mohler, and C. P. Staver. 2001. *Ecological Management of Agricultural Weeds*. Cambridge University Press, Cambridge, UK.

Lindquist, J. L., D. A. Mortensen, S. A. Clay, R. Schmenk, J. J. Kells, K. Howatt, and P. Westra. 1996. Stability of corn (*Zea mays*)-velvetleaf (*Abutilon theophrasti*) interference relationships. *Weed Sci* 44:309–313.

Lipson, M. 1997. Searching for the "O-Word": Analyzing the USDA current research information system for pertinence to organic farming. Organic Farming Research Foundation, Santa Cruz, California.

Little, C. E. 1995. *The Dying of the Trees: The Pandemic in America's Forests*. Viking, New York.

Lorio, P. L. and R. A. Sommers. 1981. Use of available resource data to rate stands for southern pine beetle risk. Pp. 75–78. In *Proceedings of the Hazard-Rating Systems in Forest Management Symposium*. General Technical Report WO–27. Forest Service, U.S. Department of Agriculture, Washington, D.C.

Lynch, S., D. Sexson, C. M. Benbrook, M. Carter, J. Wyman, P. Nowak, and J. Barzen. 2000. *Accelerating Industry-Wide Transition to Reduced-Risk Pest Management Systems: A Case Study of the Wisconsin Potato Industry Choices*. American Agricultural Economics Association, Ames, Iowa.

MacRae, I. V. 1996. IPM resources on the World Wide Web, January 17, <http://www.ipmworld. umn.edu/chapters/ macrae.htm> (13 June 2002)

MacRae, I. V. 1998. Site specific IPM, January 23, <http:// ipmworld.umn.edu/chapters/ ipmgis.htm> (13 June 2002)

Madden, J. P. 1992. *Beyond Pesticides: Biological Approaches to Pest Management in California*. ANR Publications, University of California, Oakland.

Magdoff, F. and H. van Es. 2000. *Building Soils for Better Crops*. 2nd ed. Sustainable Agriculture Network, Burlington, Vermont.

Main, C. E., T. Keever, G. J. Holmes, and J. M. Davis. 2001. Forecasting long-range transport of downy mildew spores and plant disease epidemics, April 25, <http://www.apsnet.org/online/feature/ forecast> (13 June 2002)

Mason, G. N. and C. M. Bryant. 1984. Establishing southern pine beetle hazard from aerial stand data and historical records. *For Sci* 30:375–382.

Merrigan, K. A. 2000. Politics, policy, and IPM. Pp. 497–504. In G. G. Kennedy and T. B. Sutton (eds.). *Emerging Technologies for Integrated Pest Management—Concepts, Research and Implementation*. APS Press, The American Phytopathological Society, St. Paul, Minnesota.

Merwin, I. A., S. K. Brown, D. A. Rosenberger, D. R. Cooley, and L. P. Berkett. 1994. Scab-resistant apples for the northeastern USA: New prospects and old problems. *Plant Dis* 78:4–10.

Mjelde, J. W. and H. S. J. Hill. 1999. The effect of the use of improved climate forecasts on variable costs, input usage, and production. *Agric Sys* 60:213–225.

Myers, J. H., A. Savoie, and E. van Randen. 1998. Eradication and pest management. *Annu Rev Entomol* 43:471–491.

National Academy of Sciences. 1975. *Pest Control: An Assessment of Present and Alternative Technologies*. Vol. IV. Printing and Publishing Office, Washington, D.C. 170 pp.

National Organic Program (NOP). 2002. Homepage, <http:// www.ams.usda.gov/nop> (13 June 2002)

National Potato Council. 2002. The national IPM protocol for potatoes, <http://www.npcspud.com/IPM. html> (13 June 2002)

National Research Council (NRC). 1996. *Ecologically Based Pest Management: New Solutions for a New Century*. National Academy Press, Washington, D.C.

Norman, D., R. Janke, S. Freyenberger, B. Schurle, and H. Kok. 2002. Defining and implementing sustainable agriculture (13 June 2002) <http://www.oznet.ksu.edu/sustainableag/publications/ ksas1.htm>

Oak, S. W. 1985. Adaptation of littleleaf disease hazard rating for use in forest management in South Carolina National Forests. *USDA For Serv SO Gen Tech Rep* 56:246–251.

Ohlendorf, B. L. P. and J. K. Clark. 1999. Integrated pest management for apples and pears. 2nd ed. Publication 3340. Division of Agriculture and Natural Resources, University of California, Oakland.

Pedigo, L. P. 1999. *Entomology and Pest Management*. 3rd ed. Prentice Hall, Upper Saddle River, New Jersey.

Posner, J. L., M. D. Casler, and J. O. Baldock. 1995. The Wisconsin integrated cropping systems trial: Combining agroecology with production agronomy. *Am J Altern Agric* 10:98–107.

Prokopy, R. J., J. L. Mason, M. Christie, and S. E. Wright. 1996. Arthropod pest and natural enemy abundance under second-level versus first-level integrated pest management practices in apple orchards: A four-year study. *Agric Ecosys Environ* 57:35–47.

Rabb, R. L. and F. E. Guthrie (eds.). 1970. *Concepts of Pest Management*. North Carolina State University, Raleigh.

Renner, K. A., S. M. Swinton, and J. J. Kells. 1999. Adaptation and evaluation of the WEEDSIM weed management model for Michigan. *Weed Sci* 47:338–348.

Runion, G. B., S. C. Cade, and R. I. Bruck. 1993. Effects of carbofuran and thiabendazole on incidence of pitch canker of loblolly pine. *Plant Dis* 77:166–169.

Rust, M. 1988. White pine blister rust hazard rating: An expert systems approach. *AI Appl Nat Resour Manage* 2:47–50.

Sader, S. A. and W. F. Miller. 1976. Development of a risk rating system for southern pine beetle infestations in Copiah County, Mississippi. *NASA Remote Sens Earth Res* 5:277–294.

Schaller, N. 1991. Background and status of the low-input sustainable agriculture program. Pp. 22–31. In *Sustainable Agriculture Research and Education in the Field*. National Academy Press, National Research Council, Washington, D.C.

Schweizer, E. E., P. Westra, and D. W. Lybecker. 1994. Controlling weeds in corn (*Zea mays*) rows with an in-row cultivator versus decisions made by a computer model. *Weed Sci* 42:593–600.

Schweizer, E. E., P. Westra, and D. W. Lybecker. 1998. Seedbank and emerged annual weed populations in corn (*Zea mays*) fields in Colorado. *Weed Technol* 12:243–247.

Schweizer, E. E., L. J. Wiles, D. W. Lybecker, and P. Westra. 1994. Bioeconomic modeling for weed management decisions in crops. *Proc Great Plains Agric Council Bull* 150:135–141.

Senft, D. 1997. With IPM, bigger areas are better. *Agric Res* 45:4–8.

Strand, L. L. and J. K. Clark. 1999. Integrated pest management for stone fruits. Publication 3389. Division of Agriculture and

Natural Resources, University of California, Oakland.

Sudduth, K. A., J. W. Hummel, and S. J. Birrell. 1997. Sensors for site-specific management. Pp. 183–210. In J. J. Pierce and E. J. Sadler (eds.). *The State of Site Specific Management for Agriculture*. American Society of Agronomy, Madison, Wisconsin.

Sullivan, P. July 2001. Holistic management: A whole-farm decision making framework: Fundamentals of sustainable agriculture, <http://www.attra.org/attra-pub/holistic.html> (13 June 2002)

Sutton, T. B. 1996. Changing options for the control of deciduous fruit tree diseases. *Annu Rev Phytopathol* 34:527–547.

Swanton, C. J. and S. D. Murphy. 1996. Weed Science beyond the weeds: The role of integrated weed management (IWM) in agroecosystem health. *Weed Sci* 44:437–445.

Tainter, F. H. and F. A. Baker. 1996. *Principles of Forest Pathology*. John Wiley & Sons, Incorporated, New York.

Taylor, C. R. and R. Rodriguez-Kábana. 1999a. Optimal rotation of peanuts and cotton to manage soil-borne organisms. *Agric Sys* 60:57–68.

Taylor, C. R. and R. Rodriguez-Kábana. 1999b. Population dynamics and crop yield effects of nematodes and white mold in peanuts, cotton and velvet beans. *Agric Sys* 59:177–191.

Tree Fruit Research and Extension Center. 2002. Homepage, June 11, <http://www. tfrec.wsu.edu/> (13 June 2002)

University of California. 2003. Statewide Integrated Pest Management Program. UC IPM Online, <http://www.ipm.ucdavis.edu> (30 January 2003)

University of Massachusetts Bookstore. 2002. Links, <http://www.umass.edu/umext/bookstore/ links.html> (13 June 2002)

U.S. Department of Agriculture (USDA). 2000. Integrated pest management: Reduced chemical use protects yields and the environment, April, <http://www.reeusda.gov/success/impact00/ipm.htm> (13 June 2002)

U.S. Department of Agriculture–Cooperative State Research, Extension, and Education (USDA–CSREES). 2002. Plant and animal systems, January 24, <http://www.reeusda.gov/ipm> (13 June 2002)

U. S. Environmental Protection Agency (EPA). 2003. Labeling of pesticide products under the National Organic Program (NOP), <http://www.epa.gov/pesticides/biopesticides/regtools/organic-pr-notice.htm> (21 January 2003)

VanGessel, M. J., J. Schroeder, and P. Westra. 1998. Comparative growth and development of four spurred anoda (*Anoda cristata*) accessions. *Weed Sci* 46:91–98.

VanGessel, M. J., E. E. Schweizer, K. A. Garrett, and P. Westra. 1995a. Influence of weed density and distribution on corn (*Zea mays*) yield. *Weed Sci* 43:215–218.

VanGessel, M. J., E. E. Schweizer, D. W. Lybecker, and P. Westra. 1995b. Compatibility and efficiency of in-row cultivation for weed management in corn (*Zea mays*). *Weed Technol* 9:754–760.

VanGessel, M. J., L. Wiles, E. E. Schweizer, and P. Westra. 1995c. Weed control efficacy and pinto bean (*Phaseolus vulgaris*) tolerance to early season mechanical weeding. *Weed Technol* 9:531–534.

VanGessel, M. J., E. E. Schweizer, R. G. Wilson, L. J. Wiles, and P. Westra. 1998. Impact of timing and frequency of in-row cultivation for weed control in dry beans (*Phaseolus vulgaris*). *Weed Technol* 12:548–554.

VanKirk, J. R. 1998. Database of 1995 statewide IPM needs assessments for the Northeast Region, November 20, <http://www.nysaes.cornell.edu/ipmnet/swneeds95/intro.html> (13 June 2002)

Veseth, R. 1995. Cropping systems approach to crop health management in conservation tillage, <http://pnwsteep.wsu.edu> (13 June 2002)

Washington State University (WSU). 2002. About us, April 4, <http://mtvernon.wsu.edu/frt_hort/about/us.htm > (13 June 2002)

Watkins, G. M. (ed). 1981. *Compendium of Cotton Diseases*. APS Press, The American Phytopathological Society, St. Paul, Minnesota.

Webster, T. M. and H. D. Coble. 1997. Changes in the weed species composition of the southern United States: 1974 to 1995. *Weed Technol* 11:308–317.

Whalon, M. E. and B. A. Croft. 1984. Apple IPM implementation in North America. *Annu Rev Entomol* 29:435–470.

White, D. G. (ed.). 1999. *Compendium of Corn Diseases*. 3rd ed. APS Press, The American Phytopathological Society, St. Paul, Minnesota.

Wilson, C. L. and M. E. Wisniewski (eds.). 1994. *Biological Control of Postharvest Diseases of Fruits and Vegetables: Theory and Practice*. CRC Press, Boca Raton, Florida.

Wrather, J. A., A. Y. Chambers, J. A. Fox, W. F. Moore, and G. L. Sciumbato. 1995. Soybean disease loss estimates for the southern United States, 1974 to 1994. *Plant Dis* 79:1076–1079.

Zhang, W., D. Y. Han, W. A. Dick, K. R. Davis, and H. A. J. Hoitink. 1998. Compost and compost water extract-induced systemic-acquired resistance in cucumber and arabidopsis. *Phytopathol* 88:450–455.

Zwart, K. B., J. Bloem, S. L. G. E. Burgers, L. A. Bouman, L. Brussaard, G. Lebbink, W. A. M. Didden, J. C. Y. Marinissen, J. J. Vreeken-Buys, and P. C. de Ruiter. 1994. Population dynamics in the below-ground food webs in two different agricultural systems. *Agric Ecosyst Environ* 51:187–198.

5 Rangeland and Pasture

Many noxious weeds and other pests were introduced as peoples from around the world settled throughout what is now the United States (Sheley and Petroff 1999). Weed seeds often were inadvertently introduced through the planting of contaminated crop and ornamental plant seeds, and certain deliberately introduced plants, including ornamentals, eventually escaped and became weeds. The introduction of weeds and other pests into areas devoid of natural enemies and diseases resulted in rapid population expansions and in numerous negative economic and environmental effects that have persisted to this day. Such introductions now are minimized by federal and state regulations and related activities (see Chapter 2, Section on Invasive pests for details).

In 1997, invasive rangeland weeds on federally owned land were estimated to cover more than 17 million acres (a.) and to be increasing at 14%/year (yr) (BLM 1996). At this rate, 100 a. infested today could increase to 231 a. in 5 yr and to 443 a. in 10 yr. The infestation of federal land with invasive, noxious weeds thus could extend over approximately 75 million a. by 2007. Anderson and Magleby (1997) estimated that approximately 589 million a. of pasture and rangeland exist in the contiguous United States. The challenge is to develop effective weed management strategies for containing or decreasing pest populations, to limit introduction of new pest species, and to minimize economic and ecologic effects of pests and their associated treatments.

Currently, the most effective pest management strategies are those incorporating multiple control methodologies (e.g., physical, mechanical, chemical, and biological controls) as well as public awareness campaigns, interspecific perennial plantings, and native perennial rehabilitation (Sheley and Petroff 1999). The control of invasive weeds and other pests must be viewed as a long-term commitment. This is crucial to the attainment of an acceptable pest-population level sustainable by the natural system. Rapid, often resource-intensive, responses to pest problems can slow the rate of increase of a weed or even prevent its spread. The financial resources required to sustain such a level of treatment, however, typically are great; applications, therefore, are limited to the short term.

Long-term weed control goals should focus on the establishment of natural competitors and, when needed, the introduction of highly competitive and desirable perennial plants. Although demands change over time, in the past this approach has challenged the system. Replacement plants must be managed by methods providing sustainable, competitive production year after year. Most weed problems will return if desirable competition is destroyed by disturbance or overgrazing (McNeel et al. 1999; Sheley and Petroff 1999). The ideal management strategy for insect and pathogen control in these vast but often low-value agroecosystems is the use of resistant or tolerant cultivars of grasses and legumes. Regrettably, the availability of varieties with broad-spectrum pest resistance is quite limited.

Weed Management

Education and Awareness

Individuals developing an interest in and an understanding of weeds in small or large areas must take the first steps in the weed management educational process. These individuals understand the life cycles of weeds, the reproductive techniques used for their propagation, the life expectancy of seeds in the soil, the ability of specific weeds to grow and to reproduce in a variety of geographic and ecologic zones, and the manner of weed spreading. Researchers studying invasive weeds are aware of their negative effects on the flora and the fauna of an area and can identify those plants causing health problems for animals and/or people.

The educational process includes learning about laws and regulations governing weeds in each state, county, and town. Nationally, the Federal Noxious Weed Act lists weeds regulated by the U.S. Department of Agriculture's Plant Protection and Quarantine Program. Like the federal government, each state has developed laws and regulations designating specific management practices for various weeds. Counties and cities can specify the weeds to be regulated by city weed-control districts or county governments. Each level of government emphasizes programs including identification, prevention, early detection, eradication, small-infestation management, and, finally, large-infestation management. In addition to state, county, and city regulators, federal agencies regulate invasive weed management on federal lands. In the United States, more than one-third of all western lands are owned and managed by the federal government. These lands include national forests, national wildlife refuges, national parks, federal reservations, and public lands managed by the Bureaus of Land Management and Reclamation. Throughout the western United States, federal land ownership often is intermixed with private and state lands. This pattern makes the management of noxious weeds difficult.

Because weeds and other pests have no boundaries based on land ownership, people are learning to cooperate across federal, state, and local boundaries. River basins or watersheds often make the best management areas. When a watershed management area is put in place, upper streams usually are treated first, to prevent reinvasion by water-dispersed seeds. Weed management requires a well-thought-out plan and proactive measures for identifying and acquiring necessary resources.

Land managers must be trained so that scientifically based information can be transferred and used effectively in invasive weed management. Within each state are many individuals knowledgeable about weeds and informed sufficiently to aid with management planning and project funding. Information can be obtained from university research and Extension personnel, county weed coordinators, Extension agents, federal and state agency specialists, and other informed individuals (McNeel et al. 1999).

Weed Prevention and Early Detection

Prevention is a process initiated by people aware of unwanted weed species and wishing to prepare a strategy or plan to keep species from entering certain regions. Common methods used to prevent the introduction of weed seeds follow:

1. using certified weed-free seed, forage, or hay;
2. cleaning equipment before bringing it into weed-free areas;
3. introducing gravel or soil from certified weed-free areas;
4. feeding animals certified weed-free hay and grain at least 3 weeks before introducing them into a weed-free area;
5. inspecting recreational vehicles before permitting them to enter weed-free areas;
6. educating land users about weed species that should be eliminated from an area;
7. preventing noxious weeds from producing seed, especially along roadways or streams; and
8. quarantining domesticated animals that may carry weed seeds within their digestive system, before introducing them into weed-free areas.

Prevention is the least costly method of invasive weed management. Once introductions occur, control of unwanted species can cost more than $523/a. (Taylor and McDaniel 1998).

Early detection depends on identification and location of new species in a management area. When found, invasive weed species can be eliminated or eradicated if careful follow-through and long-term monitoring methods are used. Early detection is every land management professional's responsibility. All workers on the land should be educated about weed species and should monitor infestations regularly because many hours of labor and great expense can be averted when early detection and prevention methods are applied (Sheley and Petroff 1999). Furthermore, new weed detection tools must be developed to assist large landholders in identifying weed problems in remote areas. This need is especially true for federal land managers, who often have limited resources for carrying out early weed detection programs. Early identification and assessment of weed populations are keys to allocating resources appropriately and to choosing the appropriate tool or combination of tools with which to address the problem (Figure 5.1).

Physical and Mechanical Controls

Hoeing, hand pulling, tilling, mowing, and burning are commonplace physical and mechanical weed control methods (James et al. 1991). Pulling of certain weeds when the ground is wet can remove plant crowns and prevent invasive plants from resprouting. Hoeing below active buds or crowns also can halt the resprouting of certain invasive weeds. These two methods often are used when invasive weeds are first introduced. Tillage not only stresses the problem species but can be used to prepare seed beds for new desirable perennial grasses and broadleaf plants. Establishment of new perennials is usually most successful when seeds are covered properly and spaced in one-eighth-inch- to one-fourth-inch-deep soil. In areas dominated by warm-season plants such as blue grama (*Boutela gracilis*), tillage methods such as chisel plowing have been used to renovate sodbound grasses and to increase cool-season grasses such as western wheatgrass. In rangeland, tillage is recommended only to aid in the establishment or to encourage the competitiveness of perennial grasses. In certain instances, tillage actually increases invasive species populations (McNeel et al. 1999).

A useful tool for control of annual or biennial weeds, mowing should be done at the flowering stage or before seed maturity. This practice is especially effective when expected seed viability in the soil is less than 4 yr and infestations occur in small areas.

Mowing and applying herbicides while using a new computerized, regulated wet blade mower system (Figure 5.2) have made it possible to manage invasive weeds and to decrease required control levels of herbicides. Such systems eliminate drift problems while applying concentrated herbicides in the stems of target species already cut by the mower (Figure 5.3) (Whitson 2000).

Burning is especially useful in the control of brush species (e.g., big sagebrush) whose seeds are viable

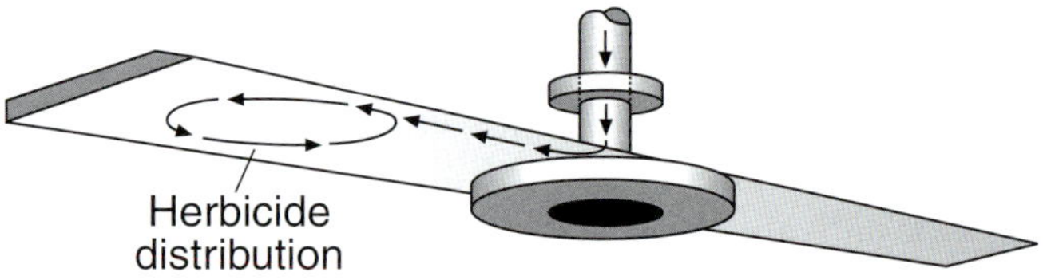

Figure 5.2. Components of a herbicide-wetblade mower system. (Source: T. D. Whitson, University of Wyoming)

Figure 5.1. Use of self-cleaning weed screens in an irrigation system to prevent the spread of weed seeds along irrigation ditches. Photo courtesy of T. D. Whitson, University of Wyoming.

Figure 5.3. Example of herbicide deposit on cut stems of greasewood by means of a wetblade mower. Photo courtesy of T. D. Whitson, University of Wyoming.

in the soil for less than 1 yr (James et al. 1991). When seeds are viable for 2 yr or longer, burning can kill desirable perennials and decrease competition. The lack of competition hampers invasive weed management. To help unify growth stages or to eliminate plant residues, plants often are burned before herbicides are applied.

Chemical Controls

For herbicides to be acceptable on rangeland, they not only should provide control of invasive weeds but also should leave perennial grasses unharmed. All rangeland herbicides providing selective weed control are classified as one of four major "modes-of-action." The modes of action, and herbicides in each category, follow:

1. **Growth regulator herbicides**—2,4-dichlorophenoxy acetic acid (2,4–D) (various brands), 4-chloro-2-methylphenoxy acetic acid (various brands), dicamba (Clarity), picloram (Grazon, Tordon, Tordon 22K), triclopyr (Remedy 4EC, Garlon 3EC, 4EC), clopyralid (Transline 3EC, Curtail 2.38EC, Reclaim 3EC)
2. **Amino acid synthesis inhibitor herbicides**—metsulfuron (Escort 60DF), trisulfuron (Amber 75DG), glyphosate (Roundup Ultra 3SC), and imazapic (Plateau 3EC)
3. **Photosynthesis inhibitor herbicides**—tebuthiuron (Spike 20% G) and hexazinone (Velpar 2EC)
4. **Cell-membrane disruptor herbicides**—paraquat (Gramoxone Extra 2.5S)

Highly selective and causing little or no damage to perennial grasses, growth regulator herbicides often cause unwanted damage to rangeland forbs (Ahrens 1994). Users must read the labels carefully regarding which species of broadleaf plants a herbicide controls. In the growth regulator family, picloram is a very effective broad-spectrum herbicide; its use is, nonetheless, restricted, and it should not be applied near surface water or on land with a shallow water table. Clopyralid is especially effective for control of the composite, or Asteraceae, family of weeds—especially knapweed and thistle species. Triclopyr is very useful in the management of unwanted brush species. The first herbicide introduced after World War II was 2,4–D, which controls annual and biennial weeds in early growth stages. This herbicide also can provide temporary control of perennial weeds, decreasing their competitive effects during the application year. Methylphenoxyacetic acid is similar in activity to 2,4–D but is much more effective against certain broadleaf weeds, which can be a component of the rangeland and pasture complex.

Amino acid synthesis herbicides, which prevent plants from producing one to three essential amino acids, are very selective (Ahrens 1994). Metsulfuron controls various rangeland broadleaf species, especially those in the mustard (Brassicaceae) family, but other grass species such as fescue (*Festuca* spp.) and creeping foxtail (*Alopecurus arundinaceus*) are killed by metsulfuron. Trisulfuron is an effective control for many annual weeds and for certain biennials and perennials in the mustard family. Glyphosate can be used selectively in perennial rangeland grasses for control of annual weeds such as downy brome. Rates higher than those recommended for pastures and rangelands will kill perennial grasses. Imazapic is a new herbicide labeled in several states. It is very selective and controls species such as leafy spurge (*Euphorbia esula* L.), houndstongue (*Cynoglossum officinale* L.), downy brome (*Bromus tectorum* L.), and members of the mustard family. Cool-season grasses are suppressed with applications made in the spring. Therefore, imazapic is most effective as a fall-applied herbicide.

Herbicides that are photosynthetic inhibitors block photosynthetic transfer of energy from sunlight. Tebuthiuron is labeled for selective control of big sagebrush (*Artemisia tridentata* L.), oak (*Quercus* spp.), and other woody species. Application rates should be low and labels followed carefully to prevent cool-season grass damage. Hexazinone is used as a spot treatment in the management of certain rangeland brush species.

Paraquat is a member of the cell membrane disruptor family. It is a contact herbicide with very rapid action and will cause immediate desiccation of all rangeland plants. Perennial plants store enough carbohydrate in their root systems to revegetate from the crowns and roots. Annuals such as downy brome (cheatgrass) and biennial rangeland weeds are killed in the contact process. Though it can be a useful method of managing unwanted annual plants germinating from dormant seed, a brownout process normally decreases total rangeland production (Ahrens 1994).

Biological Controls

The use of biological control agents to control invasive rangeland weeds is an evolving science. Many introduced weeds are major problems in the United

States but not in their native countries. Biological control methods indicate how pest problems can be addressed by the introduction of *biological agents* (i.e., pathogens or natural predators from regions in which the invasive species originated). Biological agents include insects, mites, nematodes, plant pathogens, sheep, goats, and other natural enemies. Once a potential biological agent is identified in a foreign country, it must undergo a rigorous screening process before being introduced into the United States. If the biocontrol agent attacks nontarget plants of economic or ecologic importance in the United States, then introduction will not be permitted. Weed control is especially effective when more than one biological controls for the same plant species are introduced simultaneously. Insect releases have had greater success than other types of biological controls. Currently in the United States, released insects have helped control the following weed species: diffuse and spotted knapweed (*Centaurea diffusa* and *C. maculosa*), yellow starthistle (*C. solstitialis*), leafy spurge (*Euphorbia esula*), musk thistle (*Carduus nutans*), rush skeletonweed (*Chondrilla juncea*), tansy ragwort (*Senecio jacobae*), St. Johns' Wort (*Hypericum perforatum*), and Mediterranean sage (*Salvia aethiopis*) (Whitson et al. 1996).

It takes 10 to 15 yr to measure the success of a new biological control program on invasive weed species. Insect and pathogen introductions made 25 to 30 yr ago (Morrison, Reckie, and Jensen 1998) on tansy ragwort and St. Johnswort in the Pacific Northwest have been remarkably successful. Earlier introductions are decreasing seed production and the spread of musk thistle in the Rocky Mountain region. Two flea beetles released on leafy spurge in the northern United States are showing great promise, especially on well-drained sites. In the future, other biocontrol agents, having adapted to specific sites and weed populations, may affect problem species significantly.

When weed populations expand to cover millions of acres, as many invasive rangeland species have done, the search must continue to for target-specific natural enemies that are, in the long run, able to control the infestation with little or no economic input. Releasing a biological control agent does not mean that control will be certain, straightforward, or economical. Nor does it mean that other integrated pest management (IPM) tools should not be used. Successful weed control must include a systems approach (McNeel et al. 1999) and a description of the proper uses of all available tools and must take into account the knowledge base and infrastructure of the program. The methodology used to approach a problem often is as important to success or failure as are biological interactions (Anderson et al. 2000; Sheley and Petroff 1999).

Before widespread cattle grazing, sheep and goats were important worldwide weed controllers (James et al. 1991), grazing on great amounts of broadleaf plants and being used to control plants poisonous to cattle (e.g., groundsel [*Senecio* spp.] and larkspur [*Delphinium* spp.]). Sheep generally show no ill effects upon consuming the alkaloids found in these weed species and are much more tolerant of them than are cattle. Up to half of a sheep's diet and up to 80% of a goat's diet can consist of the invasive species leafy spurge. Both animals generally prefer broadleaf plants and woody browse to perennial grasses, which are much preferred by cattle, bison, and elk.

Figure 5.4. Effectiveness of a systems approach for replacing Russian knapweed infestation with (Bozoisky) Russian wildrye, a sustainable forage species (when properly managed). Photos courtesy of T. D. Whitson, University of Wyoming.

Sheep and goats must be managed carefully when a producer concentrates them on areas with heavy populations of weeds. Without proper herding, positive control of weedy species cannot be accomplished and animal losses to predators can be great (Lym 1998; Sheley and Petroff 1999). Removal of animals from infested areas also must include a quarantine period to purge viable seeds from their digestive tracts. In many instances, use of domesticated ungulates for weed control is a long-term commitment because most weed populations will reappear once a biological control agent is removed from a system (Figure 5.4).

Plant Competition and Rehabilitation

A vegetation inventory is necessary when determinations are being made about which component of an integrated system to apply to areas infested with invasive weeds (BLM 1996). When desirable perennial grasses are in adequate supply in the infested understory, revegetation will be unnecessary after noxious weeds are removed. Often a herbicide application, fire, biocontrol agent, or grazing pattern change will shift the competitive advantage in favor of plant populations native to the site (Figure 5.5). When perennial grasses do not dominate the control site in the 2 or 3 yr after the removal of a noxious weed by means of herbicides, fire, biological control, or other management tactic, a revegetation program should be considered.

The first step in the revegetation process is to eliminate all invasive species and any other competitive vegetation preventing the establishment of perennial grass seedlings selected for a site (Whitson, Dewey, and Stougaard 1999). Sites selected for rehabilitation should be fairly free of rocks and should be located on terrain suitable for reseeding.

Figure 5.5. Combination of a herbicide and proper grazing that favor increased grass competition and sustainable weed control (left side); area on right side of fence was grazed heavily in early spring when grasses initiated new growth. Photos courtesy of T. D. Whitson, University of Wyoming.

Invasive Species Control and Site Preparation

A regional inventory should provide a basis on which to select species for rehabilitation. Inventory should include quantity and density of all plant species on a site. Vegetation can be classified as annual, biennial, or perennial (Sheley and Petroff 1999).

Each revegetation strategy should be specific to the invasive rangeland species being replaced, as well as to the site biogeographical characteristics. During the last few years, much research focusing on the elimination of various weed species and their replacement with competitive perennial grasses has been conducted according to the integrated systems approach. Areas infested with perennial species such as saltcedar (*Tamarix* spp.), Russian knapweed (*Acroptilon repens*), leafy spurge (*Euphorbia esula*), dalmatian toadflax (*Linaria genistifolia* spp. *Dalmatica*), and Canada thistle (*Cirsium ravens*) (Bottoms and Whitson 1998; Ferrell et al. 1998; Masters and Nissen 1998; Taylor and McDaniel 1998; Wilson and Kachman 1999) have been rehabilitated, and sites now are sustainable when grazed appropriately. Biennials (e.g., musk thistle) and annuals (e.g., downy brome) also have been replaced successfully with perennial grass competitors. Additional research on replacing invasive species with mixtures of forbs, grasses, and woody vegetation is needed.

Weed Monitoring and Mapping

After weed species have been found and positively identified, infested lands should be mapped and monitored carefully so that land managers can determine the direction in and the rate at which noxious weeds are spreading. Today, many data-recording methods incorporate extremely accurate global positioning and geographic information systems (Gillham, Hild, and Whitson 2001). Initial surveys establish baseline data to be used for later comparisons. Such information will aid in determinations of the spread rate on various ecologic and geographic areas. Weed demographic predictions can be made from this information. Aerial photographs and topographic maps often are used to prepare weed inventory maps (Anderson et al. 1996, 1999). Realistic objectives should be established annually, and, to facilitate identification from airplanes or satellites, maps should be prepared when plants are in their most unique phenological stage.

Monitoring includes collecting baseline data for all plant species in a management area (Sheley and Petroff 1999) and should incorporate permanent transect data with photo point data and actual clipping information. If information is recorded by species, land managers will be able to detect changes. Monitoring is especially useful when land management changes (e.g., when a herbicide application changes, a biological weed control agent or mowing is introduced, or when new weed infestations occur). Monitoring allows land managers to make decisions based on scientifically sound evidence instead of on anecdotal or informally documented experiments.

Economic and ecologic cost/benefit analyses of a specific weed-control program will be possible when production and species changes are documented thoroughly.

Integrated Systems for Weed Management

Integrated weed management is a noxious-weed management technique combining several control methods into an integrated set of tools (BLM 1996; USDA 1998). Methods include education; prevention; mechanical, physical, biological, herbicidal, and cultural controls; grazing; and land management. For example, unless the resulting ecological community is desirable and sustainable, it really does not matter whether weeds are controlled by means of a disease, an insect introduction, or a herbicide. If weeds are controlled and the area is rehabilitated with desirable plant populations, the land will regain its productivity and, it is hoped, become a healthy and sustainable plant community in spite of the continuous threat of invasion. When land managers have comprehensive information and help neighboring producers become aware of invasive weed species, a plan can be established to prevent weed introductions. When an effective management plan is in place and weeds become established, close observers of the range will find and eradicate them before they become widespread. To obviate expensive control methods for widespread infestations of invasive species, all management plans should encourage weed monitoring and early detection (Figure 5.6).

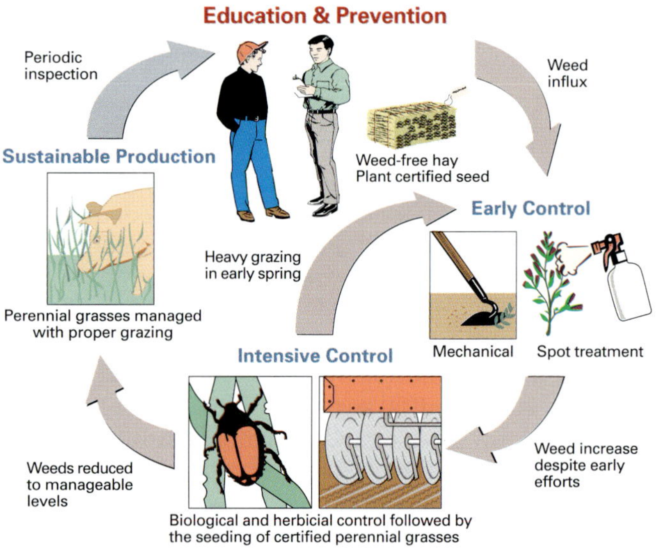

Figure 5.6. Components of an integrated system for rangeland weed control. (Source: B. Stump, University of Wisconsin.)

Rangeland management decisions should be made on the basis of information derived from a comprehensive monitoring plan. The goal of most such plans should be the development of biologically based, integrated weed management strategies that are economically and ecologically sustainable (Sheley and Petroff 1999) (Figure 5.7).

Insect and Pathogen Management

Although weeds often are the dominant pest problem in pasture and rangeland, a great number of insects and pathogens also affect range, forage, and cropping systems. Most of the two pest groups generally do not have a significant effect on overall plant productivity. Because effects can be adverse and materials and their applications costly, treatment of these pests with pesticides is usually infeasible (Center for Integrated Pest Management 1996). Insects such as the armyworm complex, chinch bugs, aphids, and weevils may cause, nonetheless, sufficient defoliation in certain regions or states to warrant monitoring and treatment.

Figure 5.7. Role of revegetation in the management system to control prolific weed infestations such as this musk thistle invasion. Photos courtesy of T. D. Whitson, University of Wyoming.

Imported fire ants (*Solenopsis invicta* and *S. richteri*) pose a new type of pest management problem in several regions, including 14 states in the South and West. In addition to the many problems they cause humans, wildlife, and even endangered species, fire ants can harm or kill cattle and other farm animals. Although nearly impossible to eradicate, this fiery pest may be controllable by means of promising insecticides and other biocontrols.

Leaf spot diseases, root rots, and plant parasitic

nematodes and associated disease complexes may contribute to poor stands and/or declines of rangeland and pasture grasses and legumes. The use of disease-resistant cultivars, where available, is the most practical management strategy for such maladies (Bernard, Gwinn, and Griffin 1998; Pederson and Quesenberry 1998). For example, Bahia grasses are resistant to most plant pathogens and insects (Seedland.com 1999a).

A potentially important nematode bacterium complex on certain grasses involves the foliage-attacking nematodes *Anguina* spp. and the bacterium *Clavibacter toxicus* (= *C. rathayi*). The bacterium, associated with these nematodes, produces corynetoxin, which induces convulsions in grazing animals; can be deadly to cattle and sheep (Bernard, Gwinn, and Griffin 1998; McKay and Ophel 1993); and has killed many thousands of animals in Australia. The problem often is encountered in annual Wimmera ryegrass (*Lolium rigidum*) and has been referred to as "annual ryegrass toxicity" (Bernard, Gwinn, and Griffin 1998; McKay and Ophel 1993). A similar disease complex can be important in paddocks of blown grass (*Agrostis avenacea*) or annual beardgrass (*Polypogon monspeliensis*). The associated disease syndrome on these grasses is known colloquially as the "flood plain staggers" (Bernard, Gwinn, and Griffin 1998).

The endophytic fungi are another group of grass-associated microbes that can affect grazing animal health. Although these fungal endophytes reside in many symptomless grasses and benefit their hosts, at times they can be very detrimental to grazing animals (Bernard, Gwinn, and Griffin 1998). Because of the toxicoses encountered in grazing cattle and horses, the endophytes *Neotyphodium coenophialum* on tall fescue and *N. lolii* on perennial ryegrass have received much study. Some 26 million a. are planted to tall fescue, and mainly to 'Kentucky 31,' in the United States, and perennial ryegrass is a major forage species in New Zealand (Bernard, Gwinn, and Griffin 1998). Although the nutrient content of endophyte-infected grasses is similar to that of endophyte-free grasses, animals reared on infected grasses grow poorly and may exhibit fescue toxicosis. This disease, which involves poor weight gain and reproductive and lactation problems, is a considerable economic problem in the United States, causing approximately $600 million in losses annually in cattle production (Bernard, Gwinn, and Griffin 1998). Horses also are affected.

In contrast to the negative effects of endophytes on animal health, their effects on tall fescue are positive. Endophytes and tall fescue, have, in fact, a mutualistic relation (Bernard, Gwinn, and Griffin 1998): endophyte-free tall fescues grow poorly, and stands may decline prematurely. Endophyte infection enhances the growth of fescue and results in the plant's becoming resistant to a number of associated pathogens and pests. Once the mechanism of this pest resistance is understood, the development of nonpesticide pest-resistance in grasses may be facilitated.

Because of associated animal toxicoses, numerous endophyte-free fescue varieties are available (Seedland.com 1999b). These grasses are not recommended for use on lawns or other recreational turf sites, however, and their utility, even in pastures, warrants further evaluation (Bernard, Gwinn, and Griffin 1998).

As this limited treatment of insects and pathogens suggests, greater research efforts are needed regarding IPM on grasslands. Research should be especially helpful as endeavors are directed increasingly at the development of sustainable crop and animal production systems.

Appendix A. Abbreviations and Acronyms

2,4–D	2,4–dichloropenoxy acetic acid
a.	acre
BLM	U.S. Bureau of Land Management
IPM	integrated pest management
yr	year

Literature Cited

Ahrens, W. H. (ed.). 1994. *Herbicide Handbook*. 7th ed. Handbook Committee, Weed Science Society of America, Champaign, Illinois. 352 pp.

Anderson, G. L., J. H. Everitt, D. E. Escobar, N. R. Spencer, and R. J. Andrasik. 1996. Mapping leafy spurge (*Euphorbia esula*) infestations using aerial photography and geographic information systems. *Geocarto Int* 11:81–89.

Anderson, G. L., C. W. Prosser, S. Hager, and B. Foster. 1999. Change detection of leafy spurge (*Euphorbia esula*) infestations using aerial photography and GIS. Pp. 223–230. In *Proceedings of the 17th Biennial Workshop on Color Photography and Videography in Resource Assessment*, 5–7 May, 1999, American Society of Photogrammetry and Remote Sensing, Bethesda, Maryland.

Anderson, G. L., E. S. Delfosse, N. R. Spencer, C. W. Prosser, and R. D. Richard. 2000. Biological control of leafy spurge: An emerging success story. Pp. 15–25. In N. R. Spencer (ed.). *Proceedings of the 10th International Symposium on Biological Control of Weeds*, 4–14 July, 1999, Montana State University, Bozeman.

Anderson, M. and R. Magleby (eds.). 1997. Agricultural resources and environmental indicators. 1996–1999. Agricultural

Handbook Number 172. Agricultural Research Service, U.S. Department of Agriculture, Washington, D.C.

Bernard, E. C., K. D. Gwinn, and G. D. Griffin. 1998. Forage grasses. Pp. 427–454. In K. R. Barker, G. A. Pederson, and G. L. Windham (eds.). *Plant and Nematode Interactions*. American Society of Agronomy, Crop Science Society of America, and Soil Science Society of America, Madison, Wisconsin.

Bottoms, R. M. and T. D. Whitson. 1998. A systems approach for the management of Russian knapweed (*Centaurea repens*). Weed technology. *Integr Syst Noxious Weed Manage Rangelands Symp* 12:363–366.

Center for Integrated Pest Management. 1996. Pests of forages and pasture, July, <http://ipmwww.ncsu. edu/AG271/forages/forages.html> (23 July 2002)

Ferrell, M. A., T. D. Whitson, D. W. Koch, and A. E. Gade. 1998. Leafy spurge (*Euphorbia esula*) control with several grass species. Weed technology. *Integr Syst Noxious Weed Manage Rangelands Symp* 12:374–380.

Gillham, J. H., A. L. Hild, and T. D. Whitson. 2001. Predicting susceptibility of invasive species using GIS. *West Soc Weed Sci Proc* 54:58.

James, L. F., J. O. Evans, M. H. Ralphs, and R. D. Child. 1991. *Noxious Weed Management Strategies in Noxious Range Weeds*. Westview Press, Boulder, Colorado.

Lym, R. G. 1998. The biology and integrated management of leafy spurge (*Euphorbia esula*) on North Dakota rangeland. Weed Technology. *Integr Syst Noxious Weed Manage Rangelands Symp* 12:367–373.

Masters, R. A. and S. J. Nissen. 1998. Revegetating leafy spurge (*Euphorbia esula*)-infested rangeland with native tallgrasses. Weed technology. *Integr Syst Noxious Weed Manage Rangelands Symp* 12:381–390.

McKay, A. C. and K. M. Ophel. 1993. Toxigenic Clavibacter/Anguina associations infecting grass seedheads. *Annu Rev Phytopathol* 31:151–167.

McNeel, H. A., R. R. Parsons, J. Sweaney, L. A. Vance, C. Henry, C. McClure, J. Free, and B. Mullin. 1999. Guidelines for coordinated management of noxious weeds: Development of weed management areas. U.S. National Park Service, U.S. Forest Service, U.S. Department of Interior, Washington, D.C. 226 pp.

Morrison, K. D., E. G. Reckie, and K. I. Jensen. 1998. Biocontrol of common St. Johnswort (*Hypericum perforatum*) with *Chrysolina hyperici* and a host-specific *Colletotrichum gloeosporiodesi*. *Weed Technol* 12:426–435.

Pederson, G. A. and K. H. Quesenberry. 1998. Clovers and other forage legumes. Pp. 399–425. In K. R. Barker, G. A. Pederson, and G. L. Windham (eds.). *Plant and Nematode Interactions*. American Society of Agronomy, Crop Science Society of America, and Soil Science Society of America, Madison, Wisconsin.

Seedland.com. 1999a. Bahiagrass pastures, <http://www.bahiagrass.com/pasture> (23 July 2002)

Seedland.com. 1999b. Fescue pastures, <http://www.fescue.com/pasture> (23 July 2002)

Sheley, R. L. and J. K. Petroff. 1999. *Biology and Management of Noxious Rangeland Weeds*. Oregon State University Press, Corvallis. 438 pp.

Taylor, J. P. and K. C. McDaniel. 1998. Restoration of saltcedar (*Tamarix* spp.)-infested flood plains on the Bosque del Apache National Wildlife Refuge. Weed Technology. *Integr Syst Noxious Weed Manage Rangelands Symp* 12:345–352.

U.S. Bureau of Land Management (BLM). 1996. Current situation. Partners against weeds, an action plan for the Bureau of Land Management. U.S. Bureau of Land Management, Washington, D.C. 7 pp.

U.S. Department of Agriculture (USDA)–Alternative Farming Systems Information Center. 1998. Alternative soil, crop, and pasture management, 28 January, <http://www.nal.usda. gov/afsic/sbjscpm.htm> (23 July 2002)

Whitson, T. 2000. Brush and noxious weed management with the Burch Wet-Blade® Mower. *West Soc Weed Sci Res Progr Rep* 53:44.

Whitson, T. D., L. C. Burrell, S. A. Dewey, D. W. Cudney, B. E. Nelson, R. D. Lee, and R. Parker. 1996. *Weeds of the West*. 5th ed. Western Society of Weed Science, Jackson, Wyoming. 630 pp.

Whitson, T. D., S. A. Dewey, and R. Stougaard. 1999. 1999–2000 weed management handbook. Utah, Montana, and Wyoming Cooperative Extension Service, Laramie.

Wilson, R. G. and S. D. Kachman. 1999. Effect of perennial grasses on Canada thistle (*Cirsium arvense*). *Control Weed Technol* 13:83–87.

6 Natural Areas Management

Natural areas, or areas set aside in national and state parks or elsewhere for management of natural resources, including native and biodiverse communities, are being invaded by nonnative plant and other pest species introduced since settlement of North America by Europeans. The movement of pests into forests and rangelands poses important problems but will not be addressed in detail in this chapter (See Chapters 4 and 5). Introductions of invasive pests have been both accidental (having occurred, for instance, as a result of the transport of seeds attached to clothing) and intentional (as a result of the introduction of commercial species). The likelihood or risk that species will be transported long distances across natural barriers such as oceans and mountain ranges has increased rapidly as transportation technology has improved; with high-speed jet travel, the potential for long-distance anthropogenic introductions is limitless: in 1990, some 330 million plants were reported to have been brought into the United States through Miami International Airport alone (OTA 1993). Although not all introduced plant species become weed problems, many weeds are of foreign origin. Of the 300 weed species of the western United States, at least 36 escaped from horticulture or agriculture (OTA 1993) (Figure 6.1).

The need to regulate the movement of noxious weeds was recognized in the United States with the passage of the Federal Noxious Weed Act of 1974. But the act identifies too few weed species as noxious weeds, and federal regulation of weeds often overlooks risks to nonagricultural areas (OTA 1993). Serious efforts to address the effects of nonindigenous invasive plants in natural areas have gained momentum only recently.

Because of shortcomings in the federal regulation of weeds in natural areas, the problem is now being addressed by cities, counties, states, and other groups. Exotic Pest Plant Councils (EPPCs), whose members include public land managers and academics in natural resource departments, recently have organized to take the lead in addressing the problem of nonindiginous plant species in natural areas. State and regional EPPCs, along with the National Park Service (NPS) and other groups, have made lists of invasive plant species occurring in natural areas (Table 6.1). Although having no statutory authority to prohibit any species, these lists educate land managers and generally heighten awareness of the problem. The number of plant species listed demonstrates the breadth of the problem: the NPS alone identifies 421 invasive species.

Passage of President Clinton's February 3, 1999 Executive Order (EO) on invasive species provided a framework for progress in the area of integrated management of invasive plant species in natural areas. This EO directed federal agencies to prevent the introduction of invasive species and to control, monitor, and restore native species to habitats affected by invasive species. It established a Federal Interagency Invasive Species Council, whose members include the Secretaries of State, Treasury, Defense, Interior, Agriculture, Commerce, and Transportation, as well as the Administrator of the Environmental Protection Agency and representatives of additional federal agencies. The council was directed to oversee imple-

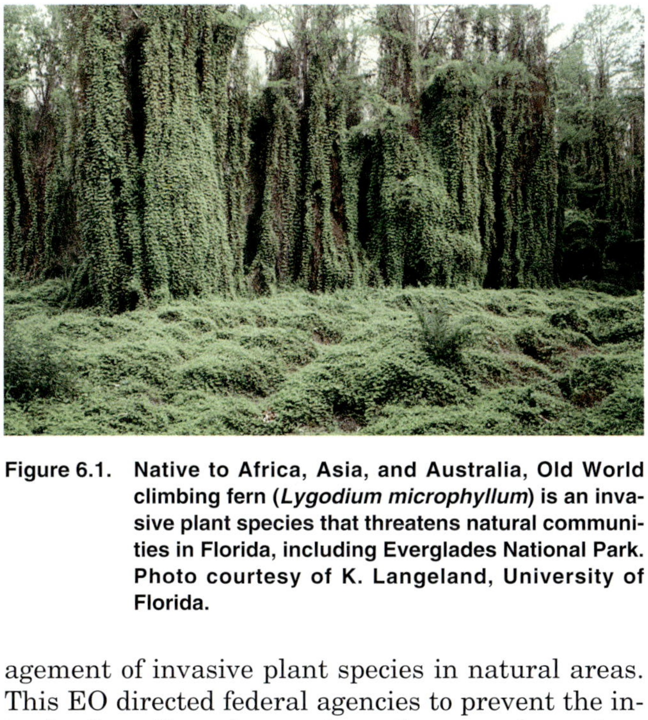

Figure 6.1. Native to Africa, Asia, and Australia, Old World climbing fern (*Lygodium microphyllum*) is an invasive plant species that threatens natural communities in Florida, including Everglades National Park. Photo courtesy of K. Langeland, University of Florida.

Table 6.1. Numbers of invasive nonindigenous plant species occurring in natural areas of the United States, as compiled on lists by various groups

California Exotic Pest Plant Council (www.caleppc.org)	
A-1: Most invasive wildland pest plants; widespread	23
A-2: Most invasive wildland pest plants; regional	20
List B: Wildland pest plants of lesser invasiveness	35
Red Alert: Species with potential to spread explosively; infestation currently restricted	16
Florida Exotic Pest Plant Council (www.fleppc.org)	
Category I: Invasive exotics that are altering native plant communities by displacing native species, changing community structures or ecological functions, or hybridizing with natives	62
Category II: Invasive exotics that have increased in abundance or frequency but have not yet altered Florida plant communities to the extent shown by Category I species	60
EPA Great Lakes National Program (www.epa.gov/grtlakes/greenacres/wildones/wo29.htm)	53
Georgia Exotic Pest Plant Council	42
National Park Service (www.nps.gov/plants/alien/sciname.htm)	421
Pacific Northwest Exotic Pest Plant Council (www.wnps.org/eppclist.htm)	
Red Alert: Great potential to spread	30
A-1: Most invasive; widespread	34
A-2: Most invasive; regional (highly to moderately invasive but still with a potential to spread)	36
B: Wildland weeds of lesser invasiveness (less aggressive)	70
Need more information	23
Tennessee Exotic Pest Plant Council (webriver.com/tneppc/)	
Rank 1: Severe threat, exotic plant species that possess characteristics of invasive species and spread easily into native plant communities and displace native vegetation; includes species that are or could become widespread in Tennessee	24
Rank 2: Significant threat, exotic species that possess some invasive characteristics, but have less impact on native plant communities; may have the capacity to invade natural communities along disturbance corridors, or to spread from stands in disturbed sites into undisturbed areas, but have fewer characteristics of invasive species than Rank 1	50
Rank 3: Lesser threat, exotic plant species that seem to principally spread and remain in disturbed corridors, not readily invading natural areas; also some agronomic weeds	47

mentation of the EO, to support field-level planning, to identify and/or to develop international recommendations, to create National Environmental Policy Act guidelines, to establish an impact-monitoring network, to develop a Web-based information network, and to prepare a National Invasive Species Management Plan. The first such plan, "Meeting the Invasive Species Challenge," was released January 18th, 2001 and is available on the Internet (USDA–NAL 2002).

Integrated management of invasive pest plants in natural areas is difficult because of the need to minimize damage to habitats. This need for caution in natural areas results in greater expenditures of time and effort than weed management in agricultural, industrial, or right-of-way settings does. Certain types of vegetation (e.g., woody or sprawling vegetation) may require removal of standing plant material even after it has been killed if its presence increases fire hazard, diminishes aesthetic appeal, or causes harm as it decays and falls. Control methods may be preventive, physical, cultural, herbicidal, or biological.

Management Methods for Invasive Weeds

Prevention

Many plant species considered by land managers to be invasive in natural areas are available in nursery and landscape trades. For example, 24 species listed as Category I invasive by the Florida Exotic Pest Plant Council (FLEPPC) are available commercially (Aylsworth 1999). Prohibiting the sale of invasive species is one way to eliminate their availability. But land managers and growers do not always agree on the invasiveness of species, and the prohibition of a commercially important species tends, of course, to be unpopular with growers. Thus, only those species that are commercially unimportant and extremely invasive tend to be prohibited at the federal or the state level. Local governments in parts of the country have passed ordinances that prohibit certain invasive plant species not controlled at higher levels or that require removal of these species from private lands—for example, before building permits will be issued.

An alternative to legislative prohibition of sale or possession of invasive species is voluntary removal from trade. For example, the Florida Nurserymen and Growers Association (FNGA) and the FLEPPC cooperated to identify 11 invasive species that were commercially available but insignificant to member nurseries. The groups now are encouraging growers to phase these species out of the market (Aylsworth 1999). The FLEPPC and the FNGA continue to work together to identify additional invasive species for phase-out of trade.

Public education is important in preventing or limiting the introduction of invasive plant species into natural areas. Many plants, including volunteer introductions and deliberately introduced landscape plants that are invasive in natural areas, are found on private lands. Some invasive species also are used as landscape plants on public lands. Educating the public not to dump yard waste containing seeds or viable vegetative materials of invasive plants and to eliminate invasive plants that can be spread by dispersal of seeds or spores should help prevent the spread of invasive plant species.

In addition to minimizing the spread of invasive plant species already present in the United States, preventing the introduction of new invasive species is important; little is being done about this problem, however. The federal Noxious Weed Act prohibits importation of listed noxious weeds, but it is applied to too few species and, as has been pointed out, overlooks risks to nonagricultural areas (OTA 1993). One difficulty attending the addition of species to the Federal Noxious Weed List is the lack of a mechanism for predicting invasiveness. Reichard and Campbell (1996) developed a decision chart designed as an aid to evaluate the invasiveness of woody plants, but no procedure has been adopted and no mechanism exists by which to limit the introduction of nonlisted plant species into the United States.

Physical Methods

Manual Removal

Although time consuming, removal by hand often constitutes a major component of effective invasive plant control. Seedlings and small saplings sometimes can be pulled from the ground, but even small seedlings of certain plants have tenacious roots that will prevent extraction or cause plants to break at the root collar. Plants breaking off at the ground will resprout; frequently, even small root-fragments left in the ground do so. Thus, repeated hand pulling or follow-up with herbicide applications often is necessary.

Figure 6.2. Specialized logging equipment is sometimes used to remove invasive tree species. Photo courtesy of F. Laroche, South Florida Water Management District.

Destruction of uprooted plant materials also is important. Stems and branches of certain species laid on the ground may sprout roots, and attached seeds may germinate. If material cannot be destroyed by methods such as burning, it should be piled in a secure area where it can be monitored and new plants killed as they appear.

Mechanical Removal

Mechanical removal can include the use of bulldozers or specialized logging equipment to remove woody plants (Figure 6.2). Because disturbance of soil creates favorable conditions for regrowth from seeds and root fragments and for recolonization by invasive nonnative plants, intensive follow-up with other control methods is essential after heavy equipment is used. Plans should be developed carefully beforehand for managing or replanting sites with native vegetation after such removal. Mechanical removal may be inappropriate in natural areas because of the potential disturbance caused by heavy equipment to soils and to nontarget vegetation.

Cultural Practices

Prescribed burning and water-level manipulation are cultural practices used in the management of pastures, rangelands, and commercial forests. Sometimes these same practices are appropriate for vegetation management in natural areas (Dehaan et al. 1993; Reid and Brooks 2000). One important consideration is the degree of degradation of the area in question. Cultural practices may affect all aspects of habitat, including native species (Stanford et al. 1996). If habitat is degraded so badly that the need to decrease the number of invasive species strongly outweighs the importance of the success of remaining native species, aggressive control strategies can be considered (Langeland and Stocker 1997). In less degraded areas, relatively careful use of integrated methods may be appropriate (Langeland and Stocker 1997).

Land-use history is essential to any understanding of the effects of fire and flood on plant species composition. Past practices affect soil structure, organic content, seed bank (both native and invasive nonnative species), and species composition. Although there is evidence that past farming and timber management practices greatly influence the outcome of cultural management, very little is known about the effects of specific historical practices (Langeland and Stocker 1997). Similar management practices conducted in areas with dissimilar histories may achieve quite different results (Duncan et al. 1999; Hedman, Grace, and King 2000). Even less is known about the effects of invasive plants entering these communities or about the subsequent effects of management on altered communities (Langeland and Stocker 1997).

An understanding of the reproductive biology of target and nontarget plant species is crucial to the effective use of control methods, and especially so with regard to methods such as fire management (Tu, Hurd, and Randall 2001). Important opportunities exist when management tools can be applied to habitats in which nonnative invasive species flower or set seed at times different from those at which native species do (Tu, Hurd, and Randall 2001).

Prescribed Burning

Fire is a normal occurrence in many ecosystems, and native species have evolved with various degrees of fire tolerance (Abrahamson and Abrahamson 1996; Kozlowski and Ahlgren 1974; Menges and Kohfeldt 1995). Throughout much of the United States, suppression of fire during this century has altered historical plant communities such as grasslands, flatwoods, and oak scrubs. As a result, the survival of fire-intolerant species has been enhanced and the coverage of species with fire adaptations has decreased. Within these communities, fire-tolerant woody species have lingered in smaller numbers, and less-fire-tolerant species have replaced ephemeral herbs (Abrahamson and Abrahamson 1996). Little is known about the amount, frequency, timing, or intensity of fire that would enhance best the historically fire-tolerant plant species, and less is known about how such a fire management regime could be used to suppress invasive species most effectively. Single fires in areas with many years of fire suppression are unlikely to restore historical species composition, and periodic fires in frequently burned areas do little to alter native species composition (Abrahamson and Abrahamson 1996).

Invasion of tree stands by exotic vines and other climbing plants has greatly increased the danger of canopy (crown) fires and, thus, of mature tree deaths (Pemberton and Ferriter 1998) (Figure 6.3). The added biomass of invasive plants can result in hotter fires and speed a fire's spread to inhabited areas. In such situations, use of fire to decrease standing biomass of invasive species may protect remaining plant populations better than no action may, although effects on nontarget native species will occur. Under these conditions, the expense of decreasing the standing biomass of invasive plant species may be justified economically on the basis of the savings resulting from subsequent fire suppression.

In general, fire can be used to suppress plant growth and even to kill certain non-fire-tolerant plants (Glitzenstein, Platt, and Streng 1995; Menges and Kohfeltd 1995). Most often, woody species populations are diminished and effects on herbaceous species are less noticeable (Glitzenstein, Platt, and Streng 1995; Menges and Kohfeltd 1995). The responses of individual plant species to fire have been studied, but very little is known about the vast majority of native plant species, and even less is known about invasive exotic species. Fire tolerance can be predicted sometimes in species with thick bark or seeds (either left in the soil or held in the canopy) that are disbursed over a wide area or adapted to fire, that is, are tolerant of high temperatures or require fire for seed release or germination (Daubenmire 1967).

The role that fire can play in suppression or elimination of invasive exotic plant species depends on many factors. The resource manager must consider, in addition to the principal factors already described,

Figure 6.3. Attack of tree stands by invasive vines and other climbing plants has increased the danger of canopy fire and thus death to mature trees. Photo courtesy of A. Ferriter, South Florida Water Management District.

potential fire effects on soil loss and water quality, historical and economic effects on buildings, harm to human life, and escape of fire to nontarget areas.

Fire has been used very successfully to manage plant species in grasslands; to maintain open *savanna* (i.e., scattered trees in herbaceous species-dominated habitats); and to promote *seral* (i.e., fire-induced or fire-tolerant) stages of forest succession (Daubenmire 1967). Very little is known about the use of fire to enhance native species while checking invasive exotic plant species. As a final caution regarding the use of fire to manage invasive plant species, frequent burning has been shown to diminish plant diversity under certain conditions, and increased fire frequency may provide opportunities for invasive plants to enter new areas (Griffis et al. 2001).

Water-Level Manipulation

A degree of success has been achieved in the regulation of water level to decrease populations of invasive plant species in aquatic and wetland habitats. The draining of aquatic sites decreases standing biomass, but little else usually is achieved unless the site is rendered less susceptible to repeated invasion when rewatered. In certain instances, planting of native species may decrease the susceptibility of aquatic and wetland sites to invasive plant species.

Carefully timed water-level increases after mechanical removal or fire management of invasive species can control subsequent germination and, with certain species, resprouting (Paveglio and Kilbride 2000; Sher, Marshall, and Gilbert 2000). Specific methods applicable to natural areas have not yet been described, however.

Reestablishment of Native Plant Species

The planting of native species can be an effective albeit expensive means of decreasing the likelihood of exotic species reinvasion after removal of nonnative species (Masters et al. 1996). Commercial plant nurseries currently provide seeds and plants of several wetland and upland species. Because species can cover a wide range of habitats and latitudes, care should be taken to obtain plant material suitable to the specific habitat under consideration. Seed collected from plants growing in northerly latitudes may do very poorly in Florida, for example. Introduction of seeds, plant parts, or whole plants should include thorough screening for unwanted plant or animal pests.

Several years often are required for plantings to become established fully, and extra care in terms of water, nutrients, and protection from fire and pests may be necessary for a time. Throughout the establishment phase, past management practices may have to be altered to avoid injury to plantings. If, for example, periodic burning or flooding is part of current management practice, it may be necessary to decrease intensity or duration until plantings are able to exhibit their inherent resistance to injury. The requirements for successful establishment of many native species are unknown, as are their tolerances to cultural invasive plant management techniques. Even when tolerances are known, responses may be affected by historical site effects; genetic strain traits; site-specific nutrition and light conditions; and soil type, hydroperiod, and microclimate interactions.

Herbicides

Herbicides are essential tools in the integrated management of invasive plant species in natural areas. Because of the need to minimize damage to nontarget species, herbicide types and application rates must be considered carefully in regard to the sensitivity of nontarget plant species, and application must be done with care. In extreme instances, nontarget damage must be weighed against both the environmental cost of nontreatment and the resulting damage that the invasive species will do to the native community.

Table 6.2. Toxicity of herbicides commonly used in natural areas of Florida (Ahrens 1994)

Herbicide	Bobwhite quail 8-day dietary LD_{50} mg/kg	Laboratory rat 96-hr oral LD_{50} mg/kg
2,4-D amine	> 5,620	> 1,000
Clopyralid	> 4,640	4,300
Glyphosate	> 4,640	> 5,000
Hexazinone	> 10,000	1,690
Imazapic	> 2,150	> 5,000
Imazapyr	> 5,000	> 5,000
Metsulfuron	> 5,620	> 5,000
Picloram	> 5,000	> 5,000
Triclopyr amine	> 10,000	2,574
Triclopyr ester	9,026	1,581

Using herbicides to control invasive plant species in natural areas results in low toxicities to wildlife (Table 6.2). Herbicides differ in terms of absorption, persistence, and selectivity characteristics and, depending on the composition of nontarget plant species, are used differently and applied by means of different techniques (Langeland and Stocker 1997).

Broadcast applications with diluted, liquid foliar-absorbed herbicide or dry formulations are used only when monocultures of invasive species exist or when nontarget vegetation is tolerant of the herbicide (Langeland and Stocker 1997). Such applications also are used when (1) the nontarget damage resulting from herbicides is more acceptable than the damage that would have occurred from the invasive species and (2) more selective application techniques are infeasible. Spot foliar application is used for small herbaceous species and sometimes for shrubs. Herbicides that are not *soil active* (absorbed by plant roots) are used if roots of sensitive species are thought to reach into the treatment area. Woody plants can be cut down and the stumps treated with herbicide, or the bark of standing trees can be treated with an oil-soluble herbicide diluted in a penetrating oil; this latter technique is referred to as "basal bark application" (Figure 6.4). Herbicides also may be placed, by means of injection equipment, directly into or in proximity to the active vascular tissue after removal of portions of the bark. This process is known as the "hack and squirt," or "frill," application.

Figure 6.4. Tedious application of herbicide to the bark of individual trees (basal bark application) is often needed to control invasive tree species with minimal effects on nontarget vegetation. Photo courtesy of K. Langeland, University of Florida.

Biological Control

Biological control has often been referred to as the only feasible approach to controlling environmentally important weeds (Waterhouse 1999), but traditional biological control agents have been released for only a small proportion of those nonindigenous species invasive in natural areas (Table 6.3; Figure 6.5). Most efforts have been directed at species that are also agricultural problems, typically in rangeland and pasture (Table 6.3). Although successes have been

Table 6.3. Status of biological control agents for invasive nonindigenous plant species in natural areas (not including aquatic weeds)

Target	Status
Carduus nutans Musk thistle	Two insects, *Rhinocullus conicus* and *Trichosirocalus horridus,* released during the 1970s have provided very successful control. Concern over nontarget effects from expanded feeding of *R. conicus* on native thistles has been expressed.
Centaurea diffusa Diffuse knapweed	Twelve insects released.
Centaurea solstitialis Yellow starthistle	Six insect species released, five of which are established to some degree in the United States.
Chondrilla juncea Rush skeletonweed	Three insects released and established.
Euphorbia esula Leafy spurge	Ten Eurasian species released, five of which have demonstrated control in the field.
Hypericum perforatum Klamathweed	Two leaf-feeding beetles established and considered a proven successful biological control program for more than 40 yr.
Lithrum salicaria Purple loosestrife	Four European species released.
Melaleuca quinquenervia Melaleuca	Melaleuca snout beetle released and established, shows promising results. Three additional insects under consideration.
Rosa multiflora Multiflora rose	An unidentified virus or MLO under consideration and an accidentally introduced seed-feeding wasp spreading.
Schinus terebinthifolius Brazilian pepper	Four insects under consideration, one with tentative approval for release.
Senecio jacobaea Ragwort	Three insects released and established.
Solanum viarum Tropical soda apple	Two insects from Brazil under consideration for release.
Tamarix spp. Saltcedar	Fifteen insect species under study, two having preliminary approval for release.
Verbascum thapsus Common mullein	One insect species released and another under consideration.

Figure 6.5. The melaleuca snout beetle (*Oxyops vitiosa*) is hoped to slow the spread of this invasive tree species by limiting flower and seed production (Source: G. Buckingham, USDA–ARS.)

reported (Deloach 1991), expectations for the success of traditional biological controls should be tempered. Only approximately 60% of released biocontrol agents have become established (Crawley 1989), only 41% have shown evidence of providing some control, and only 33% have exerted detectable control (Williamson and Fitter 1996). Although only eight examples exist, worldwide, of damage caused to nontarget plants by biological control agents (Waterhouse 1999), and although none of these examples involves serious economic or environmental damage and the majority of such damage was anticipated as a result of routine testing before release (Waterhouse 1999), scientists have expressed concerns over observed and potential damage to native species by insects released to control invasive species (Louda et al. 1997; Miller and Aplet 1993). Discovery of effective and selective biological control agents will be important in the long-term suppression of widespread invasive plant species.

Appendix A. Abbreviations and Acronyms

EO	Executive Order
EPPC	Exotic Pest Plant Council
FLEPPC	Florida Exotic Pest Plant Council
FNGA	Florida Nurserymen and Growers Association
NPS	National Park Service

Appendix B. Glossary

Natural areas. Areas set aside in national and state parks or elsewhere for management of natural resources, including native and biodiverse communities.

Savanna. Herbaceous species-dominated habitats, with scattered trees.

Seral. Fire induced or fire tolerant.

Soil active. Absorbed by plant roots.

Literature Cited

Abrahamson, W. G. and C. R. Abrahamson. 1996. Effects of fire on long-unburned Florida uplands. *J Veg Sci* 7:565–574.

Aylsworth, J. D. 1999. Invasive is out. *Ornamental Outlook* 8:40,42.

Crawley, M. J. 1989. The successes and failures of weed biocontrol using insects. *Biocontr News Inf* 19:213–223.

Daubenmire, R. F. 1967. *Plants and Environment. A Textbook of Plant Autecology*. John Wiley and Sons, Incorporated, Hoboken, New Jersey. 422 pp.

Dehaan, H., L. Vanliere, S. P. Klapwijk, and E. Vandonk. 1993. The structure and function of fen lakes in relation to water-table management in the Netherlands. *Hydrobiologia* 265:155–177.

Deloach, C. J. 1991. Past successes and current prospects in biological control of weeds in the United States and Canada. *Nat Areas* J 11:129–142.

Duncan, B. W., S. Boyle, D. R. Breininger, and P. A. Schmalzer. 1999. Coupling past management practice and historic landscape changes on John F. Kennedy Space Center, Florida. *Landscape Ecol* 14:291–309.

Glitzenstein, J. S., W. J. Platt, and D. R. Streng. 1995. Effects of fire regime and habitat on tree dynamics in north Florida longleaf pine savannas. *Ecol Monogr* 65:441–476.

Griffis, K. L., J. A. Crawford, M. R. Wagner, and W. H. Mior. 2001. Understory response to management treatments in northern Arizona ponderosa pine forests. *Forest Ecol Manage* 146:239–245.

Hedman, C. W., S. L. Grace, and S. E. King. 2000. Vegetation composition and structure of southern coastal plain pine forests: An ecological comparison. *Forest Ecol Manage* 134:233–247.

Kozlowski, T. T. and C. E. Ahlgren (eds.). 1974. *Fire and Ecosystems*. Academic Press, New York. 542 pp.

Langeland, K. A. and R. K. Stocker. 1997. Control of nonnative plants in natural areas of Florida. SP 242. Institute of Food and Agricultural Sciences, University of Florida, Gainesville. 38 pp.

Louda, S. M., D. Kendall, J. Conner, and D. Simberloff. 1997. Ecological effects of an insect introduced for the biological control of weeds. *Science* 277:1088–1090.

Masters, R. A., S. J. Nissen, R. E. Gaussoin, D. D. Beran, and R. N. Stougaard. 1996. Imidazolinone herbicides improve restoration of Great Plains grassland. *Weed Technol* 10:392–403.

Menges, E. S. and N. Kohfeldt. 1995. Life history strategies of Florida scrub plants in relation to fire. *Bull Torrey Bot Club* 122:282–297.

Miller, M. and G. Aplet. 1993. Biological control: A little knowledge is a dangerous thing. *Rutgers Law Rev* 45:285–334.

Paveglio, F. L. and K. M. Kilbride. 2000. Response of vegetation to control of reed canarygrass in seasonally managed wetlands of southwestern Washington. *Wildlife Soc Bull* 28:730–740.

Pemberton, R. W. and A. P. Ferriter. 1998. Old World climbing fern (*Lygodium microphyllum*): A dangerous invasive weed in Florida. *Am Fern J* 88:165–175.

Reichard, S. and F. Campbell. 1996. Invited but unwanted: A new decision-making process could help curb devastating effects of invasive plants on natural environments in the U.S. *Am Nurseryman* 184:39–45.

Reid, M. A. and J. J. Brooks. 2000. Detecting effects of environmental water allocations in wetlands of the Murray-Darling Basin, Australia. *Regul Rivers Res & Manage* 16:479–496.

Sher, A. A., D. L. Marshall, and S. A. Gilbert. 2000. Competition between native *Populus deltoids* and invasive *Tamarix ramosissima* and the implications for reestablishing flooding disturbance. *Conserv Biol* 14:1744–1754.

Stanford, J. A., J. V. Ward, W. J. Liss, C. A. Frissell, R. N. Williams, J. A. Lichatowich, and C. C. Coutant. 1996. A general protocol for restoration of the regulated rivers. *Regul Rivers Res & Manage* 12:391–413.

Tu, M., C. Hurd, and J. M. Randall. 2001. Weed control methods handbook, April, <http://tncweeds. ucdavis.edu/handbook.html> (24 July 2002)

U.S. Department of Agriculture–National Agricultural Library (USDA–NAL). 2002. National Management Plan: Meeting the Invasive Species Challenge, June 5, <www.invasivespecies.gov/council/nmp.shtml> (24 July 2002)

U.S. Office of Technology Assessment (OTA). 1993. Harmful non-indigenous species in the United States. OTA–F–565. Office of Technology Assessment, U.S. Government Printing Office, Washington, D.C. 390 pp.

Waterhouse, D. F. 1999. Foreword. In M. H. Julien and M. W. Griffiths (eds.). *Biological Control of Weeds: A World Catalogue of Agents and Their Target Weeds*. 4th ed. CABI Publishing, New York. 223 pp.

Williamson, M. and A. Fitter. 1996. The varying success of invaders. *Ecology* 77:1661–1666.

7 Integrated Management of Aquatic Vegetation

Management of aquatic vegetation is essential to navigation, irrigation, flood control, and natural and recreational resources protection (Figure 7.1). Both native and nonnative plant species may be targets for management. Most research related to integrated management of aquatic plants, however, has been directed toward those nonnative invasive species that most severely impact water use (see Murphy and Pieterse [1990] for a review of research related to the integrated management of invasive aquatic species).

Figure 7.1. Management of aquatic vegetation is essential to navigation, irrigation, flood control, and natural and recreational resources. Photo courtesy of K. Langeland, University of Florida.

Aquatic plants have three basic growth forms: (1) *submersed plants* are attached to the bottom of a water body and grow into the water column; (2) *emersed plants* are attached to the bottom and extend above the water surface; and (3) *floating plants* are unattached to the bottom and float on the water surface. All three growth forms are represented as weeds. Management strategies for key target species will be discussed individually later in this chapter.

The integrated pest management (IPM) of aquatic weeds must be viewed in a context different from that of pathogens and arthropod pests of crops. Aquatic vegetation, especially the most problematic species, should be managed at the lowest possible level. This guideline is based on (1) the reproductive potentials and rapid growth rates of invasive aquatic plant species under most growth conditions found in U.S. waters and (2) the difficulties associated with bringing uncontrolled populations back to manageable levels (Hoyer and Canfield 1997; Schardt 1997). Management at the lowest possible level is often referred to as *maintenance control*.

The common practice in integrated control of aquatic weeds is oriented toward combining management procedures to achieve cost effectiveness (Murphy and Pieterse 1990). The goal of a management plan is to protect the water body from the detrimental effects of invasive aquatic plant species and to minimize the ecologic effects of management practices (Langeland 1991). In addition to cost, the biologic importance of aquatic plants to aquatic habitats and the complexity of aquatic ecosystems must be considered in the development of an integrated aquatic plant management plan. Public perceptions and public uses of water bodies also must be taken into account. Consideration of all these factors and their interactions can be a complex task.

Principles of a Successful Integrated Aquatic Plant Management Program

Hoyer and Canfield (1997) suggested six principles of a successful integrated aquatic plant management program for lakes. These principles, of various degrees of importance, can be applied to management of aquatic vegetation in ponds, lakes, rivers, and ditches.

1. **Identify the uses of the water body and determine whether any of these uses is impaired or benefited by aquatic vegetation.** Water uses and characteristics potentially affected by nuisance aquatic vegetation include specific wildlife habitats, fishing, swimming, navigating (recreational or commercial), conveying flood or irrigation water, and enjoying the outdoors.

Intended water use(s) influence(s) the necessary management level (e.g., whole-lake control or specific-area control) and the management strategies or tactics. For example, if the only purpose of a reservoir is to hold irrigation water or to cool a power plant, grass carp can be used for cost-effective, whole-lake vegetation control. If vegetation interferes with swimming only in certain areas of a lake and vegetation is to be maintained for wildlife habitat in other parts, small areas of vegetation may be controlled with bottom covers, harvesters, or contact herbicides.

2. **Understand plant ecology and water-body ecology**. Knowledge of the ecology of a water body is important in the making of IPM decisions regarding lake systems but is less important for making decisions related to managing vegetation in drainage or irrigation conveyance systems. Understanding plant and water-body ecology is important when vegetation is to be managed for fishing and wildlife habitat. Fish populations and wildlife assemblages respond to changes in amount and type of vegetation in a water body. Depending on the trophic status of the water body, water chemistry also can be affected by vegetation management (e.g., where sufficient nutrients are available, managing large amounts of aquatic macrophytes can lead to dense phytoplankton blooms).
3. **Set management goals**. Management goals can be thought of as water-body conditions to be achieved (Hoyer and Canfield 1997). These goals depend on water use and influence choice of method. For an irrigation canal in a citrus grove, management goals likely will be as simple as elimination of all vegetation for an extended period. In contrast, management goals in large, multiple-use, public water-bodies can be quite complex. For example, boaters and swimmers may desire a minimal level of vegetation, whereas anglers for largemouth bass may desire abundant vegetation. In the latter instance, use of selective herbicides, spot treatment with contact herbicides, or mechanical harvesting of noxious vegetation, integrated with planting of tolerant, more-beneficial vegetation, may be necessary. Management or, if possible, eradication of invasive nonnative aquatic species always should be a priority. These plant species will be discussed in detail.
4. **Consider a range of management techniques**. Management methods range from hand removal to herbicide use. As has been discussed, the methods chosen for an integrated aquatic plant management plan will depend largely on water uses and management goals. Monetary considerations also are a factor. Methods such as hand pulling and using bottom barriers are adaptable only to small ponds or beach sites, whereas herbicides or grass carp are required for large-scale or whole-lake control. The following are management techniques to consider in the development of an integrated aquatic plant management plan.

Hand Removal. Removal by hand of small amounts of vegetation interfering with beach areas or boat docks may be the only vegetation control necessary. Hand removal is labor intensive and must be conducted routinely. Frequency and practicality of continued hand removal will depend on availability of labor, regrowth or reintroduction potential of vegetation, and desired control level (Wade 1990). Floating booms can be used to minimize reintroduction of unwanted vegetation. To minimize regrowth, hand removal of aquatic vegetation may be used in combination with other methods such as herbicides or benthic barriers and has the added advantage of being very selective in removing undesired vegetation and maintaining desired plants.

Mechanical Removal. Specialized machines in a wide variety of sizes and with various accessories are available for removing aquatic vegetation in different situations (Figure 7.2). There are small machines that can be used for

Figure 7.2. Although expensive to operate, aquatic weed harvesters provide an important method of aquatic weed management. Photo courtesy of W. T. Haller, University of Florida, Institute of Food and Agricultural Sciences.

limited areas, as well as large machines that can be used, in combination with transports and shore conveyors, in extensive or whole-lake operations. Such machines commonly are called *mechanical harvesters* or *weed harvesters*, and the process they facilitate is called *mechanical harvesting* or *removal*.

Because it possesses several advantages, mechanical removal is in certain circumstances an important method of aquatic plant management (Wade 1990). Immediate control can be achieved, and because plant matter is removed from the water, the problem of decaying vegetation—which can be associated with herbicide-killed vegetation—is minimized. Water can be used immediately, whereas there may be restrictions on irrigation and domestic water use after the application of certain herbicides.

Mechanical removal has several disadvantages. It usually costs more and is slower and less efficient than other methods, and associated maintenance and repair costs are higher. Certain bodies of water are unsuitable for mechanical removal because of the depth of water and the presence of obstructions. Plant fragments may drift to infest new areas. Temporary increases in turbidity may result from disturbance of sediments during aquatic plant harvest. A suitable area for disposal of harvested plants must be available. Wildlife (e.g., small fish, snakes, turtles, and desirable vegetation) are removed with harvested weeds. This process must be repeated during the growth season.

Water-Level Manipulation. Water-level manipulation refers to (1) the raising of water levels to control aquatic vegetation by drowning and (2) the lowering of water levels to control aquatic vegetation by exposing it to freezing, drying, or heating. Manipulation of water level in aquatic plant management is limited to lakes and reservoirs with adequate water control structures.

In aquatic plant management, lowering of water level, or *drawdown*, is used more commonly than raising of water level (Figure 7.3). For many years, drawdown has been used to oxidize and to consolidate flocculent sediments, to alter fish populations, and to control aquatic weeds in lakes. Use of drawdown in aquatic plant management also may need to be restricted because of considerations such as water use patterns and water rights (e.g., disruption of recreational or agricultural use) or a predictable source of water for refilling. Additional considerations include electrical power generation and minimal stream flows for protecting endangered species.

Drawdown usually is conducted during winter months so that plants are exposed to both drying and freezing. Summer drawdown also can be effective although it usually affects agricultural and recreational water use more negatively, stresses fish populations more significantly, and has a greater potential to enhance the spread of emergent plants such as cattails, rushes, and willows. This strategy generally alters the composition of aquatic vegetation while not necessarily producing desirable changes. Responses of various aquatic plant species differ widely and at times unpredictably. Submersed aquatic plants tend to respond in various ways to drawdown, whereas emergent plants tolerate or are stimulated by it (Langeland 1991).

The main advantage of drawdown as a method of aquatic plant management is its low cost—unless recreational income or power generation is lost. Secondary benefits include sediment oxidation and consolidation and fisheries enhancement (Hoyer and Canfield 1997). Potential undesirable effects include decreases in the number of desirable species, increases in the number of undesirable tolerant species such as the invasive, nonnnative hydrilla (*Hydrilla verticillata*); expansion of undesirable species to deeper areas; creation of floating islands (Hoy-

Figure 7.3. ***Drawdown*, the lowering of water level, sometimes is used to control aquatic vegetation by exposing it to freezing, drying, or heating. Photo courtesy of K. Langeland, University of Florida.**

er and Canfield 1997); and loss of storage water and recreational benefits when insufficient water is available for refill (Hoyer and Canfield 1997).

Light Penetration. Submersed aquatic plants sometimes can be controlled or suppressed by decreasing light penetration into the water. Light penetration can be restricted by the use of special dyes, special fabric bottom covers, or fertilization.

Although dyes are not pesticides, only those approved for such a purpose should be used in water. These specially produced dyes block the light that plants need for photosynthesis and are nontoxic to aquatic organisms, humans, or animals that might drink the treated water. Dyes are effective only in ponds with little to no flow, however, and generally are effective only in water deeper than 3 feet.

Although gases produced in bottom sediments accumulate under nonpermeable bottom covers such as plastic and eventually cause the covers to float to the surface, various materials, including black plastic and specially manufactured bottom covers, have been used to prevent rooted aquatic plants from growing. Specially made bottom covers also can prevent submersed aquatic plant growth. In addition to preventing light from reaching the bottom, these materials also prevent rooted aquatic plants from becoming established. These special materials are expensive and must be maintained to prevent sediment accumulation on top of the cover. Thus, their use is restricted generally to ornamental ponds, swimming areas, or around boat docks.

Nutrient Limitation. Plant growth can be limited if even one nutrient critical for growth is in short supply. Because nitrogen, phosphorus, or carbon usually are the nutrients limiting plant growth in lakes (Wetzel 1975), intentional limiting of one or more of these nutrients can be used to decrease the growth of floating plants that derive nutrients from the water column. Rooted aquatic vegetation derives nutrients from the lake bottom (Wetzel 1975) and thus will not respond to decreased nutrients in the water column.

In certain geographic regions, nutrients are in such short supply that aquatic plants cannot grow to problem levels. Where inputs with a human source have accelerated plant growth, nutrients can be limited by identifying and abating the nutrient source(s). If the lake has received external phosphorus inputs for a long period, it also may be necessary to affect internal nutrient availability by precipitation with agents such as alum. Although nutrient limitation is theoretically possible, no examples are available in the scientific literature of the use of nutrient limitation to manage nuisance rooted macrophyte populations in lakes (Hoyer and Canfield 1997).

Biological Control. Classical biocontrol with fungal pathogens as well as microbial herbicides has been considered for aquatic vegetation management. *Classical biocontrol* refers to the introduction of a pathogen from the geographic origin of a weed into the area where control is desired. Experience in the southeastern United States indicates that in heavily used and intensively managed waterways a microbial herbicide strategy is more suitable than a classical approach (Charudattan 1990). The potential to deploy certain pathogens as traditional agents in some areas and as microbial herbicides in others and to combine traditional and microbial herbicide agents deserves additional study. According to Charudattan (1990), underwater systems impose severe technological and ecological constraints rendering attempts at biological control of undesirable submersed plants impractical; currently, fungal control is feasible only for emergent species. Fungal pathogens of water hyacinth (*Eichhornia crassipes*), hydrilla, Eurasian watermilfoil (*Myriophylum spicatum*), parrotsfeather (*M. aquaticum*), water lettuce (*Pistia stratiotes*), giant water fern (*Salvinia molesta*), and alligatorweed (*Alternanthera philoxeroides*) have been identified (Charudattan 1990). No pathogen, however, has been developed into a commercially available microbial herbicide.

Much effort has been expended in the identification of arthropods as classical biological control agents for aquatic weeds, and greater success has been achieved through the use of such agents than through the use of microbial agents. Thirteen insects have been identified as biological control agents for six of the most invasive aquatic plants in the United States (Table 7.1). As yet, none has obviated the use of other control methods. These insects will be discussed later in reference to their respective host plants.

Several plant-feeding, or *phytophagous*, fish

Table 7.1. Insect biological control agents on invasive aquatic plant species in the United States (Langeland 1991)

Biocontrol agent	Target plant species
Agasicles hygrophyla *Amynothrips andersoni* *Vogtia malloi*	Alligatorweed (*Alternanthera philoxeroides*)
Neochotina eichhorniae *N. bruchi* *Sameodes albigutalis*	Water hyacinth (*Eichhornia crassipes*)
Baguous affinis *B. hydrillae* *Hydrellia balciunasi* *Parapoynx diminutalis*	Hydrilla (*Hydrilla verticillata*)
Euthrychiopsis lecontei (native)	Eurasian watermilfoil (*Myriophyllum spicatum*)
Neohydronomous affinis	Water lettuce (*Pistia stratiotes*)
Cyrtabagous salvineae	Giant water fern (*Salvinia molesta*)

species have been considered biological control agents (Opuszcynski and Shireman 1995). The grass carp, now produced as a sterile triploid, has found widespread use in nonselective aquatic weed control and in hydrilla management programs (Langeland 1996). Its integration as a hydrilla biological control agent will be discussed in greater detail later in this chapter. Used mainly for algae control, *Tillapia* spp. have played a limited role in aquatic vegetation control in the United States and will not be discussed further in this report.

Herbicides. The market for aquatic herbicides is small compared with markets such as those for agronomic crops. Moreover, additional testing for registration and reregistration is required to label products for use in water. Thus, the monetary incentive for manufacturers to register aquatic herbicides is low. Only six herbicide active ingredients are registered in aquatic plant management (Table 7.2). Of the many known herbicidal active ingredients with potentials against nuisance aquatic plants, only those ingredients that can be shown to be safe to use in the aquatic environment and that have sufficient market potential are registered or reregistered. Herbicides registered for aquatic use are nonpersistent and of low environmental toxicity. Even though aquatic herbicides must meet the rigorous registration standards

Table 7.2. Herbicide active ingredients used for managing invasive aquatic plant species in the United States (Langeland 1991)

Herbicide active ingredient	Target species
2,4–D	Water hyacinth (*Eichhornia crassipes*)
Cuprous ion	Hydrilla (*Hydrilla verticillata*) Water hyacinth (*Eichhornia crassipes*) Eurasian watermilfoil (*Myriophyllum spicatum*)
Diquat dibromide	Hydrilla (*Hydrilla verticillata*) Water hyacinth (*Eichhornia crassipes*) Eurasian watermilfoil (*Myriophyllum spicatum*) Water lettuce (*Pistia stratiotes*)
Endothall	Hydrilla (*Hydrilla verticillata*) Eurasian watermilfoil (*Myriophyllum spicatum*)
Fluridone	Hydrilla (*Hydrilla verticillata*) Eurasian watermilfoil (*Myriophyllum spicatum*) Torpedo grass (*Panicum repens*)
Glyphosate	Torpedo grass (*Panicum repens*)

of the U.S. Environmental Protection Agency, opposition to the use of aquatic herbicides by the public is often encountered. The educational component of an integrated aquatic plant management program often requires teaching individuals about the pesticide registration process and the environmental safety of aquatic herbicides. Even with the few available registered aquatic herbicides and the public skepticism regarding their environmental safety, most integrated aquatic plant management programs rely on the use of herbicides. This situation results from their cost effectiveness and environmental safety, both of which are recognized by aquatic plant managers and regulators.

5. **Develop an action plan and a program to monitor the success or failure of management activities**. Once the crucial information identified earlier in this report is available to stakeholders of the integrated aquatic plant management plan, decisions can be made to synthesize a plan that is satisfactory to all parties. A comprehensive written plan provides continuity as participants change (Hoyer and Canfield 1997). It allows for monitoring of control methods and is flexible so that changes can be made when new information becomes available or change becomes necessary for other reasons. And it is long term so that vegetation can be managed on a continuing basis. Such management plans are important for private lakes so that all waterfront property owners or members of a lake owners' association can agree on management goals, methods, and outcomes and are even more important for multiple-use public waters. In the latter instance, regulatory agencies will oversee or assist in plan development and implementation.

   Florida's use of multi-agency/public task forces or committees to develop work plans for aquatic plant management is a helpful example of how an integrated aquatic plant management plan can be developed for public water bodies (Allen 1993). Committees are formed for individual water bodies/systems because each situation is unique. Task forces consider the multiple use characteristics of the water body, its hydrology, location, economics, fisheries, wildlife populations, and relevant historical trends and public concerns. An example is the Crystal and Homosassa Rivers Aquatic Plant Management Interagency Committee. Because this public water system is highly populated and heavily used for a great variety of recreational activities and at the same time provides a refuge for the endangered West Indies manatee, it is an especially complex system for which to develop an aquatic plant management plan. The Florida multi-agency/public task force set up to develop its work plan is composed of representatives from agencies and public user groups that may have conflicting interests in the management of aquatic plants. Agencies and groups represented include the U.S. Fish and Wildlife Service (FWS), the U.S. Army Corps of Engineers, the Florida Department of Environmental Protection, Citrus County, the University of Florida Center for Aquatic and Invasive Plants, the Save the Manatee Club, and the Crystal River Chamber of Commerce. (Other task forces typically include members of fishing and hunting associations.) This specific committee's efforts have resulted in an aquatic plant management plan integrating (1) both application of contact and systemic herbicides and mechanical harvesting of the invasive submersed plants hydrilla and Eurasian watermilfoil, (2) plantings of beneficial native aquatic plants to protect the West Indian manatee, and (3) provision of optimized recreational benefits.

6. **Establish a long-term aquatic plant management educational program**. A long-term aquatic plant management educational plan is essential so that all stakeholders understand the information provided for developing the plan. This insight must encompass ecology as well as the potential and observed ecological responses of various control methods. Education almost always is needed in regard to the environmental and the human-health safety of herbicides and biological controls. Educational commitment must be long term because of the changes in population and the limits of human memory. Once the plant problem is removed, it generally is a matter of only a few years before users begin to question why the maintenance control of aquatic plants is being done (Hoyer and Canfield 1997). This is a crucial time for educational efforts explaining why weeds should be maintained at a low level instead of allowed to run rampant intermittently.

Integrated Management of Specific Invasive Plants

Alligatorweed

First identified in the United States in 1890, alligatorweed is believed to have been brought in ship ballast from its native South America to Charleston Harbor (Gangstadt 1977). By the 1960s, this pest occupied more than 160,000 acres (a.) of water from Florida to Virginia (Gangstadt 1977). Difficult to control with herbicides, alligatorweed was an early target for classical biological control research. Three insects—*Agasicles hygrophila*, the alligatorweed flea beetle; *Amynothrips andersoni*, the alligatorweed thrips; and *Vogtia malloi*, the alligatorweed stem borer—were approved and released between 1964 and 1971 (Coulson 1977; Gangstadt 1977). In warm regions and in large bodies of water, alligatorweed is a minor problem as long as flea-beetle populations are sufficient (Coulson 1977).

The farther north along the Atlantic coast that alligatorweed populations occur, however, the more completely management programs must integrate herbicide (usually glyphosate) applications and augment flea-beetle populations. Overwintering of flea-beetle populations depends on latitude and on winter severity. The Jacksonville District of the U.S. Army Corps of Engineers undertakes a yearly collection of alligatorweed flea-beetles from local waters for shipment to representatives of state or federal agencies in locations whose ecologies can benefit from augmentation of insect populations (U.S. Army Corps of Engineers 2002) (Figure 7.4). In North Carolina and Virginia, which make up the extreme northern boundaries of alligatorweed on the Atlantic coast, insects released for biological control of alligatorweed are ineffective (Langeland 1986). Here, a maintenance program depending exclusively on applications of glyphosate is necessary. In small ponds and ditches throughout that region, the suppression level offered by insects often is unacceptable and herbicide application, usually of glyphosate, is the predominant control method.

Water Hyacinth

Native to the Amazon Basin, water hyacinth (*Eichhornia crassipes*) was introduced into the United States in 1884. By 1948, 63,000 a. of this species existed in Florida alone (Hammack 1948). Effects of unmanaged water hyacinth range from flooding to degraded aquatic habitats (Gowanlach 1944; Penfound and Earle 1948). Aggressive management programs are necessary to prevent water hyacinth from having significant detrimental effects on water resources. Because it has a population doubling time of as little as 6 to 18 days (Mitchell 1976), water hyacinth should be managed at the lowest possible level (Figure 7.5). Maintenance control results in both diminished detrimental effects of water hyacinth growth and decreased herbicide use (Joyce 1985).

The water hyacinth weevils *Neochetina eichhorniae* and *N. bruchi* were introduced as water hyacinth biocontrol agents in the United States in 1972 and 1974, respectively (Langeland 1991). The water

Figure 7.4. **The Jacksonville District of the U.S. Army Corps of Engineers undertakes a yearly collection of alligatorweed flea-beetles (*Agasicles hygrophyla*) for shipment to representatives of state or federal agencies. Photo courtesy of K. Langeland, University of Florida.**

hyacinth moth, *Sameodes albigutalis*, was introduced in 1977. Although established and capable of decreasing the vigor of plants or controling them in growth-limited situations, these insects have not decreased measurably the need for operational water hyacinth control programs where the plant has been a problem historically (Haller 1996).

The fungal pathogen *Cercospera rodmanii* has received attention because of its potential integration into water-hyacinth control programs (Charudattan 1986). Because of both technical difficulties in ensuring efficacy and uncertainties in the marketplace, *C. rodmanii* has, however, never been registered as a bioherbicide (Charudattan 1990).

Spray programs incorporating 2,4–dichlorophenoxy acetic acid (2,4–D), diquat dibromide, or complexed copper as herbicides are used to maintain water hyacinth at maintenance control levels (Schardt 1997). 2,4–dichlorophenoxy acetic acid, the most widely used herbicide for water hyacinth control, is also the most cost effective and is selective against water hyacinth at use rates below which nontarget native plant populations such as spatterdock (*Nuphar luteum*) are not damaged (Hanlon and Haller 1991). Diquat dibromide is used when water hyacinth is growing in bulrush (*Scirpus validus, S. californicus*) communities, for the nontarget bulrush plants contacted by overspray quickly recover, unlike those affected by systemic 2,4–D (Thayer and Joyce 1990). Copper-containing herbicides are used in proximity to domestic water supply intakes because they do not have the label-required setback distances from intakes that diquat and 2,4–D do. Diquat dibromide and 2,4–D do not harm water hyacinth weevils directly, and it has been recommended that areas of water hyacinth be left unsprayed to provide refugia for the weevils (Haag 1986a,b). In practice, this concept is not used because spray programs, being imperfect, themselves leave areas of refugia; intentionally leaving large areas for refugia would be contrary to the principles of maintenance control and detrimental to the effectiveness of spray programs. Thus, although established biological control agents may slow the growth rate of water hyacinths, extensive scouting to identify small populations and spraying these small populations with appropriate herbicides to maintain them at the lowest possible level are keys to successful management (Schardt 1997).

A typical integrated water hyacinth control program integrates growth suppression by means of traditional biological controls, depending on the situation. Tactics include extensive scouting to identify small water-hyacinth populations and spraying small numbers of plants frequently with the appropriate herbicide to minimize nontarget effects while maintaining plants at the lowest possible level.

Figure 7.5. When not managed at maintenance levels, water hyacinth (*Eichhornia crassipes*) will displace native plant communities in lakes and rivers. Photo courtesy of J. Schardt, Florida Department of Environmental Protection.

Hydrilla

Native to the warmer regions of Asia, hydrilla was discovered in U.S. waters in 1960 and now is a common problem in the Atlantic coastal states, southeastern states, Gulf Coastal states, California, and Washington (Langeland 1996). Hydrilla is a problem in many aquatic environments, e.g., agricultural water conveyance canals and natural lakes, where it interferes with recreational activities, impedes water flow, displaces native-plant communities, and alters fisheries populations and water chemistry (Langeland 1996).

Four insects—*Bagous affinis*, the hydrilla tuber weevil; *B. hydrillae*, the hydrilla stem borer; *Hydrellia balciunasi*, the Australian leaf mining fly; and *H. pakistanae*, a leaf mining fly—have been approved as biocontrol agents in the United States and released (Langeland 1996). Another insect, *Parapoynx diminutalis*, a small aquatic moth, was released accidentally. None of these insects has had measurable or widespread effects on hydrilla populations (Cuda, J. 2001. Personal communication).

Hydrilla is high on the food preference list of the grass carp, which has been used for biological control of hydrilla since the early 1970s (Opuszcynski and Shireman 1995). Genetically sterile triploid grass carp now are used instead of diploid fish, which have the potential to escape stocking sites to suitable spawning habitats (Malone 1984). Initial stocking rates of more than 50 grass carp/a. proved too high for selective hydrilla control and usually resulted in removal of all vegetation, including nontarget vegetation, from the stocked water body because once preferred foods were consumed, the fish moved on to less-preferred foods (Van Dyke, Leslie, and Nall 1984).

Optimal grass carp stocking rate depends on many interacting factors unique to individual water bodies. Computer-based stocking rate simulation models have been developed to help aquatic-plant managers (Stewart and Boyd 1994). Because poststocking mortality cannot be predicted and results depend on actual numbers of fish present (not on initial stocking rate), these models do not reliably determine with precise prediction of outcome the stocking rates for individual water bodies. Low stocking rates of fewer than 10 fish/a. are used and integrated now with initial and perhaps subsequent herbicide applications (Eggeman 1994; Jaggers 1994) (Figure 7.6). In a recent survey of grass carp-stocked lakes, Hanlon (1999) concluded that stocking lakes with more than 3 to 4 fish/a. resulted in complete control of vegetation, with the exception of a few species that grass carp have difficulty consuming. Stockings of 3 to 4 fish/a. could result in control of nuisance levels of submersed vegetation, such as hydrilla, while some predominantly unpalatable submersed, floating leafed, and emergent vegetation could be maintained. At stocking rates below 3 to 4 fish/a., little control may be achieved. Fish populations of 3 to 4 fish/a. are difficult to achieve, however, because of the unknown mortality rates of grass carp.

The contact herbicides endothall and diquat (sometimes tankmixed with a copper-based herbicide) and the systemic herbicide fluridone can be integrated with the low grass carp stocking rates described already (Eggeman 1994; Jaggers 1994). But these herbicides often are used alone to avoid the potential for unpredictable results associated with the stocking of grass carp. The amount of herbicide needed for control of large areas of hydrilla has been decreased greatly by application of fluridone in low doses over extended periods (Fox, Haller, and Shilling 1994). Because of the great expense, mechanical harvesting rarely is used for hydrilla management; it is used, however, when herbicides or grass carp cannot be used, for example, in open systems with high rates of water exchange (Langeland 1996).

Eurasian Watermilfoil

Eurasian watermilfoil, native to Europe, Asia, and northern Africa, first was documented in the United States in 1942 (Couch and Nelson 1985) and now occurs throughout the country (U.S. Geological Survey 2001).

A native weevil (*Euthrychiopsis lecontei*) has been implicated in the natural declines of Eurasian watermilfoil, but its effects are insufficiently predictable to make application practical. Integration of mechanically harvested areas of great human activity with partitioned, less-used areas into no-harvest zones has been suggested as a feasible approach to Eurasian watermilfoil control (Sheldon and O'Bryan 1996). But this practice has been neither adopted widely nor proved. Grass carp and herbicides are used for Eurasian watermilfoil control with integration characteristics and limitations similar to those of hydrilla. Mechanical harvesting is used more often as a stand-

Figure 7.6. **Genetically sterile triploid grass carp are used alone or in combination with herbicide application to control hydrilla (*Hydrilla verticuillata*) and other submersed aquatic vegetation. Photo courtesy of R. K. Stocker, University of Florida, Institute of Food and Agricultural Sciences.**

alone management for Eurasian watermilfoil in northern states, which have a shorter growing season, than for hydrilla in southern states.

Torpedo Grass

Torpedo grass was introduced from the Old World into the United States before 1876 (Beal 1896) and now occurs from Florida to Texas, northward along the Atlantic coast to North Carolina, and in California and Hawaii (Langeland and Burks 1998). Torpedo grass quickly forms monocultures displacing native vegetation in water bodies and is a significant pest in citrus grove irrigation canals and on golf courses.

No biological controls have been developed for torpedo grass, and no methods other than herbicides are effective for controlling it (Langeland and Smith 1998). The only herbicide registered for aquatic use and effective against torpedo grass is glyphosate, which achieves at best incomplete control with single or repeat applications. Because of the extensive number of dormant buds on rhizomes, repeat applications are necessary to bring an established torpedo grass population under control. Smith, Langeland, and Hanlon (1998) provided a treatment regime that decreases herbicide use and cost by decreasing glyphosate rate in successive applications.

Water Lettuce

Water lettuce was seen in the United States as early as 1774 and may have been introduced by natural means or by humans (Langeland and Burks 1998). It is a tropical plant native to Africa or to South America, and in the United States is limited to warm regions. It occurs throughout peninsular Florida and Westward to Texas and also is found persisting in coastal South Carolina. It forms vast mats, disrupting submersed plant communities and interfering with water movement and navigation.

The water lettuce weevil (*Neohydronomous affinis*) was released in Florida in 1986 and 1987 (Gallagher and Haller 1990). Having formed self-perpetuating populations and being dispersed throughout the peninsular portion of the state, the weevils now are affecting water lettuce populations. Weevil populations and their effects are unpredictable, however. Water lettuce is very sensitive to low doses of diquat herbicide, and management is integrated heavily toward herbicide application. In monocultures, diquat is used alone. In mixed populations of water lettuce and water hyacinth, a mixture of diquat and 2,4–D is used.

Giant Water Fern

A recent introduction to the United States from Southeastern Brazil, giant water fern first was found in the United States in South Carolina in 1995 and now has invaded Florida, Georgia, Alabama, Louisiana, Mississippi, Texas, Arizona, California, and Hawaii (U.S. Geological Survey 2001). Its expected range in the United States is in fresh waters, from areas experiencing frost but not ice, to warmer regions (Whiteman and Room 1991). Effects are similar to those described for water hyacinth and water lettuce.

Giant water fern has been an important weed in much of both the tropical and the subtropical regions of the world. After evaluations of numerous insects for its biological control, all of which were unsuccessful, *Cyrtobagous salviniae* was found to be very successful in controlling giant water fern (Room 1986). This weevil should be able to control giant salvinia in most locations (Room 1986). *Cyrtobagous salviniae* exists on populations of the common water fern (*Salvinia minima*) in Florida (Carter, T. 2001. Personal communication). The effectiveness of *C. salviniae* in controlling giant water fern in the United States is unknown.

A national Giant Salvinia Task Force (GSTF) was formed to develop an integrated management program for giant water fern (Helton and Chilton 2001). The GSTF, a cooperative effort among the U.S. Department of Agriculture (USDA), FWS, and state and local governments, proposed that affected agencies respond using public education, aquatic herbicides, biological control, and mechanical removal where feasible (Helton and Chilton 2001). Even before formation of the task force, the U.S. Geological Survey had produced a public information fact sheet and Web page (U.S. Geological Survey 2001). More than 80,000 copies of the fact sheet have been distributed. The herbicides diquat, chelated copper, glyphosate, and fluridone are being used in emergency efforts to contain populations, and a team of USDA scientists is conducting, under quarantine, studies of the Australian strain of *Cyrtobagous salviniae*, studies that must be undertaken before release is feasible (Helton and Chilton 2001).

Appendix A. Abbreviations and Acronyms

2,4–D amine	2,4-dichlorophenoxy acetic acid
a.	acre
FWS	U.S. Fish and Wildlife Service

GSTF	Giant Salvinia Task Force
ha	hectare
USDA	U.S. Department of Agriculture

Appendix B. Glossary

Classical biological control. The introduction of an organism from the geographic origin of a weed into the area where control is desired.

Drawdown. Intentional lowering of water level.

Emersed plants. Plants attached to the bottom of a water body and extending above the water surface.

Floating plants. Plants unattached to the bottom of a water body and floating on the water surface.

Maintenance control. The lowest possible level of management.

Mechanical removal. Removal of unwanted vegetation by means of specialized machines.

Phytophagus. Feeding on plants.

Submersed plants. Plants attached to the bottom of a water body and growing into the water column.

Water-Level manipulation. (1) The raising of water levels to control aquatic vegetation by drowning and (2) the lowering of water levels to control aquatic vegetation by exposing it to freezing, drying, or heating.

Literature Cited

Allen, N. P. 1993. The role of the task force in today's aquatic plant management. *Aquatics* 15:11–18.

Beal, W. J. 1896. *Grasses of North America*. Volume 2. Holt and Company, New York. 675 pp.

Charudattan, R. 1986. Integrated control of waterhyacinth (*Eichhornia crassipes*) with a pathogen, insects and herbicides. *Weed Sci* 34:26–30.

Charudattan, R. 1990. Biological control of aquatic weeds by means of fungi. Pp. 141–142. In A. H. Pieterse and K. J. Murphy (eds.). *Aquatic Weeds: The Ecology and Management of Nuisance Aquatic Vegetation*. Oxford University Press, New York. 593 pp.

Couch, R. and E. Nelson. 1985. Myriophyllum in North America. Pp. 8–18. In L.W. J. Anderson (ed.). *Proceedings of the 1st International Symposium on Watermilfoil (*Myriophyllum spicatum*) and Related Haloragaceae Species*, July 23–24, Vancouver, British Columbia. The Aquatic Plant Management Society, Incorporated, Vicksburg, Mississippi. 223 pp.

Coulson, J. R. 1977. Biological control of alligatorweed, 1959–1972. A review and evaluation. Technical Bulletin Number 1547. U.S. Department of Agriculture, Washington, D.C. 98 pp.

Eggeman, D. 1994. Integrated hydrilla management plan utilizing herbicides and triploid grass carp in Lake Istokpoga. Pp. 164–165. In *Proceedings of the Grass Carp Symposium*, March 7–9, 1994, University of Florida, Gainesville. U.S. Army Corps of Engineers Waterways Experiment Station, Vicksburg, Mississippi. 242 pp.

Fox, A. M., W. T. Haller, and D. G. Shilling. 1994. Use of fluridone for hydrilla management in the Withlacoochee River, Florida. *J Aquat Plant Manage* 32:47–55.

Gallagher, J. E. and W. T. Haller. 1990. History of development of aquatic weed control in the United States. *Rev Weed Sci* 5:115–191.

Gangstadt, E. O. 1977. Potential growth of aquatic plants in Republic of the Philippines and projected methods of control. *J Aquat Plant Manage* 14:10–14.

Gowanlach, J. N. 1944. The economic status of the waterhyacinth in Louisiana. *Louisiana Conservationist* 1:6,8.

Haag, K. A. 1986a. Effective control of waterhyacinth using Neochetina and limited herbicide application. *J Aquat Plant Manage* 24:70–75.

Haag, K. A. 1986b. Effects of herbicide application on mortality and dispersive behavior of the waterhyacinth weevils, *Neochetina eichhorniae* and *Neochetina bruchi* (Coleoptera:Curculionida). *Environ Entomol* 15:1192–1198.

Haller, W. T. 1996. Operational aspects of chemical, mechanical, and biological control of waterhyacinth in the United States. P. 137. In R. Charudattan, R. Labrada, T. D. Center, and C. Kelly-Begazo (eds.). *Strategies for Waterhyacinth Control: Report of a Panel of Experts Meeting, 11–14 September 1995, Fort Lauderdale, Florida*. Food and Agricultural Organization, United Nations, Rome. 217 pp.

Hammack, J. A. 1948. Distribution of waterhyacinth in Florida. Pp. 21–28. In *Proceedings of the Soil Science Society of Florida*. E. O. Painter Printing Company, De Land, Florida.

Hanlon, C. and B. Haller. 1991. The effect of 2,4–D amine on the growth of Spatterdock. *Aquatics*13:18–20.

Hanlon, S. G. 1999. Evaluation of macrophyte control in 41 Florida lakes using triploid grass carp (*Ctenopharyngodon idella*) at different stocking rates. Master's Thesis, University of Florida, Gainesville. 112 pp.

Helton, R. J. and E. Chilton. 2001. Giant salvinia (*Salvinia molesta* D. S. Mitchell) in Texas: A destructive menace. *Aquatics* 23:12–18.

Hoyer, M. V. and D. E. Canfield. 1997. Aquatic plant management in lakes and reservoirs. North American Lake Management Society, and Aquatic Plant Management Society, U.S. Environmental Protection Agency, Washington, D.C. 101 pp.

Jaggers, B. V. 1994. Economic considerations of integrated hydrilla management. A case history of Johns Lake, Florida. Pp. 151–163. In *Proceedings of the Grass Carp Symposium*, March 7–9, 1994, University of Florida, Gainesville. Waterways Experiment Station, U.S. Army Corps of Engineers, Vicksburg, Mississippi. 242 pp.

Joyce, J. C. 1985. Benefits of maintenance control of waterhyacinth. *Aquatics* 7:11–13.

Langeland, K. A. 1986. Management program for alligatorweed in North Carolina. Report Number 224. Water Resources Research Institute, University of North Carolina, Raleigh. 34 pp.

Langeland, K. A. 1991. Aquatic pest control applicator training manual. University of Florida Cooperative Extension Service, Institute of Food and Agricultural Sciences, Gainesville. 107 pp.

Langeland, K. A. 1996. *Hydrilla verticillata* (L.F.) Royle (Hydrocharitaceae), The perfect aquatic weed. *Castanea* 61:293–304.

Langeland, K. A. and K. C. Burks. 1998. Identification and biology of non-native plants in Florida's natural areas. University of Florida, Gainesville. 165 pp.

Langeland, K. A. and B. Smith. 1998. Torpedograss: Forage gone wild. *Wildland Weeds* 1:4–6.

Malone, J. M. 1984. Triploid white amur. *Fisheries* 9:36.

Mitchell, D. S. 1976. The growth and management of *Eichhornia crassipes* and *Salvinia* spp. in their native environment and in alien situations. Pp. 165–176. In C. K. Varshney and J. Rzoska (eds.). *Aquatic Weeds SE Asia*. W. Junk Publishers, The Hague.

Murphy, K. J. and A. H. Pieterse. 1990. Present status and prospects of integrated control of aquatic weeds. Pp. 222–227. In A. H. Pieterse and K. J. Murphy (eds.). *Aquatic Weeds: The Ecology and Management of Nuisance Aquatic Vegetation*. Oxford University Press, New York. 593 pp.

Opuszcynski, K. and J. V. Shireman. 1995. *Herbivorous Fishes: Culture and Use for Weed Management*. CRC Press, Boca Raton, Florida. 223 pp.

Penfound, W. T. and T. T. Earle. 1948. The biology of the waterhyacinth. *Ecol Monogr* 18:449–472.

Room, P. M. 1986. Biological control in solving the world's *Salvinia molesta* problems. Pp. 271–276. In *Proceedings of the European Weed Research Society/Association of Applied Biologists 7th Symposium on Aquatic Weeds*. Quorn Selective Reproduction Limited, Loughborough, Leicester, United Kingdom.

Schardt, J. D. 1997. Maintenance control. Pp. 229–243. In D. Simberloff, D. C. Schmitz, and T. C. Brown (eds.). *Strangers in Paradise*. Island Press, Washington D.C. 467 pp.

Sheldon, S. P. and L. M. O'Bryan. 1996. The effects of harvesting Eurasian watermilfoil on the aquatic weevil *Euhrychiopsis lecontei*. *J Aquat Plant Manage* 34:76–77.

Smith, B., K. Langeland, and C. Hanlon. 1998. Comparison of various glyphosate application schedules to control torpedograss. *Aquatics* 20:4–9.

Stewart, R. M. and W. A. Boyd. 1994. Simulation model evaluation of sources of variability in grass carp stocking requirements. Pp. 85–92. In *Proceedings of the Grass Carp Symposium*, March 7–9, 1994, University of Florida, Gainesville. Waterways Experiment Station, U.S. Army Corps of Engineers, Vicksburgh, Mississippi. 242 pp.

Thayer, D. D. and J. C. Joyce. 1990. The impact of herbicides on bulrush communities: An evaluation of waterhyacinth control efforts within native plants. *Aquatics* 12:12–19.

U.S. Army Corps of Engineers. 2002. Biological control of exotic aquatic and wetland plants, December 7, <http://www.saj.usace.army.mil/conops/apc/weed_bio.html> (29 July 2002)

U.S. Geological Survey. 2001. Nonindigenous aquatic species, September 19, <http://nas.er.usgs.gov/ ferns> (29 July 2002)

Van Dyke, J. M., A. J. Leslie, and L. E. Nall. 1984. The effects of grass carp on the aquatic macrophytes of four Florida lakes. *J Aquat Plant Manage* 22:84–86.

Wade, P. M. 1990. Physical control of aquatic weeds. Pp. 93–135. In A. H. Pieterse and K. J. Murphey (eds.). *Aquatic Weeds: The Ecology and Management of Nuisance Aquatic Vegetation*. Oxford University Press, New York.

Wetzel, R. G. 1975. *Limnology*. W. B. Saunders Company, Philadelphia, Pennsylvania. 473 pp.

Whiteman, J. B. and P. M. Room. 1991. Temperatures lethal to *Salvinia molesta* Mitchell. *Aquat Bot* 42:27–35.

8 Food Animal Integrated Pest Management

The livestock and poultry industries are characterized by production and marketing systems ranging from the large, vertically integrated companies that dominate poultry and swine production to the small, independent growers who sell directly to consumers. Each segment of the industry faces a unique set of pest management challenges, in terms of both the pests and the nuances of management approaches that best suit the production environment.

The basic elements of integrated pest management (IPM) for livestock and poultry are the same as those for other agricultural commodities. Cultural practice, biological control, monitoring, and chemical control all play important roles in effective IPM programs for livestock and poultry pests. New regulations governing pesticide use have decreased the availability of pesticides over the past decade. Although these actions have caused some difficulty for the animal industry, they have made producers more receptive to the application of a full range of pest management options. Amendments to the Federal Insecticide, Fungicide and Rodenticide Act in 1988 and passage of the Food Quality Protection Act in 1996 have limited both the availability of existing pesticides to the animal industry and the introduction of new pesticide chemistries for minor-use commodities. The needs to manage pesticide resistance, to confront nuisance issues, to decrease the risk of arthropod-borne disease to man and animals, and to continue producing food profitably will define, collectively, the future path of IPM in the livestock and poultry industries.

Livestock

Dairy Cow and Confined Cattle Integrated Pest Management

The house fly (*Musca domestica*) (Figure 8.1) and the stable fly (*Stomoxys calcitrans*) (Figure 8.2) are the primary insect pests affecting confined cattle in dairies and feedlots in the United States. Public health and nuisance issues are major economic incentives for managing house flies. Milk must be produced, han-

Figure 8.1. The house fly—a primary pest of confinement livestock and poultry production systems. Photo courtesy of M. Stringham, North Carolina State University.

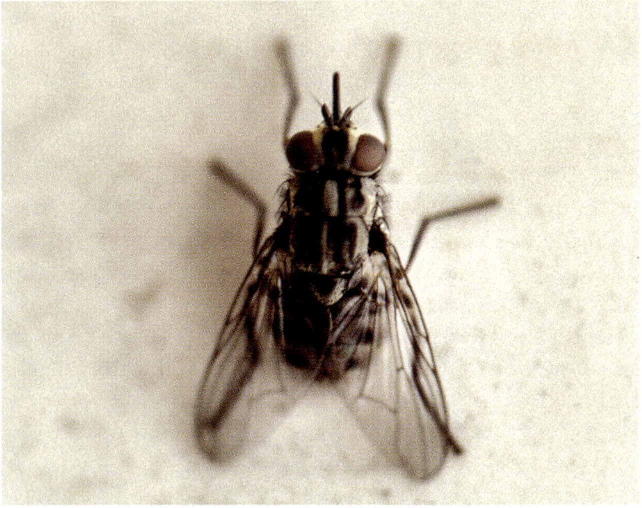

Figure 8.2. The stable fly—a biting muscoid fly commonly associated with livestock production. Photo courtesy of D. W. Watson, North Carolina State University.

dled, and stored in a clean and sanitary environment, and public health authorities consider numerous flies in and around a dairy to be evidence of unsatisfactory sanitation (Rutz, Geden, and Pitts 1993). Public nuisances caused by flies dispersing from dairies and feedlots into suburban areas are a growing concern.

Stable flies are biting-insects that may take several blood meals a day from humans and other animals. Because the bite of the stable fly is quite painful, cattle will avoid feeding and bunch up to avoid this pest. The stable fly is capable of transmitting several livestock pathogens, notably those causing anthrax, brucellosis, swine erysipelas, and equine infectious anemia (Knapp, Charron, and Burg 1992).

Bovine responses to stable fly feeding range from behaviors denoting pain and irritation such as bunching, head throwing, and foot stamping, to physiological changes. The latter includes increased body temperature, increased urinary nitrogen output, and elevated cortisol levels (Wieman et al. 1992).

Integrated pest management is recognized in the industry as a preferred best management practice to minimize pest problems in dairy cow and feedlot cattle systems and to help protect farm workers, consumers, and the environment (Rutz and Watson 1998). Traditionally, dairy farmers have relied heavily on insecticides, a management strategy that has led to resistance in the fly population (Rutz et al. 1999). Integrated pest management in dairy systems spans several years of research and Extension efforts.

To manage fly populations, the dairy IPM plan relies on accurate pest identification and monitoring and on cultural, biological, and chemical control strategies (Rutz and Watson 1998). Among these strategies, cultural control (i.e., manure management by eliminating the habitat conducive to growth and development of the immature stages of flies) is essential to the fly management program (Rutz, Geden, and Pitts 1993). Prompt removal of manure, soiled bedding, and spoiled feed limits the opportunity for flies to breed in the barn. Land application of manure and bedding increases mortality of existing larvae through drying and chilling during the winter. Natural enemies of flies also play important roles in on-farm fly management. For instance, parasitoids attack the pupal stage of the fly. The female parasitoid deposits an egg in the fly puparium, the larval parasitoid kills and consumes the fly pupa, and the newly developed parasitoid emerges from the fly pupa in 2 to 3 weeks (Rutz, Geden, and Pitts 1993). Although ten species of parasitoids are associated commonly with dairy systems, one species, *Muscidifurax raptor*, has been used extensively in IPM studies (Rutz and Watson 1998). An indigenous species, *M. raptor* is found throughout the Midwest, mid-Atlantic states, and northeastern United States. *M. raptorellus* (Figure 8.3) has shown promise for IPM programs (Petersen and Cawthra 1995). A gregarious pupal parasitoid, female *M. raptorellus* deposits three to five eggs in each fly pupa, increasing its own reproductive potential. Many species of parasitoids are available commercially through insectaries (Hunter 2002).

Well-informed producers benefit most by purchasing parasitoids suitable to their region and needs. Adult fly populations are monitored based on the spot card technique, which helps producers establish an action threshold based on their tolerances for flies; certain producers may tolerate the presence of some, whereas other producers may accept no flies. The spot card technique estimates the adult fly population by determining the number of fly specks on ten cards distributed throughout a barn. Cards are replaced weekly, and running averages tallied. If tolerance thresholds are exceeded, a low-residual pyrethrin space spray is recommended to control adult flies while minimizing parasitoid mortality (Rutz, Geden, and Pitts 1993).

Figure 8.3. ***Muscidifurax raptorellus*** **(and other parasitoids)—effective biological control agents for fly management in livestock and poultry production. Photo courtesy of D. W. Watson, North Carolina State University.**

Many IPM efforts in feedlot cattle systems have concentrated on manure management, biological control, migration, and feedlot design (Campbell et al. 1999). Demonstrating the IPM principle that no single strategy will provide satisfactory control, Thomas et al. (1996) proved that frequent cleanings of feedlots decreased stable fly populations by only 51%. But the addition of natural enemies to the management plan enhanced fly management further. The house fly parasitoid, *Muscidifurax zaraptor*, was the most abundant species in Nebraska feedlots (Campbell et al. 1999); a second species, *Spalangia nigroaenea*, was the dominant parasitoid attacking the stable fly (Greene, Guo, and Chen 1998; Jones and Weinzierl 1997). Augmentation of naturally occurring parasitoid populations has been practiced through frequent releases of commercially available selected strains of these antagonists (Campbell et al. 1999; Weinzierl and Jones 1998).

Dispersion and migration of adult house flies and stable flies are strategies of great importance to their management (Campbell et al. 1999). Although resident populations of flies are suspected of overwintering on the farm, unpublished reports of massive migrations of large numbers of flies have suggested that these insects may move with changing weather patterns (Hogsette and Ruff 1985; Petersen and Meyer 1983). Subsequent analysis of isozymes from dispersing stable flies indicated that these insects may move with frontal systems over great distances (Jones, C. J. et al. 1991). Stable flies collected from Minnesota and Iowa showed that substantial gene flow occurs among populations in these contiguous states (Krafsur 1967).

Figure 8.4. The horn fly (a biting muscoid fly)—a significant pest of pastured cattle worldwide. Photo courtesy of D. W. Watson, North Carolina State University.

Range Cattle Integrated Pest Management

American cattle producers have not faced severe pest-control problems since the screwworm fly was eradicated from North America (Graham and Hourrigan 1977). This well-recognized eradication program relied on the inundative release of sterile male (SM) screwworm flies. Sterile flies successfully mated with wild female flies, and the result was a precipitous decline in offspring. Screwworm fly populations currently are limited to isolated regions of Central and South America. Research efforts are concentrated in Panama and include remote sensing to establish ecological habitats, population dynamics, and behavior (Kunz et al. 1999).

Important external pests of range cattle in the United States include the biting flies (horn flies, mosquitoes, gnats, horse flies, stable flies, and black flies), face flies, cattle grubs, lice, mites, and ticks. These pests cost U.S. livestock producers an estimated $3 billion annually (Kunz et al. 1999).

The horn fly (*Haematobia irritans*) (Figure 8.4) is a serious problem: estimated losses due to this pest exceed $800 million annually and result from decreased feed efficiency, weight gain, and milk production (Kunz et al. 1999). After taking frequent blood meals from the host, female horn flies deposit their eggs in dung pats, where the larvae develop (Bruce 1964).

At first, horn flies were fairly easy to control with the introduction of pyrethroid-impregnated ear tags, which were a convenient, inexpensive, long-lasting method, but widespread use of the ear tag placed heavy selection pressure on the horn fly population, which in turn caused resistance to the pyrethroids (Kunz et al. 1999).

Several resistance management strategies have been suggested for the horn fly. Among those strategies that have been investigated are insect growth regulators and ivermectin as a feed through (insecticide feed additives are incorporated into the feed, pass through the animal, and prevent insect larval development in the manure) or a bolus (insecticide is formulated into a large pill or tablet that, when administered to an animal, slowly releases insecticide into the digesting feed, preventing insect development in the manure). Concern for their effects on nontarget species, and difficulty of administration to animals are factors that have limited acceptance of these methods (Floate 1998; Kunz et al. 1999). Resistance management also may include alternating active ingredients to delay the onset of pyrethroid resistance and to allow pyrethroid-susceptible populations to enter the gene pool. Rotational tagging requires producers to change active ingredients every 1 or 2 years to delay resistance onset. Ear tags were used by 39.5% of the Texas beef cattle producers, and organophosphate (OP) ear tags were used predominantly in a Kansas pesticide use survey (Cress et al. 1995; Hall, Holloway, and Hoelscher 1994). Of the 20 types of ear tags now on the market, half are OP tags. Resistance to

OP tags is developing in horn fly populations in certain regions of the country, but resistance onset may be postponed by means of proper management and tag rotation (Barros et al. 2001).

The mechanisms of diapause have been investigated and simulation models developed (Kunz et al. 1999; Lysyk 1999). It is thought that the manipulation of *diapausing* (overwintering as pupae) strains and nondiapausing strains may have management potential by decreasing overwintering hardiness.

The potential for biological control of the horn fly has not been evaluated fully. For example, the microbial ecology of the horn fly may provide a means of control. The presence of endophytes in forage grasses may have an effect on horn fly survival (Dougherty et al. 1998). Dung pat and midgut microflora influence the growth, development, and survival of the horn fly larva (Perotti et al. 2000).

Similarly, competition and predation by other arthropods within the dung pat are suspected to affect horn fly survival (Kunz et al. 1999). For example, dung beetles have great potential for inclusion as a biological control component of IPM programs. Dung beetles destroy the dung pat, thus decreasing the survival of horn fly larvae. The imported dung beetle *Onthophagus gazella* has been established in Texas, California, and Georgia (Kunz et al. 1999). *Onthophagus taurus* has become established throughout much of the South, from North Carolina to California (Fincher 1990). To date, the effect of these beetles on horn fly populations is unknown. Surprisingly, no parasites or predators specific to horn flies have been identified.

As with the screwworm project, the SM release technique was demonstrated to affect horn fly reproduction (Kunz et al. 1999). The combined use of an insect growth regulator, methoprene, and released sterile flies eliminated horn flies from a portion of an island. As in the screwworm project, the released SM flies mated with the native population, and infertile eggs resulted.

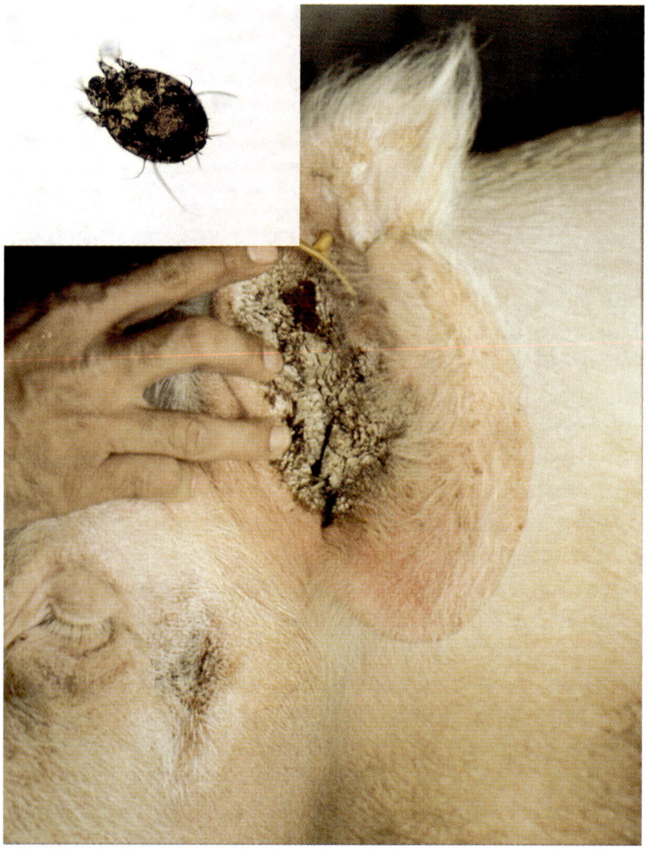

Figure 8.5. Sarcoptic mange mite infestations adversely affect the performance of hogs and other livestock. Photo courtesy of J. J. Arends, JABB, Inc.

Swine Integrated Pest Management

Arthropods of importance to swine production include the itch mite, the hog louse, the house fly, the stable fly, and the cockroach (Watson 2002). Management of these pests decreases disease risk and production losses resulting from poor growth and development and inefficient feed conversion (Holscher et al. 1999).

Mange is an inflammatory skin condition caused by the itch mite *Sarcoptes scabiei*, an obligate parasite that tunnels in the epidermis of pigs (Figure 8.5). The hog louse, *Haematopinus suis*, is an obligate blood-feeding parasite. Infested animals scratch vigorously and rub against objects for relief. Mites and lice can be found on pigs throughout the year but are most noticeable during the winter (Davis and Moon 1990). The primary means of transmission is direct contact with infested pigs (Davis and Moon 1990).

Strict biosecurity, through restricted movement of people and equipment and through rodent management, prevents the establishment of itch mites and hog lice (Holscher et al. 1999). Isolation and prophylactic treatment of all new stock ensures the introduction of pest-free animals into the herd. Injectable endectocides and pour-on or spray formulations of parasiticides are very effective in the management of itch mites and lice on swine (Arends, Skogerboe, and Ritzhaupt 1999). Biological control, host resistance, and insect growth regulators are unavailable for the control of mites and lice on swine.

The house fly and the stable fly are relatively minor pests in liquid-waste management systems. Largely unsuitable for fly development, lagoons provide relatively inexpensive, simple swine-waste treatments that decrease nutrient content and biological activity before land application (Rutz and Axtell 1978). Fly management in solid-waste systems is difficult, especially when pigs are kept on straw bedding. Manure and urine mixed with bedding promote fly growth and development. In solid-waste systems, IPM programs such as those in dairies and feedlots provide satisfactory control of house and stable flies.

The cockroach has become an increasingly important pest in swine production in recent years (Waldvogel et al. 1999). Occasionally, predators or parasitoids attack the Oriental cockroach, *Blatta orientalis*, and the German cockroach, *Blattella germanica* (Figure 8.6). Nondiscriminating cockroaches feed on various foods, including animal feeds and feces. The economic importance of cockroach infestations in swine production remains to be documented. The cockroach population has the potential to expand from hundreds to thousands of individuals, thereby posing problems to both workers and swine (Waldvogel et al. 1999). German and Oriental cockroaches are recognized as possible disease vectors, and hypersensitivity to cockroach allergens is quite common among humans (Brenner, Koehler, and Patterson 1987; Kopanic, Sheldon, and Wright 1994).

Figure 8.6. The German cockroach—an emergent structural pest in confinement swine production. Photo courtesy of M. Stringham, North Carolina State University.

Although biosecurity may be in force, cockroaches undermine it by moving readily between barns and nurseries (Waldvogel et al. 1999). Swine suffer from numerous mechanically transmitted disease agents, including mycotoxins, bacteria, and viruses (Taylor 1995). Porcine reproductive and respiratory syndrome, gastroenteritis, and endoparasites severely affect the young growing pig (Waldvogel et al. 1999).

Control and management of Oriental and German cockroaches can be accomplished through the application of conventional insecticides as residual surface sprays and baits (Wickham 1995). Pyrethroid and OP insecticides are used commonly in swine production. Although insecticide resistance has developed in the urban setting, it has not been documented in swine production. Inorganic insecticides such as boric acids and diatomaceous earth also have been used to decrease pesticide exposures (Cochran 1995; Kaakeh and Bennett 1997).

The entomopathogenic fungus *Metarhizium anisopliae* has been used to help manage cockroaches. Parasitoids targeting the oothecae have been cultured and released to help decrease cockroach populations (Pawson and Gold 1993). Parasitoids are least effective against the German cockroach, which carries the ootheca for several weeks.

Poultry

Fly Management

Muscoid flies (*Musca domestica* and *Fannia* spp.) are the primary pests of dry-waste systems for layer, broiler breeder, and other types of production systems in which manure is allowed to accumulate beneath cages or slats. Other species often associated with poultry house manure are dump flies (*Hydrotaea* = *Ophyra aenescens* and *H. ignava* = *leucostoma*), soldier flies (*Hermetia illucens*), fruit flies (*Drosophila repleta*), and, on rare occasions, the stable fly (*Stomoxys calcitrans*) (Arends and Stringham 1992). Blowflies also may be present where broken eggs accumulate or daily mortality is handled improperly. On the whole, however, none of these species is as pestiferous in the poultry house environment as house flies (*M. domestica*) or lesser house flies (*Fannia* spp.). Efforts to manage house flies invariably decrease the incidence of associated fly species.

Monitoring is encouraged so that changes in the manure environment or house fly population can be detected. Decisions about timing and targeting of pesticide treatments and other interventions depend on consistent monitoring. Manure condition, "drinkers" (water 'fountains'), and larval activity are monitored visually at regular intervals to detect wet spots and leaks and to discover concentrations of fly larvae. Baited traps, spot cards, and sticky cards or ribbons are used to monitor fly populations (Hogsette, Jacobs, and Miller 1993; Lysyk and Axtell 1986). Action thresholds, namely, fly or speck counts that warrant intervention, depend on the needs of the specific site.

The principle objectives of fly IPM in poultry houses are to decrease manure moisture and to effect biological control. Drinker management, ventilation, sanitation, site drainage, building maintenance, dietary adjustments, and other cultural practices that minimize the introduction of excess moisture and encourage drying are essential. Manure moisture below 50% inhibits fly breeding (Fatchurochim, Geden, and Axtell 1989; Mullens, Hinkle, and Szijj 1996) and promotes colonization by a variety of beneficial arthropods by improving their ability to find prey (Axtell 1999; Legner 1971). Parasitoids (*Muscidifurax* spp. and *Spalangia* spp.), predaceous beetles (*Carcinops pumilio*) (Figure 8.7), and predatory mites (*Macrocheles muscaedomesticae*) (Figure 8.8) are the principal beneficial arthropods present, and they play major roles in the suppression of house fly populations in places where manure remains moderately dry (Geden, Stinner, and Axtell 1998; Wills and Mullens 1991). Parasitoids are available commercially and may be released in poultry houses to augment natural populations (Axtell 1999; Hunter 2002; Rutz and Patterson 1990). Releases are not always successful, however. Gaps in our knowledge of host-parasite interactions in various manure and production environments, and the need to improve overall quality of mass-reared parasites, are two factors hampering efforts to use parasitoids reliably in house fly control (Geden, Smith, and Rutz 1992; Geden et al. 1999; Kaufman et al. 2000).

Figure 8.7. The hister beetle (*Carcinops pumilio*)—a valuable predator of muscoid flies in poultry houses. Photo courtesy of R. C. Axtell, North Carolina State University.

Figure 8.8. Macrochelids (and other predaceous mites) feed on fly eggs deposited in the manure of poultry houses. Photo courtesy of R. C. Axtell, North Carolina State University.

Sticky ribbons, baited traps, and electrical grids can be used when fly populations are low to moderate. Their effectiveness is limited, however, by the need for continual maintenance or replacement—a time-consuming task considering that multiple devices are needed for each poultry house to maintain adequate fly control. The high cost of devices such as the specialized electrical grids ($400 to $600 each) tested for use in caged layer houses is another limiting factor that must be considered (Rutz and Scoles 1988).

Decreased reliance on pesticides is a goal of fly IPM in poultry production. Fewer pesticides minimize environmental and human risks and allow for more effective management of pesticide resistance. Another, less obvious, reason—that of managing pesticide use—is equally important. Although pesticides continue to play a role in managing house fly populations, when applied inappropriately they often contribute to the severity and frequency of fly outbreaks (Axtell and Arends 1990; Farrell 2001; Mandeville, Mullens, and Yu 1990). Most pesticides registered for use in poultry production are broad-spectrum materials and must be used with care to preserve the community of beneficial arthropods found in manure. Understandably, to prevent such an occurrence, IPM programs discourage broadcast treatments of manure. The insect growth regulator cyromazine is the only exception and may be applied to manure as a larvicide with little effect on beneficial parasites and predators (Axtell and Edwards 1983; Wills, Mullens, and Mandeville 1990). Its use by the poultry industry is restricted to layer, broiler breeder, and caged pullet production. The use of this important fly management tool has not been trouble free, however. Cyromazine resistance, presumably due to overreliance or application of sublethal concentrations, via feed through or spray applications, is known to occur (Popischil et al. 1996; Scott et al. 2000; Shen and Plapp 1990; Sheppard et al. 1989). Pyriproxifen, an insect growth regulator labeled for general use in poultry, may be a fly-management option (Zhang and Shono 1997). It is being promoted as having minimal effects on beneficial arthropods in poultry houses, but additional field research is needed to evaluate that claim thoroughly.

The future of fly IPM in poultry operations likely will include several options in addition to those discussed already. The fungal pathogens (Figure 8.9) *Entomophthora muscae*, *Beauveria bassiana*, and *Metarhizium anisopliae* and the bacterium *Bacillus thuringiensis* show promise as fly management tools, but much work remains to be done before they can be deployed successfully (Geden, Rutz, and Steinkraus 1995; Mullens, Rodriguez, and Meyer 1987; Watson and Peterson 1993; Watson et al. 1995). Interestingly, two of the previously mentioned fly species also may be useful control agents. Larvae of the dump fly (*Hydrotaea* spp.) are predators of other fly species' larvae. Refinements in ways to establish and to encourage *Hydrotaea* populations may provide a way of excluding houseflies (Hogsette and Washington 1995; Turner and Carter 1990). The black soldier fly (*Hermetia illucens*), whose larvae are exceptionally large and active, may provide yet another way to decrease house fly populations significantly in poultry houses (Sheppard 1983; Sheppard and Tomberlin 1999). Great numbers of soldier fly larvae churn manure into a more fluid state than is acceptable to ovipositing house flies. Additionally, the larvae have high nutritional value and can be harvested easily for use in animal feeds.

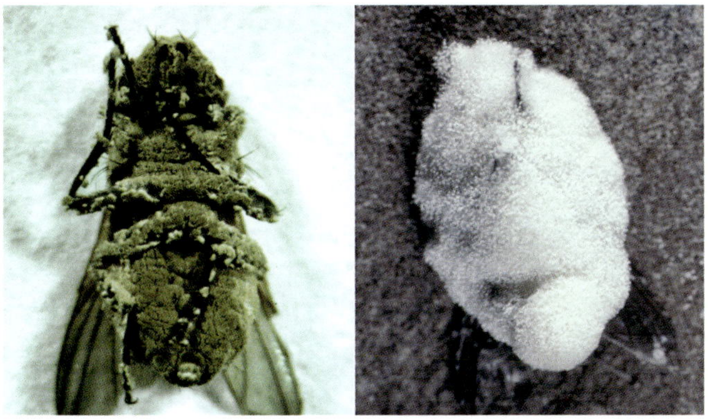

Figure 8.9. Fungal pathogens such as *Metarhizium anisopliae* (left) and *Beauveria bassiana* (right) hold promise as biocontrol agents for house flies and other manure-inhabiting pests of poultry and livestock. Photo courtesy of D. W. Watson, North Carolina State University.

Beetle Management

The lesser mealworm (*Alphitobius diaperinus*) (Figure 8.10), the hide beetle (*Dermestes maculatus*), and the larder beetle (*D. lardarius*) are the primary beetle pests of poultry (Axtell 1999). The lesser mealworm is known to harbor a number of pathogens and parasites affecting birds (Arends 1997; Despins et al. 1994; Watson, Guy, and Stringham 2000). Its potential to become a nuisance in infested litter spread on land near residences (Hinchey 1997) and in the mechanical transmission of pathogens such as *Escherichia coli* and *Salmonella* serovars to humans also is a concern (Jones, F. T. et al. 1991; McAllister, Steelman, and Skeeles 1994; McAllister et al. 1996). It has been suggested that lesser mealworms are an effective biological control for manure-breeding flies (Wallace, Winks, and Voestermans 1985). The adults and larvae may help dry manure with their tunneling activity, and adults are minor predators of fly larvae, but the negative aspects of this pest limits its poten-

Figure 8.10. The lesser mealworm—the predominant beetle pest of poultry production. Photo courtesy of D. W. Watson, North Carolina State University.

tial for biological control.

Both dermestid beetles and lesser mealworms are predominantly structural pests in most types of poultry housing. Lesser mealworms are most abundant in deep-litter houses such as those used to grow turkeys and broilers and in high-rise poultry houses (Pfieffer and Axtell 1980; Rueda and Axtell 1997). Dermestid beetles are less numerous in general but may become serious pests in high-rise layer houses (Axtell 1999). Lesser mealworm populations can be huge, often covering more than 70% of the manure surface in high-rise layer houses or visibly churning the litter throughout an infested broiler or turkey house (Geden et al. 1999). Lesser mealworm larvae and adults are noted for insulation damage caused by their tunneling. Dermestid beetles and larvae also tunnel into insulation, but are better known for the damage they cause structural timbers in high-rise layer houses.

Management options for the control of these beetle pests are limited and restricted mainly to the use of pesticides (Arends and Stringham 1992; Weaver 1996). Monitoring (visual or with sampling devices) is possible but impractical in many situations (Safrit and Axtell 1984; Stafford et al. 1988). Cultural practices are for the most part designed to augment pesticide use. Cold weather, especially freezing, is sometimes an effective method of control. Between flocks, poultry houses can be opened and the beetle population exposed to subfreezing temperatures for at least a week during the winter (Arends and Stringham 1992). Where litter is left in place, it may be tilled to dissipate heat and to bring beetle life stages to the surface, where many will freeze. Poultry growers in locations where prolonged freezing temperatures are less predictable must rely on alternative cultural practices to decrease beetle populations. Frequent manure removal is one cultural practice that helps decrease beetle populations in caged layer houses. Similarly, total removal of litter from broiler and turkey houses after each flock helps suppress lesser mealworm numbers by physically removing them from the poultry house (Geden et al. 1999). The downside of these practices is that fresh litter may not be available and old litter must be reused. The cost of equipment for more-frequent manure removal or of automated systems also may be prohibitive for many producers. Recently, barriers against beetle movement have been used successfully in high-rise layer houses (Geden and Carlson 2001). Although beetles are not controlled, barriers prevent damage to insulation and to wood above the manure collection area.

Pesticide treatments to suppress lesser mealworm populations are sprayed on the litter after each flock in broiler and turkey production (Weaver 1996). The soil surface also may be treated if old litter is cleaned out completely after the removal of birds. Such prescription treatments sometimes decrease beetle populations sufficiently over the course of several production cycles to eliminate the need for treatment between subsequent flocks. Unfortunately, beetle infestations often rebound after two or more flocks once treatments are halted. Recycling litter for multiple flocks in broiler production makes beetle control more difficult as well. Although intended to decrease production costs, this practice eliminates a useful cultural practice (the physical removal of beetles with the litter) from the beetle management equation. Pesticides are used in a similar manner to decrease lesser mealworm and dermestid beetle infestations in breeder and caged-layer production; a single application can be applied to bare soil before a new flock is placed. Because flock cycles for these types of poultry often exceed 12 months, treatments for beetle control sometimes are applied to the manure beneath slats or cages while birds are present. This practice seldom is recommended, however. In addition to the difficulty of application, the broad-spectrum insecticides used often decimate the beneficial arthropods that play an important role in fly suppression (Axtell 1999).

Prospects for advances in beetle management are limited. Natural enemies of the lesser meal-worm exist, but only a few have been investigated for their potential to aid in management. The fungus *Beauveria bassiana* shows promise as a biocontrol agent, and the development of *Bacillus thuringiensis* for beetle control also may be possible (Axtell 1999; Steinkraus, Geden, and Rutz 1991). The protozoan *Gregarina alphitobii* may predispose lesser mealworms to infection with a second, more virulent, protozoan, *Farinocystis tribolii*, or the fungus *B. bassiana* (Apuya et al. 1994; Steinkraus, Brooks, and Geden 1992). Still, it likely will be some time before a commercial product is available.

Ectoparasite Management

In general, ectoparasite infestations are restricted to layer and breeder flocks remaining in production for 12 months or longer. Broilers and market turkeys seldom are affected because they are sold to market long before ectoparasite populations become well established (Axtell 1999). Northern fowl mites (*Ornithonyssus sylviarum*) (Figure 8.11), chicken mites (*Dermanyssus gallinae*), bedbugs (*Cimex lectularius*), and chicken lice (mainly *Menacanthus* spp.)

Figure 8.11. The northern fowl mite—the most common ectoparasite of layer and breeder birds. Photo courtesy of D. W. Watson, North Carolina State University.

are the ectoparasites most often reported in the poultry industry (Arends and Stringham 1992). Of these, the northern fowl mite occurs most frequently (Hogsette et al. 1991). Most of the common ectoparasites of poultry are blood feeders believed to diminish bird performance although economic studies (primarily of the northern fowl mite) have produced conflicting results in this regard (Arends, Robertson, and Payne 1984; Loomis et al. 1970). Poultry mites and bedbugs also have been reported to cause transitory skin irritations and other allergic reactions in poultry workers (Schofield and Dolling 1993; Varma 1993). Three additional ectoparasites of poultry—turkey chiggers (*Neoschoengastia americana*), sticktight fleas (*Echidnophaga gallinacea*), and fowl ticks (*Argas* spp.)—commonly reported in the review literature (Arends 1997) are no longer important pests of modern commercial production because of changes in standard flock-management practices. There may be a resurgence of these ectoparasites, however, as the practice of free-range production becomes more prevalent in response to the demands of a growing organic market.

Prevention is the most effective pest management tool for the control of ectoparasite infestations. Wild birds are a common source of ectoparasites in poultry flocks (Arends 1997; Kells and Surgeoner 1997). The shift to closed, environmentally controlled housing in layer production has done much to decrease the incidence of ectoparasite infestation associated with wild birds over the years (Geden et al. 1999). Building design and exclusion practices, such as closure of entry points with screening or other material, properly fitted entry doors, and construction techniques that protect exposed edges of insulation beneath eaves and at building corners will help keep wild birds out of poultry houses. Strict biosecurity decreases the likelihood of moving an identified ectoparasite problem between poultry houses or between poultry farms and helps prevent the spread of infestations to workers' homes. Coverall changes for workers as they move between poultry houses and farms is an important first step toward stopping the spread of infestation. Sanitizing egg flats and other transportable poultry- or egg-handling equipment, sanitizing transport vehicles, and routing service vehicles from clean to infested farms are other practices used to decrease or to eliminate the spread of ectoparasites (Arends and Stringham 1992; Axtell and Arends 1990). The elimination of ectoparasite infestations in pullet flocks before they are moved to layer houses is necessary to contain outbreaks (Geden et al. 1999). Similar cultural practices are used in pullet production to prevent the introduction of infested replacement birds into production flocks and to ensure that new flocks have no pre-existing infestations.

The detection of different ectoparasites requires different monitoring strategies. Bedbugs and chicken mites are found feeding on host birds only at night. Although skin lesions caused by the bites of chicken mites and bedbugs may be seen on individual birds, careful inspection of the premises is the most practical approach to detecting these pests (Arends and Stringham 1992). Northern fowl mites and chicken lice are obligate parasites that live on their hosts. Monitoring relies on the capture of individual birds and their examination for the presence of mites or lice. Bird and premise inspections are suited best to smaller production flocks in which daily management duties can be adjusted to accommodate monitoring efforts. Larger flocks often require that a minimum of 50 birds be examined (Arends and Stringham 1992)—a time-consuming task that often is neglected when a single employee is responsible for overseeing the operation of an automated layer house containing approximately 100,000 laying hens (Geden et al. 1999). Alternative approaches to ectoparasite monitoring have been evaluated and may prove more satisfactory once developed fully (Mullens, Hinkle, and Szijj 2000).

Pesticides remain the most effective method of control for established ectoparasite infestations. Carbamates, OPs, and pyrethroids are generally effective for direct application to birds and as premise treatments, but the total number of active ingredients available for use is limited (Fletcher and Axtell 1991, 1993). Permethrin-impregnated strips have been used to control northern fowl mites in layers and breeders with a degree of success (Axtell 1991; Hall et al. 1984). Pyrethroid resistance in chicken mites raises questions, however, about the continued efficacy of pyrethroids in general and the wisdom of using slow-release devices in particular (Beugnet et al. 1997). Light oils and insecticidal soaps have been shown to control poultry mites and may prove useful as alternatives to traditional pesticides if concerns about associated air sac infections in birds are addressed (Geden et al. 1999; McKeen, Loomis, and Dunning 1983; Schering-Plough 1995).

Only when applied correctly do sprays and dusts provide satisfactory control as premise treatments (Arends and Stringham 1992). Litter and manure must be removed and houses thoroughly cleaned before such treatments are applied to control chicken mites, bedbugs, and lice. Because chicken mites and bedbugs can live a considerable period without feeding and are likely to remain hidden in protected locations throughout the house, multiple treatments are necessary to eliminate infestations before birds are reintroduced (Stringham 1994). Spray or dust treatments of birds present additional problems. To be effective, treatment must penetrate the feathers to reach ectoparasites on the skin (Arther and Axtell 1982). Additionally, because the residual life of registered pesticides applied to birds is shorter than the life cycle of most ectoparasites, a second application often is required for acceptable control.

The use of biological control for ectoparasites is largely undeveloped, but future research may lead to effective biopesticides. Three fungi and a single bacterium have been shown to exhibit activity against mites and other ectoparasites. *Trenomyces histophtorus* is a fungus originally isolated from chicken lice that has potential for use (Meola and DeVaney 1976). Mycotoxins produced by the fungi *Beauveria bassiana* and *Metarhizium anisopliae* have insecticidal activity that may prove useful in the control of mites and other poultry pests (Grove and Pople 1980). The bacterium *Bacillus thuringiensis* was shown to be toxic to at least one species of sheep louse and may prove equally effective against the chicken-biting louse and the northern fowl mite (Drummond, Miller, and Pinnock 1992; Mullens et al. 1988).

Future Needs for Food Animal Integrated Pest Management

Waste management continues to be a serious concern for livestock and poultry producers. Essential to the survival of the industry is the development of alternative manure handling systems that alleviate environmental concerns, decrease manure solids, recycle nutrients, and yet provide value-added products such as feed supplements, compost, or vermiculture substrate (Lo, Lau, and Liao 1993; Mikkelsen 1999; Sheppard et al. 1994).

Basic studies of dispersal, population dynamics, and behavior under diverse production or waste-handling systems are lacking for flies, beetles, and other pests of food animals. Crucial to developing IPM strategies is an understanding of the biology and the ecology of naturally occurring biological control agents (Axtell 1999; Rutz and Watson 1998). Nuisance and public health hazards, including cockroaches and flies involved in the transmission of foodborne and antibiotic-resistant pathogens, must be studied. More practical and accurate ways of monitoring ectoparasites and other common pests are needed to improve the timing and effectiveness of IPM interventions. Alternatives for conventional insecticide applications are needed, as are safe, effective, environmentally compatible IPM strategies.

Appendix A. Abbreviations and Acronyms

IPM	integrated pest management
OP	organophosphate
SM	sterile male

Appendix B. Glossary

Diapausing. Overwintering as pupae.

Mange. Mange is an inflammatory skin condition caused by the itch mite, *Sarcoptes scabiei*, an obligate parasite that tunnels in the epidermis of pigs

Literature Cited

Apuya, L. C., S. M. Stringham, J. J. Arends, and W. M. Brooks. 1994. Prevalence of protozoan infections in darkling beetles from poultry houses in North Carolina. *J Invertebr Pathol* 63:255–259.

Arends, J. J. 1997. External parasites and poultry pests. Pp.785–813. In B. W. Calnek (ed.). *Diseases of Poultry*. 9th ed. Iowa State University Press, Ames.

Arends, J. J. and S. M. Stringham. 1992. *Poultry Pest Management*. Publication AG–474. Cooperative Extension Service, North Carolina State University, Raleigh. 30 pp.

Arends, J. J., S. H. Robertson, and C. S. Payne. 1984. Impact of northern fowl mite on broiler breeder flocks in North Carolina. *Poult Sci* 63:1457–1461.

Arends, J. J., T. L. Skogerboe, and L. K. Ritzhaupt. 1999. Persistent efficacy of doramectin and ivermectin against experimental infestations of *Sarcoptes scabiei* var. *suis* in swine. *Vet Parasitol* 82:71–79.

Arther, F. H. and R. C. Axtell. 1982. Comparisons of permethrin formulations and application methods for northern fowl mite control on caged laying hens. *Poult Sci* 61:4879–4884.

Axtell, R. C. 1991. Control of northern fowl mites in poultry. Pp. 639–646. In F. Dusbabek and V. Bukva (eds.). *Modern Acarology: Proceedings of the VIII International Congress of Acarology*, Ceské Budekpvoce, Czechoslovakia, August 6–11, 1990.

Axtell, R. C. 1999. Poultry integrated pest management: Status and future. *IPM Rev* 4:53–73.

Axtell, R. C. and J. J. Arends. 1990. Ecology and management of arthropod pests of poultry. *Annu Rev Entomol* 35:101–126.

Axtell, R. C. and T. D. Edwards. 1983. Efficacy and non-target effects of Larvadex®, as a feed additive for controlling house flies in caged-layer poultry manure. *Poult Sci* 62:2371–2377.

Barros, A. T. M., J. Ottea, D. Sanson, and L. Foil. 2001. Horn fly (Diptera: Muscidae) resistance to organophosphate insecticides. *Vet Parasitol* 2021:1–14.

Beugnet, F., C. Chauve, M. Gauthey, and L. Beert. 1997. Resistance of the red poultry mite to pyrethroids in France. *Vet Rec* 140:577–579.

Brenner, R. J., P. G. Koehler, and R. S. Patterson. 1987. Health implications of cockroach infestations. *Infect Med* 4: 349–360.

Bruce, W. G. 1964. The history and biology of the horn fly, *Haematobia irritans* (Linnaeus): With comments on control. Technical Bulletin Number 157. Agricultural Experiment Station and Entomology Research Division, Agricultural Research Service, U.S. Dept. of Agriculture, Raleigh, North Carolina. 32 pp.

Campbell, J. B., I. L. Berry, J. Chirico, B. C. Clymer, G. L. Greene, R. D. Hall, S. R. Skoda, G. D. Thomas, R. A. Weinzierl, and J. B. Welch. 1999. Confined cattle summary. Pp. 120–126. In J. A. Hogsette and C. J. Geden (eds.). *Research and Extension Needs for Integrated Pest Management for Arthropods of Veterinary Importance*. Proceedings of a workshop in Lincoln, Nebraska, April 12–14, 1994. Center for Medical, Agricultural, and Veterinary Entomology, Agricultural Research Service, U.S. Department of Agriculture, Gainesville, Florida.

Cochran, D. G. 1995. Toxic effects of boric acid on the German cockroach. *Experientia* 51:561–563.

Cress, D. C, D. E. Mock, G. L. Greene, T. J. Byram, E. J. Theissen, C. Barnes, and B. Schmitz. 1995. 1994 Cattle Pesticide Summary. Kansas agricultural chemical usage. Cooperative Extension Service, Kansas State University, Manhattan. 54 pp.

Davis, D. P. and R. D. Moon. 1990. Dynamics of swine mange: A critical review. *J Med Entomol* 27:727–737.

Despins, J. L., R. C. Axtell, D. A. Rives, J. S. Guy, and M. D. Ficken. 1994. Transmission of enteric pathogens of turkeys by darkling beetle larva (*Alphitobius diaperinus*). *J Appl Poult Res* 3:1–5.

Dougherty, C. T., F. W. Knapp, L. P. Bush, J. E. Maul, and J. Van Willigen. 1998. Mortality of horn fly (Diptera: Muscidae) larvae in bovine dung supplemented with Loline alkaloids from tall fescue. *J Med Entomol* 35:798–803.

Drummond, R. O., K. Miller, and D. E. Pinnock. 1992. Toxicity of *Bacillus thuringiensis* against *Damalinia ovis* (Phthiraptera: Mallophaga). *J Invert Pathol* 60:102–103.

Farrell, J. 2001. "Complaints about farm fester." *St. Petersburg Times*, May 19, pg. 1.

Fatchurochim, S., C. J. Geden, and R. C. Axtell. 1989. Filth fly (Diptera) oviposition and larval development in poultry manure of various moisture levels. *J Entomol Sci* 24:224–231.

Fincher, G. T. 1990. Biological control of dung-breeding flies: Pests of pastured cattle in the United States. Pp. 137–151. In D. A. Rutz and R. S. Patterson (eds.). *Biocontrol of Arthropods Affecting Livestock and Poultry*. Westview Press, Boulder, Colorado. 316 pp.

Fletcher, M. G. and R. C. Axtell. 1991. Susceptibilities of northern fowl mite, *Ornithonyssus sylviarum* (Acarina: Macronyssidae), and chicken mite, *Dermanyssus gallinae* (Acarina: Dermanyssidae), to selected acaricides. *Exp & Appl Acarol* 13:137–142.

Fletcher, M. G. and R. C. Axtell. 1993. Susceptibility of the bedbug, *Cimex lectularius* L., to selected insecticides and various treated surfaces. *Med Vet Entomol* 7:69–72.

Floate, K D. 1998. Off-target effects of ivermectin on insects and on dung degradation in southern Alberta, Canada. *Bull Entomol Res* 88:25–35.

Geden, C. J. and D. A. Carlson. 2001. A mechanical barrier for preventing climbing by lesser mealworm (Coleoptera: Tenebrionidae) and hide beetle (Coleoptera: Dermestidae) larvae in poultry houses. *J Econ Entomol* 94:1610–1617.

Geden, C. J., R. E. Stinner, and R. C. Axtell. 1988. Predation by predators of the house fly in poultry manure: Effects of predator density, feeding history, interspecific interference, and field conditions. *Environ Entomol* 17:320–329.

Geden, C. J., L. Smith, and D. A. Rutz. 1992. Rapid deterioration of searching behavior, host destruction, and fecundity of the parasitoid *Muscidifurax raptor* (Hymenoptera: Pteromalidae) in culture. *Ann Entomol Soc Am* 85:179–187.

Geden, C. J., D. A. Rutz, and D. C. Steinkraus. 1995. Virulence of different isolates and formulations of *Beauveria bassiana* for house flies and the parasitoid *Muscidifurax raptor*. *Biol Control* 5:615–621.

Geden, C. J., J. J. Arends, R. C. Axtell, D. R. Barnard, D. M. Gaydon, L. A. Hickle, J. A. Hogsette, W. F. Jones, B. A. Mullens, M. P. Nolan, Jr., M. P. Nolan, III, J. J. Petersen, and D. C. Sheppard. Poultry summary. 1999. Pp. 2–22. In J. A. Hogsette and C. J. Geden (eds.). *Proceedings of the Research and Extension Needs for Integrated Pest Management for Arthropods of Veterinary Importance Workshop*. Proceedings of a workshop in Lincoln, Nebraska, April 12–14, 1994. Center for Medical, Agricultural, and Veterinary Entomology, Agricultural Research Service, U.S. Department of Agriculture, Gainesville, Florida.

Graham, O. H. and J. L. Hourrigan. 1977. Eradication programs for the arthropod parasites of livestock. *J Med Entomol* 13:629–658.

Greene, G. L., Y. J. Guo, and H. Y. Chen. 1998. Parasitization of house fly pupae (Diptera: Muscidae) by *Spalangia nigroaenea* (Hymenoptera: Pteromalidae) in cattle feedlot environments. *Biol Control* 12:7–13.

Grove, J. F. and M. Pople. 1980. The insecticidal activity of beauveracin and enniatin complex. *Mycopathologia* 70:103–105.

Hall, K. D., R. L. Holloway, and C. E. Hoelscher. 1994. Texas Beef Cattle Pest Control Practices. Agricultural and Environmental Safety, Texas Agricultural Extension Service, Texas A & M, College Station. 80 pp.

Hall, R. D., J. M. Vandepopuliere, W. Jaynes, and F. J. Fischer. 1984. A new in-cage treatment system for control of northern fowl mites on laying hens. *Poult Sci* 63:628–632.

Hinchey, F. 1997. "Bugged residents sue over beetles." *The Columbus Dispatch*, October 2, Pp. 1–2.

Hogsette, J. A. and J. P. Ruff. 1985. Stable fly (Diptera: Muscidae) migration in northwest Florida. *Environ Entomol* 14:170–175.

Hogsette, J. A. and F. Washington. 1995. Quantitative mass production of *Hydrotaea aenescens* (Diptera: Muscidae). *J Econ Entomol* 88:1238–1242.

Hogsette, J. A., J. F. Butler, W. V. Miller, and R. D. Hall. 1991. Annotated bibliography of the northern fowl mite, *Ornithonyssus sylviarum* (Canestrini & Fanzago) (Acari: Macronyssidae). *Entomol Soc Am Misc Pub* 76:1–62.

Hogsette, J. A., R. D. Jacobs, and R. W. Miller. 1993. The sticky card: A device for studying the distribution of adult house fly (Diptera: Muscidae) populations in closed poultry houses. *J Econ Entomol* 86:450–454.

Holscher, K. H., R. E. Williams, C. E. Hoelscher, G. R. Strother, D. E. Bay, J. L. Riner, W. F. Lyon, R. D. Moon, and I. Hagsten. 1999. Swine IPM summary. Pp. 127–143. J. A. Hogsette and C. J. Geden (eds.). *Research and Extension Needs for Integrated Pest Management for Arthropods of Veterinary Importance*. Proceedings of a workshop in Lincoln, Nebraska, April 12–14, 1994. Center for Medical, Agricultural, and Veterinary Entomology, Agricultural Research Service, U.S. Department of Agriculture, Gainesville, Florida.

Hunter, C. D. 2002. Suppliers of beneficial organisms in North America, May 3, <http://www.cdpr.ca. gov/docs/dprdocs/goodbug/organism.htm> (5 August 2002)

Jones, C. J. and R. A. Weinzierl. 1997. Geographical and temporal variation in pteromalid (Hymenoptera: Pteromalidae) parasitism of stable fly and house fly (Diptera: Muscidae) pupae collected from cattle feedlots in Illinois. *Environ Entomol* 26:421–432.

Jones, C. J., J. A. Hogsette, R. S. Patterson, D. E. Milne, G. D. Propp, J. F. Milio, L. G. Rickard, and J. P. Ruff. 1991. Origin of stable flies (Diptera: Muscidae) on west Florida beaches: Electrophoretic analysis of dispersal. *J Med Entomol* 28:787–795.

Jones, F. T., R. C. Axtell, D. V. Rives, S. S. Scheideler, F. R.Traver, Jr., R. L. Walker, and M. J. Wineland. 1991. A survey of *Salmonella* contamination in modern broiler production. *J Food Protect* 54:502–507.

Kaakeh, W. and G. W. Bennett. 1997. Evaluation of trapping and vacuuming compared with low-impact insecticide tactics for managing German cockroaches in residences. *J Econ Entomol* 90:976–982.

Kaufman, P. E., S. J. Long, D. A. Rutz, and C. S. Glenister. 2000. Prey- and density-mediated dispersal in *Carcinops pumilio* (Coleoptera: Histeridae), a predator of house fly (Diptera: Muscidae) eggs and larvae. *J Med Entomol* 37:929–932.

Kells, S. A. and G. A. Surgeoner. 1997. Sources of northern fowl mite (*Ornithonyssus sylviarum*) infestation in Ontario egg production facilities. *J Appl Poult Res* 6:221–228.

Knapp, F. W., A. E. Charron, and J. G. Burg. 1992. Disease transmission by the stable fly. Pp. 25–38. In G. D. Thomas and S. R. Skoda (eds.). *The Stable Fly: A Pest of Humans and Domestic Animals*. MP 64. University of Nebraska, Lincoln. 148 pp.

Kopanic, R. J., B. W. Sheldon, and C. G. Wright. 1994. Cockroaches as vectors of *Salmonella*: laboratory and field trials. *J Food Prot* 57:125–132.

Krafsur, E. S. 1967. Allozyme variation in stable flies (Diptera: Muscidae). *Biochem Genet* 31:231–240.

Kunz S. E, A. B. Broce, J. F. Butler, R. L. Byford, W. F. Fisher, L. D. Foil, J. E. George, R. R. Gerhardt, F. R. Holbrook, H. G. Kinzer, J. E. Lloyd, L. Meek, R. J. Meyer, J. A. Miller, G. R. Mullen, C. D. Steelman, and P. D. Teel. 1999. Range cattle IPM summary. Pp. 70–120. In J. A. Hogsette and C. J. Geden (eds.). *Research and Extension Needs for Integrated Pest Management for Arthropods of Veterinary Importance*. Proceedings of a workshop in Lincoln, Nebraska, April 12–14, 1994. Center for Medical, Agricultural, and Veterinary Entomology, Agricultural Research Service, U.S. Department of Agriculture, Gainesville, Florida.

Legner, E. F. 1971. Some effects of the ambient arthropod complex on the density and potential parasitization of muscoid flies in poultry wastes. *J Econ Entomol* 64:111–115.

Lo, K. V., A. K. Lau, and P. H. Liao. 1993. Composting of separated solid swine wastes. *J Agric Eng Res* 54:307–317.

Loomis, E. E., E. L. Bramhall, J. A Allen, R. A. Ernst, and L. L. Dunning. 1970. Effects of the northern fowl mite on white Leghorn chickens. *J Econ Entomol* 63:1885–1889.

Lysyk, T. J. 1999. Effect of temperature on time to eclosion and spring emergence of postdiapausing horn flies (Diptera: Muscidae). *Environ Entomol* 28:387–397.

Lysyk, T. J. and R. C. Axtell. 1986. Comparison of baited jug-trap and spot cards for sampling house fly, *Musca domestica* (Diptera: Muscidae), populations in poultry houses. *Environ Entomol* 14:815–819.

Mandeville, J. D., B. A. Mullens, and D. S. Yu. 1990. Impact of selected pesticides on field population dynamics of parasitic Hymenoptera (Pteromalidae) in caged-layer poultry manure in southern California, U.S.A. *Med Vet Entomol* 4:261–268.

McAllister, J. C., C. D. Steelman, and J. K. Skeeles. 1994. Reservoir competence of the lesser mealworm (Coleoptera: Tenebrionidae) for *Salmonella typhimurium* (Eubacteriales: Enterobacteriaceae). *J Med Entomol* 31:369–372.

McAllister, J. C., C. D. Steelman, J. K. Skeeles, L. A. Newberry, and E. E. Gbur. 1996. Reservoir competence of *Alphitobius diaperinus* (Coleoptera: Tenebrionidae) for *Escherichia coli* (Eubacteriales: Enterobacteriaceae). *J Med Entomol* 33:983–987.

McKeen, W. D., E. C. Loomis, and L. L. Dunning. 1983. Effectiveness of Ectiban, egg oil, Rabon or Sevin for control of northern fowl mites on laying hens. *Poult Sci* 62:2343–2346.

Meola, S. and J. DeVaney. 1976. Parasitism of Mallophaga by *Trenomyces histophtorus*. *J Invert Pathol* 28:195–201.

Mikkelsen, R. L. 1999. Vermicomposting for the treatment of swine waste. Pp. 60–64. In G. B. Havenstein (ed.). *Animal Waste Management Symposium*. North Carolina State University, Raleigh.

Mullens, B. A., J. L. Rodriguez, and J. A. Meyer. 1987. An epizootiological study of *Entomophthora muscae* in muscoid fly populations on southern California poultry facilities, with emphasis on *Musca domestica*. *Hilgardia* 55:1–41.

Mullens, B. A., W. D. McKeen, J. L. Rodriquez, and J. D. Mandeville. 1988. Evaluation of ABG–6280 (thuringiensin) for control of northern fowl mite. *Insectic & Acaracide Tests* 13:405.

Mullens, B. A., N. C. Hinkle, and C. E. Szijj. 1996. Role of the poultry manure pad in manure drying and its potential relationship to filth fly control. *J Agric Entomol* 13:331–337.

Mullens, B. A., N. C. Hinkle, and C. E. Szijj. 2000. Monitoring northern fowl mites (Acari: Macronyssidae) in caged laying hens: Feasibility of an egg-based sampling system. *J Econ Entomol* 93:1045–1054.

Pawson, B. M. and R. E. Gold. 1993. Evaluating different release strategies for the control of American cockroaches using the egg parasitoid, *Aprostocetus hagenowii*. Pp. 407–413. In K. B. Wildey and W. H. Robinson (eds.). *Proceedings of the 1st Inter-*

national Conference on Insect Pests in the Urban Environment. ICIPUE, Milton Keynes, United Kingdom.

Perotti, M. A., T. J. Lysyk, L. D. Kalischuk-Tymensen, L. J. Yanke, and B. Selinger. 2000. Growth and survival of immature *Haematobia irritans* (Diptera: Muscidae) is influenced by bacteria isolated from cattle manure and conspecific larvae. *J Med Entomol* 38:180–187.

Petersen, J. J. and J. K. Cawthra. 1995. Release of a gregarious *Muscidifurax* species (Hymenoptera: Pteromalidae) for the control of filth flies associated with confined beef cattle. *Biol Control* 5:279–284.

Petersen, J. J. and J. A. Meyer. 1983. Characterization and seasonal distribution of breeding sites of stable flies and house flies (Diptera: Muscidae) on eastern Nebraska feedlots and dairies. *J Econ Entomol* 76:103–108.

Pfieffer, D. G. and R. C. Axtell. 1980. Coleoptera of poultry manure in caged-layer houses in North Carolina. *Environ Entomol* 9:21–28.

Popischil, R., K. Szomm, M. Londershausen, I. Schroeder, A. Tuber, and R. Fuchs. 1996. Multiple resistance in the larger house fly *Musca domestica* in Germany. *Pestic Sci* 48:333–341.

Rueda, L. M. and R. C. Axtell. 1997. Arthropods in litter of poultry (broiler chicken and turkey) houses. *J Agric Entomol* 14:81–91.

Rutz, D. A. and R. C. Axtell. 1978. Factors affecting production of the mosquito, *Culex quinquefasciatus*, from anaerobic animal waste lagoons. Technical Bulletin 256. Agricultural Experiment Station, North Carolina State University, Raleigh. 32 pp.

Rutz. D. A. and R. S. Patterson (eds.). 1990. *Biocontrol of Arthropods Affecting Livestock and Poultry*. Westview Press, Boulder, Colorado. 316 pp.

Rutz, D. A. and G. A. Scoles. 1988. Evaluation of fly-electrocuting black light devices in caged-layer poultry facilities. *Poult Sci* 67:871–877.

Rutz, D. A. and D. W. Watson. 1998. Parasitoids as a component in an integrated fly-management program on dairy farms. Pp. 185–201. In R. L. Ridgeway, M. Hoffman, C. S. Glenister, and M. Inscoe (eds.). *Mass-Reared Natural Enemies: Applications, Regulation, and Needs*. Entomological Society of America Monograph, Annapolis, Maryland.

Rutz, D. A., C. J. Geden, and C. W. Pitts. 1993. Pest management recommendations for dairy. Pennsylvania State and Cornell University Cooperative Extension Publication 10M493. Pennsylvania State University, University Park. 11 pp.

Rutz, D. A., T. Fincher, J. George, R. Gerhardt, C. J. Jones, T. Lysyk, R. W. Miller, C. Pitts, E. T. Schmidtman, and D. W. Watson. 1999. Dairy cattle IPM Summary. Pp. 21–70. In J. A. Hogsette and C. J. Geden (eds.). *Research and Extension Needs for Integrated Pest Management for Arthropods of Veterinary Importance*. Proceedings of a workshop in Lincoln, Nebraska, April 12–14, 1994. Center for Medical, Agricultural, and Veterinary Entomology, Agricultural Research Service, U.S. Department of Agriculture, Gainesville, Florida.

Safrit, R. D. and R. C. Axtell. 1984. Evaluation of sampling methods for darkling beetles (*Alphitobius diaperinus*) in litter of turkey and broiler houses. *Poult Sci* 63:2369–2375.

Schering-Plough Animal Health. 1995. *Accide™ Backgrounder Technical Bulletin*. American Scientific Laboratories, Union, New Jersey. 30 pp.

Schofield, C. J. and W. R. Dolling. 1993. Bedbugs and kissing-bugs (bloodsucking Hemiptera). Pp. 483–516. In R. P. Lane and R. W. Crosskey (eds.). *Medical Insects and Arachnids*. Chapman & Hall, Norwell, Massachusetts.

Scott, J. G., T. G. Alefantis, P. E. Kaufman, and D. A. Rutz. 2000. Insecticide resistance in house flies from caged-layer poultry facilities. *Pest Manage Sci* 56:1–7.

Shen, J. and F. W. Plapp, Jr. 1990. Cyromazine resistance in the house fly (Diptera: Muscidae): Genetics and cross-resistance to diflubenzuron. *J Econ Entomol* 83:1689–1697.

Sheppard, D. C. 1983. House fly and lesser house fly control utilizing the black soldier fly in manure management systems for caged laying hens. *Environ Entomol* 12:1239–1242.

Sheppard, D. C. and J. Tomberlin. 1999. Adverse effects on house fly larvae when reared with black soldier fly larvae. In *Proceedings of 43rd Annual Livestock Insect Workers Conference*, Chattanooga, Tennessee, June 27–30, 1999.

Sheppard, D. C., N. C. Hinkle, J. S. Hunter, III, and D. M. Gaydon. 1989. Resistance in constant exposure livestock insect control systems: A partial review with some original findings on cyromazine resistance in house flies. *Fl Entomol* 72:360–369.

Sheppard, D. C., G. L. Newton, S. A. Thompson, and S. Savage. 1994. A value-added manure management system using the black soldier fly. *Biores Tech* 50:275–279.

Stafford, K. C., III, C. H. Collison, J. G. Burg, and J. A. Cloud. 1988. Distribution and monitoring lesser mealworms, hide beetles, and other fauna in high-rise, caged-layer poultry houses. *J Agric Entomol* 5:89–101.

Steinkraus, D. C., C. J. Geden, and D. A. Rutz. 1991. Susceptibility of lesser mealworm (Coleoptera: Tenebrionidae) to *Beauvaria bassiana*: Effects of host stage, formulation, substrate, and host passage. *J Med Entomol* 28:314–321.

Steinkraus, D. C., W. M. Brooks, and C. J. Geden. 1992. Discovery of the neogregarine *Farinocystis tribolii* and an eugregarine in the lesser mealworm, *Alphitobius diaperinus*. *J Invert Pathol* 59:203–205.

Stringham, S. M. 1994. Ectoparasite and rodent management. Pp. 70–76. In *Proceedings of the 1994 Georgia Poultry Conference*, Gainesville. October 11–12, 1994.

Taylor D. J. 1995. *Pig Diseases*. 6th ed. St. Edmundsbury Press, Bury St. Edmunds, Suffolk, United Kingdom. 367 pp.

Thomas, G. D., S. R. Skoda, D. R. Berkebile, and J. B. Campbell. 1996. Scheduled sanitation to reduce stable fly (Diptera: Muscidae) populations in beef cattle feedlots. *J Econ Entomol* 89:411–414 .

Turner, E. C., Jr. and L. Carter. 1990. Mass rearing and introduction of *Ophyra aenescens* (Wiedemann) (Diptera: Muscidae) in high-rise caged layer houses to reduce house fly populations. *J Agric Entomol* 7:247–257.

Varma, M. R. G. 1993. Ticks and mites (Acari). Pp. 597–658. In R. P. Lane and R. W. Crosskey (eds.). *Medical Insects and Arachnids*. Chapman & Hall, Norwell, Massachusetts.

Waldvogel, M. G., C. B. Moore, G. W. Nalyanya, S. M. Stringham, D.W. Watson, and C. Schal. 1999. Integrated cockroach (Dictyoptera: Blattellidae) management in confined swine production. Pp. 183–188. In W. H. Robinson, F. Rettich, and G. W. Rambo (eds.). *Proceedings of the 3rd International Conference on Urban Pests*, Grafické Závody Hronov, Czech Republic, July 19–22, 1999.

Wallace, M. M. H., R. G. Winks, and J. Voestermans. 1985. The use of the beetle, *Alphitobius diaperinus* (Panzer) (Coleoptera: Tenebrionidae), for the biological control of poultry dung in high-rise layer houses. *J Austr Inst Agric Sci* 51:214–219.

Watson, D. W. 2002. Swine pest management. Pp. 815–817. In David Pimentel (ed.). *Encyclopedia of Pest Management*. Mar-

cel Dekker Publishing, New York.
Watson, D. W. and J. J. Petersen. 1993. Seasonal activity of *Entomophthora muscae* (Zygomycetes: Entomophthorales) in *Musca domestica* L. (Diptera: Muscidae) with reference to temperature and humidity. *Biol Control* 3:182–190.
Watson, D. W., C. J. Geden, S. J. Long, and D. S. Rutz. 1995. Efficacy of *Beauveria bassiana* for controlling the house fly and stable fly (Diptera: Muscidae). *Biol Contr* 5:405–411.
Watson, D. W., J. S. Guy, and S. M. Stringham. 2000. Limited transmission of turkey coronavirus in young turkeys by adult *Alphitobius diaperinus* (Coleoptera: Tenebrionidae). *J Med Entomol* 37:480–483.
Weaver, J. E. 1996. The lesser mealworm, *Alphitobius diaperinus*: Field trials for control in a broiler house with insect growth regulators and pyrethroids. *J Agric Entomol* 13:93–97.
Weinzierl, R. A. and C. J. Jones. 1998. Releases of *Spalangia nigroaenea* and *Muscidifurax zaraptor* (Hymenoptera: Pteromalidae) increase rates of parasitism and total mortality of stable fly and house fly (Diptera: Muscidae) pupae in Illinois cattle feedlots. *J Econ Entomol* 91:1114–1121.
Wickham, J. C. 1995. Conventional insecticides. Pp. 109–148. In M. K. Rust, J. M. Owens, and D. A. Reierson (eds.). *Understanding and Controlling the German Cockroach*. Oxford University Press, United Kingdom.
Wieman, G. A., J. B. Campbell, J. A. Deshazer, and I. L. Berry. 1992. Effect of stable flies (Diptera: Muscidae) and heat stress on weight gain and feed efficiency of feeder cattle. *J Econ Entomol* 85: 1835–1842.
Wills, L. E. and B. A. Mullens. 1991. Vertical distribution of dipterous larvae and predatory arthropods in accumulated caged layer poultry manure in southern California. *J Agric Entomol* 8:59–66.
Wills, L. E., B. A. Mullens, and J. D. Mandeville. 1990. Effects of pesticides on filth fly predators (Coleoptera: Histeridae, Staphylinidae; Acarina: Macrochelidae, Uropodidae) in caged layer poultry manure. *J Econ Entomol* 83:451–457.
Zhang, L. and T. Shono. 1997. Toxicities of pyriproxifen to susceptible and resistant strains of houseflies. *Appl Entomol Zool* 32:373–378.

Further Reading

Axtell, R. C. 1986. Fly control in confined livestock and poultry production. Technical Monograph CGA 690 00034–A, Ciba–Geigy Corporation, Agricultural Division, Greensboro, North Carolina. 59 pp.
Axtell, R. C. 1999. Poultry integrated pest management: Status and future. *Integr Pest Manage Rev* 4:53–73.
Axtell, R. C. 2002a. *Fly Management Simulator (FMS)*, <http://ipmwww.ncsu.edu/vetent/expert.html> (5 August 2002)
Axtell, R. C. 2002b. *Livestock Pest Expert System (LPES)*, <http://ipmwww.ncsu.edu/vetent/expert. html> (5 August 2002)
Axtell, R. C. 2002c. *Poultry Pest Expert System (PPES)*, <http://ipmwww.ncsu.edu/vetent/expert.html> (5 August 2002)
DeVaney, J. A. 1986. Symposium: Arthropods of economic importance to the poultry industry. Ecotoparasites. *Poult Sci* 65:649–656.
Fletcher, M. G., R. C. Axtell, R. E. Stinner, and L. R. Wilhoit. 1991. Temperature-dependent development of immature *Carcinops pumilio* (Coleoptera: Histeridae), a predator of *Musca domestica* (Diptera: Muscidae). *J Entomol Sci* 26:99–108.
Kettle, D. S. 1990. *Medical and Veterinary Entomology*. 2nd ed. CAB International, Wallingford, United Kingdom. 658 pp.
Rueda, L. M. and R. C. Axtell. 1985. Guide to common species of pupal parasites (Hymentoptera: Pteromalidae) of the house fly and other muscoid flies associated with poultry and livestock manure. Bulletin 278. Agricultural Research Service, North Carolina State University, Raleigh. 88 pp.
Rutz, D. A. and R. S. Patterson (eds.). 1990. *Biocontrol of Arthropods Affecting Livestock and Poultry*. Westview Press, Boulder, Colorado. 316 pp.
Wilhoit, L. R., R. E. Stinner, and R. C. Axtell. 1991. Computer simulation model of house fly management in confined animal production systems. Technical Bulletin 296. Agricultural Research Service, North Carolina State University, Raleigh. 81 pp.
Williams, R. E., R. D. Hall, A. B. Broce, and P. J. Scholl (eds.). 1985. *Livestock Entomology*. John Wiley & Sons, Hoboken, New Jersey. 335 pp.

9 Wildlife-Damage Control

Wildlife damage management is an important aspect of integrated pest management (IPM). Contemporary managers recognize the importance of practices that prevent problems and those that address existing ones. Cultural practices coupled with exclusion, repellents, toxicants, and other lethal control measures provide tools to address most mammalian wildlife damage situations. Public education and legal and social considerations are becoming a major component of wildlife damage efforts. Professionals are becoming increasingly aware of considerations that go beyond the biological. Effective allocation of resources to solve problems caused by wildlife is hampered by a lack of knowledge concerning the magnitude of such problems (Conover et al. 1995).

Wildlife management is often thought of in terms of protecting, enhancing, and nurturing wildlife populations and the habitat needed for their well-being. But many species require, at one time or another, management actions to decrease conflicts with other wildlife species or with people.

Wildlife damage control is an increasingly important aspect of wildlife management because of expanding human populations and intensifying land-use practices. Concurrent with this growing need to limit the conflicts between wildlife and humans, public attitudes and environmental regulations are restricting use of some traditional tools of control, such as toxicants and traps (Hygnstrom, Timm, and Larson 1994). Agencies and individuals carrying out control programs are being scrutinized more carefully to ensure that their actions are justified, environmentally safe, and in the public interest. Thus, wildlife damage control activities must be based on sound economic, ecological, and sociological principles and must be carried out as positive, necessary components of overall wildlife management programs (Hygnstrom, Timm, and Larson 1994).

Wildlife damage control programs can be thought of as incorporating four stages: defining problems, understanding problem species ecology, applying control methods, and evaluating results. *Defining problems* refers to determining the species and numbers of animals causing the problem, the amount of loss, the nature of the conflict, and other biological and social factors related to the problem. *Understanding problem species* refers to understanding the life history of the species, especially in relation to the problem at hand. *Applying control methods* refers to taking the information gained from stages 1 and 2, to develop an appropriate management program to alleviate or to decrease the conflict between human and pest. *Evaluating results* allows an assessment of the decrease in damage relative to the costs of the control and its effects on target and nontarget populations and on the environment. Increasingly, emphasis is being placed on the implementation of IPM systems in which several control methods are combined, coordinated, and used simultaneously with other management practices.

Ungulates

Damage Assessment

Ungulate (hoofed-animal) damage to various agricultural, forestry, and ornamental crops that is caused by feeding, trampling, and antler rubbing is increasing. Deer browsing in winter on buds of apple or on other fruit trees can decrease yields for the next year (Austin and Urness 1989) or alter the growth pattern of tree limbs adversely (Harder 1970). Similar browsing on nursery plants and in Christmas tree plantations can decrease or eliminate product value (Scott and Townsend 1985). Browsing of hardwood saplings and young fir trees in regenerating forests can decrease growth rates, deform trees, and even cause plantation failures (Crouch 1976; Tilghman 1989).

Tree damage caused by antler rubbing can be a problem (Scott and Townsend 1985). Smooth-barked, small trees (0.5 to 1 inch [in.] [1.3 to 2.5 centimeters {cm}] in diameter, at 6 in. [15 cm] above ground) such as green ash, plum, and cherry, were preferred for antler rubbing by white-tailed deer invading an Ohio nursery (Nielsen, Dunlap, and Miller 1982).

Accurate estimates of economic loss from ungulate browsing and rubbing in orchards, nurseries, and re-

Figure 9.1. Deer damaging hay. Photo courtesy of J. Knight, Montana State University.

forestation projects are difficult to obtain. Losses in yield or tree value may accumulate for many years after damage occurs and depend on other stresses, including rodent damage, inflicted on the plants. In 1983, Ohio growers reported average losses to deer of $82/acre (a.) ($204/hectare [ha]) for orchards, $89/a. ($219/ha) for Christmas tree plantings, and $108/a. ($268/ha) for nursery plantings (Scott and Townsend 1985). Losses are estimated to be in the millions of dollars annually in certain U.S. states (Black et al. 1979; Connelly, Decker, and Wear 1987; Craven 1983).

Deer also feed on various agricultural crops, especially young soybean plants and ripening ears of corn. Hygnstrom and Craven (1988) estimated a mean loss of 2,397 pounds of corn/a. (2,680 kilograms/ha) for 51 unprotected corn fields in Wisconsin. Yield losses in soybean fields are most severe when feeding occurs during the first week of sprouting (DeCalesta and Schwendeman 1978). Elk and deer in certain areas raid haystacks and cattle feedlots (Eadie 1954) (Figure 9.1).

Damage Identification

Ungulates do not have an upper set of incisors. Thus, twigs or plants nipped by these hoofed species do not show the neat, sharp-cut edge left by most rodents and rabbits, but instead show a rough, shredded edge and usually a square or ragged break. Pearce (1947) observed that deer in the Northeast seldom browse higher than 6 feet (ft) (1.8 m) from a standing position, but are able to reach up to 8 ft (2.5 m) by rearing up on their hind legs. Elk and moose browse to a height of approximately 10 ft (3 m). Deer seldom browse on branches more than 1 in. (2.5 cm) in diameter. Moose and elk will gnaw the bark of aspen trees. When male ungulates rub the velvet from their antlers, the scarring is confined generally to the trunk area up to 3 ft (1 m) high (Pearce 1947).

Damage Prevention and Control

Various exclusion methods often are crucial for damage prevention. Where ungulates are abundant or crops particularly valuable, fencing may be the only way to minimize damage. Several fencing designs are available to meet specific needs (Craven and Hygnstrom 1994). Temporary electric fences are useful in protecting garden and field crops during snow-free periods (Figure 9.2). Deer and elk are attracted to these fences by smell (such as a coating of peanut butter) and are lured into contacting the fence with their noses. The resulting shock is a very strong negative stimulus, and ungulates learn to avoid the

Figure 9.2. Electrical elk fence. Photo courtesy of J. Knight, Montana State University.

fenced area. Permanent high-tensile electric fences provide year-round protection and are suited best to high-value specialty or orchard crops. The electric shocking power and the unique fence designs present both psychological and physical barriers to deer and elk. This type of electric fence can be used even after the ground is frozen (Schmidt and Knight 2000). Permanent woven-wire fences provide the ultimate barrier. They require little maintenance but are very expensive to build.

Tree protectors such as plastic tree wrap or woven-wire cylinders protect young trees from deer and elk. Four-foot (1.2-m) woven-wire cylinders can keep deer from rubbing tree trunks with their antlers (Craven and Hygnstrom 1994).

Wooden panels have been used traditionally to exclude deer and elk from haystacks. Stockyards also have been protected by welded-wire panels and woven wire. More recently, haystacks have been protected by wrapping with a plastic snow fence (Craven and Hygnstrom 1994). The material comes in 8-foot width rolls and is fairly light and easy to use.

Appropriate cultural methods and habitat modifications will depend on habitat. Damage to ornamental plants can be minimized by selecting landscape and garden plants that are less preferred by deer and elk. In many instances, original landscape objectives can be met by planting species that have a degree of resistance to deer and elk damage (Hill and Knight 1998).

Crops should be harvested as early as possible to limit the period of vulnerability to deer. Susceptible crops should be planted as far from wooded cover as practical, so as to decrease the potential for severe damage. Habitat modification is not recommended. Destruction of wooded or brushy cover as a means of decreasing deer use would destroy valuable habitat for other wildlife. And because deer and elk forage over a large area, it is unlikely that all available cover would be on the same farm or ranch that is sustaining damage.

Lure crops have been planted to attract deer away from highways and crop fields where deer traditionally have caused damage (Figure 9.3). Their effectiveness has been variable, and concern has been raised that an artificial food source eventually may increase deer densities and concomitant problems. Specific recommendations are not yet available regarding plant selection and timing or proximity of lure crops.

Promising research on the use of chemosterilants and immunocontraception to decrease or to eliminate reproduction is under way. Because specificity, efficacy, and delivery of contraceptive agents continue to be problems, contraception is unlikely to be applied in rural/agricultural landscapes in the foreseeable future. The use of contraception for herd control is suited best to urban parks, refuges, and other discrete areas.

One of the keys to success with frightening devices and repellents is to take action at the first sign of a problem. It is difficult to change ungulates' movements or behavioral patterns once established. Additionally, frightening devices and repellents should be used when crops are most susceptible to damage, for example, during the silking to tasseling stages for field corn or during the blossom stage for soybeans (Craven and Hygnstrom 1994).

Frightening devices such as gas exploders, shell crackers, fireworks, and gunfire can provide rapid but temporary relief from ungulate damage. Ungulates tend to develop strategies to avoid or to build tolerance of frightening devices.

Figure 9.3. White-tailed deer near a road in West Virginia. Photo by Ken Hammond, U.S. Department of Agriculture.

Repellents are suited best for use in orchards and gardens and on ornamental plants. High cost, use limitations, and variable effectiveness make most repellents impractical on row crops, pastures, or other large areas. Success with repellents is evidenced by decrease, and not by the elimination, of damage.

Repellents are described in terms of mode of action (i.e., "contact" or "area"). *Contact repellents*, applied directly to plants, repel by taste. They are most effective when applied to trees and shrubs during the dormant period. New growth that appears after treatment is unprotected. Almost all contact repellents may diminish the palatability of forage crops and should not be used on plant parts destined for human consumption (Craven and Hygnstrom 1994).

Applied near the plants to be protected, *area repel-*

lents repel deer and elk by odor alone. Such repellents usually are less effective than contact repellents but can be used in perimeter applications and in other situations in which contact repellents cannot (Hygnstrom 1994).

No toxicants are registered for deer or elk control. Poisoning with any product for any reason is illegal and unlikely to be tolerated by the public.

In special environments such as city parks, refuges, or suburban neighborhoods, it may be necessary or desirable to capture deer alive and to move them to other areas. Deer can be captured safely with rocket nets, drop-door box traps, or tranquilizer guns, but these techniques are expensive and time consuming and require the expertise of professional wildlife biologists (Craven and Hygnstrom 1994). Live capture and relocation is seldom a practical alternative unless delicate public relations problems mandate live removal as the only choice.

Figure 9.4. Elk grazing. Photo by Ken Hammond, U.S. Department of Agriculture.

Effective harvest and use of the legal deer and elk season are probably the most effective means of controlling their populations (Craven and Hygnstrom 1994) (Figure 9.4). By permitting hunting, landowners provide public access to a public resource while decreasing damage problems. Because of the daily and seasonal movements of deer and elk, only rarely does a single landowner control all the land a deer uses. As a result, neighboring landowners should cooperate. Landowners, the state wildlife agency, and local hunters should reach a consensus about a desirable population level for an area before deer are removed. In urban areas and parks where hunting is generally prohibited, it may be necessary to use professional sharpshooters to remove deer.

Rodents

Damage Assessment

Rodents seldom are observed in the act of causing damage, and frequently their damage is difficult to measure. Nonetheless, assessments of damage have indicated that rodents and other small mammals cause tremendous annual losses of food and fiber in the United States. In the late 1970s, forest animal damage to Douglas fir and ponderosa pine in Washington and Oregon was estimated to total $60 million annually, and potential decreases in the total value of forest resources was estimated at $1.83 billion (Black et al. 1979; Brodie et al. 1979). Although these figures included losses attributable to ungulates, rodents and hares were responsible for much of the damage.

Miller (1987) surveyed forest managers and natural resource agencies in 16 southeastern states and estimated annual wildlife-caused losses, primarily by beavers, to be $11.2 million on 70 million a. (28.4 million ha). An additional $1.6 million was spent to control wildlife damage on this land. Arner and Dubose (1982) estimated that economic loss to beavers exceeded $4 billion over a 40-year (yr) period on 988,000 a. (400,000 ha) in the southeastern United States. Annual loss in Mississippi to nonimpounded timber was estimated at $215 million over a period of at least 10 yr (Bullock and Arner 1985).

Rats cause substantial losses to sugarcane. Lefebvre, Ingram, and Yang (1978) estimated annual losses to be approximately $6 million ($95/a. [$235/ha]) in one-third of the area producing sugarcane in Florida. Hawaiian losses were reported to be in excess of $20 million/yr (Seubert 1984). Ferguson (1980) estimated that voles caused losses that approached $50 million to apple growers in the eastern United States. Losses of forage on rangelands to rodents, rabbits, and hares also are known to be extensive; accurate estimates of the monetary losses are difficult to obtain, however, because of the nature of the damage and the wide area over which it occurs (Marsh 1985).

Pearson and Forshey (1978) compared yields of apple trees visibly damaged by voles to those not showing damage to determine the dollar losses in gross return/tree. Richmond et al. (1987) determined decreases in growth, yield, and fruit size of apple trees damaged by pine vole populations of known size maintained in enclosures around the trees.

An index of rodent damage to sugarcane was developed through sampling at harvest to determine the percentage of stalks damaged (Lefebvre, Ingram, and

Yang 1978). Forage losses have been estimated by means of comparing production on areas with and without rodents (Foster and Stubbendieck 1980; Luce, Case, and Stubbendieck 1981; Turner 1969). Sauer (1977) used exclusion cylinders to determine losses of forage to ground squirrels. Alsager (1977) described a method of determining forage production decreases from pocket gopher damage.

These methods are useful in evaluating the efficacy of control techniques. But loss estimates must be converted to accurate assessments of dollar loss to enable cost-benefit analysis of control programs. This conversion is difficult, given the vast acreages involved and the variability in rodent populations. In other situations, control may be necessary irrespective of cost (for example, when rats or mice are present in homes or when diseases such as rabies or hantavirus are present).

These examples illustrate the complexity of damage situations and the need for more-accurate damage assessment methods, which constitute an area of high priority for future research. Lack of methods for determining damage levels has been a noteworthy impediment to the development of cost-effective control strategies (Dolbeer, Holler, and Hawthorn 1994).

Damage Identification

Most rodents are secretive and not observed easily; many are nocturnal (Dolbeer, Holler, and Hawthorn 1994). Often the investigator must rely on various signs such as tracks, trails, tooth marks, droppings, or burrows to determine the species doing the damage. Trapping may be necessary to make a positive identification of small rodents; often, more than one species is involved.

Characteristics of the damage also may provide clues to the species involved. In orchards, for example, major stripping of roots usually is caused by pine voles whereas damage at the root collar or on the trunk, up to the extent of snow depth, is caused most often by meadow voles (Dolbeer, Holler, and Hawthorn 1994). In sugarcane, various species of rats gnaw stalks so that they are hollowed out between the internodes but usually not severed completely. Rabbits, in contrast, usually gnaw through the stalks, leaving only the ring-shaped internodes.

Damage to plants can be grouped generally as follows: root damage—pocket gophers and pine voles; trunk debarking—meadow voles, squirrels, porcupines, wood rats, rabbits, and mountain beavers; stem and branch cutting—beavers, rabbits, meadow voles, mountain beavers, porcupines, and rabbits; debudding—red squirrels and chipmunks (Dolbeer, Holler, and Hawthorn 1994). These characteristics can aid in identification of the species responsible, but positive identification should be made on the basis either of species-specific signs (tracks, hair, droppings) or individual capture.

Damage Prevention and Control

Because of the great number of potential rodent pest-species, a comprehensive description of prevention and control techniques is beyond the scope of this chapter. General statements can be made regarding the usual considerations attending application of IPM principles to rodents. Detailed, species-specific information can be found in the publication *Prevention and Control of Wildlife Damage*, by Hygnstrom, Timm, and Larson (1994).

With commensal rodents, the only certain solution is usually exclusion or elimination of all entry points to protected structures (Dolbeer, Holler, and Hawthorn 1994). It should be noted that these rodents are capable of chewing their own entry; thus management requires ongoing effort. Barriers of small mesh hardware cloth can be used to exclude field rodents from young trees or gardens (Dolbeer, Holler, and Hawthorn 1994). Beaver and muskrats also have been excluded by use of fencing. Exclusion of mice or voles from fields rarely is practical.

Cultural practices that modify structures or habitat probably represent the greatest potential to minimize rodent problems. Sanitation practices, weed control, grazing practices, crop rotation, grain buffer strips, and flood irrigation can affect rodent populations. Control of tap-rooted plants and selection of damage-resistant varieties of alfalfa or tree seedlings also will decrease rodent damage (Dolbeer, Holler, and Hawthorn 1994).

Other methods such as trapping and toxicants commonly are used to control rodents. The ability to control with repellents or ultrasonic devices has not been established.

Mammalian Predators

Damage Assessment

Mammalian predators always have been a concern to livestock producers. Wade and Bowns (1982) estimated that the direct loss of sheep and goats to coyotes in the United States ranged from $75 million to

$150 million annually. Pearson (1986), using a summary of other studies and surveys, estimated the loss of sheep, lambs, and goats to predators (primarily coyotes) at $68,160,000 in the 17 western states in 1984. Terrill (1988), using data from all 50 states, reported that from 1985 to 1987 annual losses of sheep and lambs to coyotes and other predators ranged from $69 million to $83 million. In 1990, 490,000 sheep and lambs valued at $21.7 million and 129,400 goats valued at $5.6 million were lost to mammalian predators in the United States (NASS 1991). In 1991, the National Agricultural Statistics Service estimated that mammalian predators in the United States killed 106,000 cattle and calves valued at $41.5 million (NASS 1992). Losses of poultry to predators, although not well documented, also are thought to be substantial (Dolbeer, Holler, and Hawthorn 1994).

Mammalian predators, especially red foxes, striped skunks, raccoons, and mink, significantly affect waterfowl nesting success in small wetland areas surrounded by agricultural lands. A study in North Dakota indicated nesting success of only 8% for mallards on such wetlands, half that needed to sustain the population (Cowardin, Gilmer, and Shaiffer 1985). Adept at catching nesting hens as well as destroying eggs, the red fox evidently is the most important waterfowl predator (Sargeant, Allen, and Eberhardt 1984).

Damage Identification

Because predation is observed rarely, accurate assessment of losses to specific predators often requires careful investigative work. The first action in determining the cause of animal death is to check for signs on the animal and around the kill site. Size and location of tooth marks often will indicate predatory species. Extensive bleeding usually is characteristic of predation. Where external bleeding is not apparent, the hide can be removed from the carcass—particularly around the neck, throat, and head—and the area can be checked for tooth holes, subcutaneous hemorrhage, and tissue damage (Dolbeer, Holler, and Hawthorn 1994). Hemorrhage occurs only if skin and tissue damage occurs while the animal is alive. Animals that die from causes other than predation do not normally show external or subcutaneous bleeding although bloody fluids may be lost from body openings (Bowns 1976). Animal losses are easiest to evaluate if examination is conducted when the carcass is fresh (Wade and Bowns 1982).

Animals are not always killed by a throat attack; some are pulled down from the side or rear. In these situations, blood often appears on the sides, hind legs, and tail areas. A calf can have its tail chewed off, and its nose may have tooth marks or be completely chewed by the predator when the tongue is eaten (Bowns 1976).

Tracks and droppings alone are proof neither of depredation nor of the species responsible for any depredation (Dolbeer, Holler, and Hawthorn 1994). Rather, when combined with other characteristics of depredation, tracks and droppings can constitute evidence that a particular predator is in the area and can aid determinations of which species is causing the problem.

Damage Prevention and Control

For exclusion, heavy woven-wire fencing at least 10 ft (3 m) high is required to discourage predators such as coyotes, raccoons, fox, bears, and mountain lions. Overhead fencing also is necessary for permanent and predictable protection (Knight 1994). Fencing usually is practical only for high-value livestock and poultry. When practical, night fencing under lights or in sealed buildings is useful.

Electric fencing with alternating charged and grounded wires can exclude coyotes, raccoons, fox, bears, and mountain lions. Wires should be 10 ft (3 m) high, spaced 4 in. (10 cm) apart, and charged with at least 5,000 volts (Knight 1994).

Because most predators prefer to hunt and thus stay where escape cover is close at hand, the utility of cultural practices is limited. Although often impractical, removal of brush and trees within one-quarter mile (0.4 km) of building and livestock concentrations may result in decreased predation (Knight 1994).

Chronic predation has led to certain ranchers' shifting from sheep to cattle production. And in certain areas with high levels of predation, ranchers have changed from cow/calf to steer operations.

Frightening devices such as bright lights, flashing white lights, blaring music, barking dogs, and changes in the placement of scarecrow objects in livestock depredation areas may repel predators temporarily (Knight 1994), as may a strobe light/siren device. Guard dogs, llamas, and donkeys have been successful means of protecting sheep from coyotes (Green, Henderson, and Collinge 1994).

No chemical repellents have shown sufficient effectiveness to be registered for use. Moreover, no chemical toxicants are registered for mountain lion or bear control. M–44 ejector devices for use with sodium cyanide-loaded plastic capsules are registered for coyote control and for fox control in certain states, as are livestock protection collars containing compound 1080 (Green, Henderson, and Collinge 1994).

Mountain lions and coyotes are extremely strong and require very well-made traps (Knight 1994). These lions are trapped easily, however, along habitual travel ways, in depredation areas, and at kill sites. Although blind sets usually are made in narrow paths frequented by lions, baits made of fish products, poultry, porcupine, rabbits, or deer parts, as well as curiosity lures such as catnip, rhodium oil, and house cat urine and gland materials are effective attractants. Mountain lions are very curious and respond to hanging and moving flags of skin, feathers, or bright objects.

There are many effective methods for trapping coyotes. Like mountain lions, coyotes follow regular paths and crossings. They also prefer high hills or knolls from which they can view terrain. They establish regular scent posts along their paths. Leghold traps can be used along these routes as effective and versatile coyote control tools (Green, Henderson, and Collinge 1994) (Figure 9.5).

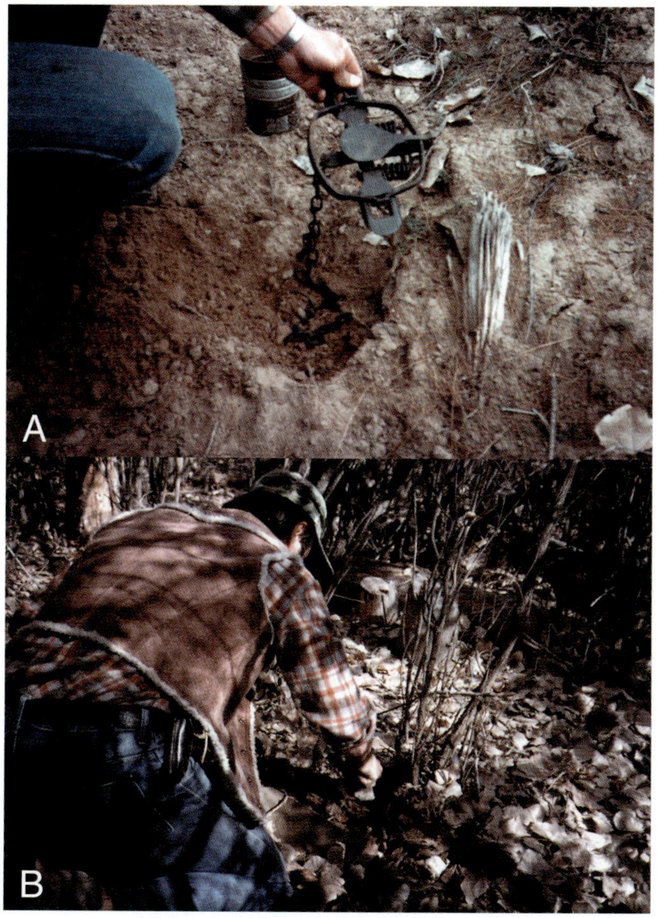

Figure 9.5. A. Setting a coyote leghold trap. B. Camouflaging a leghold trap. Photos courtesy of J. Knight, Montana State University.

Other predators require specifically designed trapping techniques. Culvert and barrel traps are effective tools for trapping bears, for instance. The Aldrich-type foot snare can be used to catch bear and mountain lions (Hygnstrom 1994). The set is made on trails frequented by lions or at baits used by bears. Stones or sticks are used to direct foot placement over the triggering device. Snares suspended in runways and trails or set with baits in cubby arrangement can be set to kill mountain lions or coyotes.

If coyotes, bears, or mountain lions return to a fresh kill to feed, they can be shot. They also can be called into shooting range with predator calls, particularly sounds that simulate the distress cry of deer or rabbits (Green, Henderson, and Collinge 1994). Trained dogs can be used to pursue predators. Hunting of predators as big-game animals should be encouraged in areas of predation when possible. Aerial gunning from fixed-wing aircraft and helicopters sometimes is used to kill coyotes (Green, Henderson, and Collinge 1994).

Acknowledgments

Much of the information in this chapter was taken from *Prevention and Control of Wildlife Damage*, by Hygnstrom, Timm, and Larson, published by the University of Nebraska Cooperative Extension in 1994.

Appendix A. Abbreviations and Acronyms

a.	acre
cm	centimeter
ft	foot
ha	hectare
in.	inch
yr	year

Appendix B. Glossary

Area repellents. Chemicals that repel deer and elk by odor alone.

Contact repellents. Chemicals that repel deer and elk by contact.

Ungulate. Hoofed animal.

Literature Cited

Alsager, D. E. 1977. Impact of pocket gophers (*Thomomys talpoides*) on the quantitative productivity of rangeland vegetation in southern Alberta: A damage assessment tool. Pp. 47–57. In W. B. Jackson and R. E. Marsh (eds.). *Vertebrate Pest Control and Management Materials*. American Society for Testing Materials Standard Testing Procedures, Philadelphia, Pennsylvania.

Arner, D. H. and J. S. Dubose. 1982. The impact of the beaver on the environment and economics in the southeastern United States. *Trans Int Congr Game Biol* 14:241–247.

Austin, D. D. and P. J. Urness. 1989. Evaluating production losses from mule deer depredation in apple orchards. *Wildl Soc Bull* 17:161–165.

Black, H. C., E. J. Dimock, II, J. Evans, and J. A. Rochelle. 1979. Animal damage to coniferous plantations in Oregon and Washington. Part I. A survey, 1963–75. Corvallis Research Bulletin 25. Forest Research Laboratory, Oregon State University.

Bowns, J. E. 1976. Field criteria for predator damage assessment. *Utah Sci* 37:26–30.

Brodie, J. D., H. C. Black, E. J. Dimock, II, J. Evans, C. Kao, and J. A. Rochelle. 1979. Animal damage to coniferous plantations in Oregon and Washington. Part II. An economic evaluation. Corvallis Research Bulletin. Forest Research Laboratory, Oregon State University.

Bullock, J. F. and D. H. Arner, 1985. Beaver damage to nonimpounded timber in Mississippi. *South J Appl For* 9:137–140.

Connelly, N. A., D. J. Decker, and S. Wear. 1987. Public tolerance of deer in a suburban environment: Implications for management and control. *Proc East Wildl Damage Contr Conf* 3:207–218.

Conover, M. R., W. C. Pitt, K. K. Kesler, T. J. DuBow, and W. A. Sanborn. 1995. Review of human injuries, illnesses, and economic losses caused by wildlife in the United States. *Wildl Soc Bull* 23:407–414.

Cowardin, L. M., D. S. Gilmer, and C. W. Shaiffer. 1985. Mallard recruitment in the agricultural environment of North Dakota. *Wildl Monogr* 92:1–37.

Craven, S. R. 1983. New directions in deer damage management in Wisconsin. *Proc East Wildl Damage Contr Conf* 1:65–67.

Craven, S. R. and S. E. Hygnstrom. 1994. Deer. *Prevention and Control of Wildlife Damage*. Cooperative Extension, University of Nebraska, Lincoln.

Crouch, G. L. 1976. Deer and reforestation in the Pacific northwest. *Proc Vertebr Pest Conf* 7:298–301.

DeCalesta, D. S. and D. B. Schwendeman. 1978. Characterization of deer damage to soybean plants. *Wildl Soc Bull* 6:250–253.

Dolbeer, R. A., N. R. Holler, and D. W. Hawthorn. 1994. Identification and assessment of wildlife damage: An overview. *Prevention and Control of Wildlife Damage*. University of Nebraska Cooperative Extension, Lincoln.

Eadie, W. R. 1954. *Animal Control in Field, Farm and Forest*. The Macmillan Company, New York. 257 pp.

Ferguson, W. L. 1980. Rodenticide use in apple orchards. *Proc East Pine Meadow Vole Symp* 4:2–8.

Foster, M. A. and J. Stubbendieck. 1980. Effects of the Plains pocket gopher (*Geomys bursarius*) on rangeland. *J Range Manage* 33:74–78.

Green, J. S., F. R. Henderson, and M. D. Collinge. 1994. Coyotes. *Prevention and Control of Wildlife Damage*. Cooperative Extension, University of Nebraska, Lincoln.

Harder, J. D. 1970. Evaluating winter deer use of orchards in western Colorado. *Trans North Am Wildl Res Conf* 35:35–47.

Hill, C. and J. Knight. 1998. Minimizing deer damage to residential plantings. MontGuide MT 1998–14. Extension Service, Montana State University, Bozeman.

Hygnstrom, S. E. 1994. Black bears. *Prevention and Control of Wildlife Damage*. University of Nebraska Cooperative Extension, Lincoln.

Hygnstrom, S. E. and S. R. Craven. 1988. Electric fences and commercial repellents for reducing deer damage in cornfields. *Wildl Soc Bull* 16:291–296.

Hygnstrom, S. E., R. Timm, and G. Larson. 1994. *Prevention and Control of Wildlife Damage*. University of Nebraska Cooperative Extension, Lincoln.

Knight, J. E. 1994. Mountain lions. *Prevention and Control of Wildlife Damage*. Extension Service, University of Montana, Bozeman.

Lefebvre, L. W., C. R. Ingram, and M. C. Yang. 1978. Assessment of rat damage to Florida sugarcane in 1975. *Proc Am Soc Sugar Cane Technol* 7:75–80.

Luce, D. G., R. M. Case, and J. L. Stubbendieck. 1981. Damage to alfalfa fields by Plains pocket gophers. *J Wildl Manage* 45:258–260.

Marsh, R. E. 1985. Competition of rodents and other small mammals with livestock in the United States. Pp. 485–508. In S. M. Gaafar, W. E. Howard, and R. E. Marsh (eds.). *Parasites, Pests and Predators*. Elsevier Science Publishers B.V., Amsterdam, The Netherlands.

Miller, J. E. 1987. Assessment of wildlife damage on southern forest. Pp. 48–52. In J. Dickson and O. Maughan (eds.). *Managing Southern Forests for Wildlife and Fish*. U.S. Forest Service, U.S. Department of Agriculture, New Orleans, Louisiana.

National Agricultural Statistics Service (NASS). 1991. Sheep and goat predator loss. National Agricultural Statistics Board, U.S. Department of Agriculture, Washington, D.C. 23 pp.

National Agricultural Statistics Service (NASS). 1992. Cattle and calves death loss. National Agricultural Statistics Board, U.S. Department of Agriculture, Washington, D.C. 23 pp.

Nielsen, D. G., M. J. Dunlap, and K. V. Miller. 1982. Pre-rut rubbing by white-tailed bucks: Nursery damage, social role, and management options. *Wildl Soc Bull* 10:341–348.

Pearce J. 1947. Identifying injury by wildlife to trees and shrubs in northeastern forest. Wildlife Research Report 13. U.S. Fish Wildlife Service, Bethesda, Maryland. 29 pp.

Pearson, E. W. 1986. A literature review of livestock losses to predators in western US. Unpublished final report. U.S. Fish and Wildlife Service, Denver Wildlife Research Center, Colorado. 20 pp.

Pearson, K. and C. G. Forshey 1978. Effects of pine vole damage on tree vigor and fruit yield in New York orchards. *Hort Sci* 13:56–57.

Richmond, M. E., C. G. Forshey, L. A. Mahaffy, and P. N. Miller. 1987. Effects of differential pine vole populations on growth and yield of McIntosh apple trees. *Proc East Wildl Damage Control Conf* 3:296–304.

Sargeant, A. B., S. H. Allen, and R. T. Eberhardt. 1984. Red fox predation on breeding ducks in midcontinent North America. *Wildl Monogr* 89:1–41.

Sauer, W. C. 1977. Exclusion cylinders as a means of assessing losses of vegetation due to ground squirrel feeding. Pp. 14–21. In W. B. Jackson and R. E. Marsh (eds.). *Test Methods for Vertebrate Pest Control and Management Materials*. American Society for Testing and Materials, Philadelphia, Pennsylvania.

Schmidt, L. and J. Knight. 2000. Electric fencing to control deer and elk on Montana's farms and ranches. Montguide MT 2000–10. Extension Service, Montana State University, Bozeman.

Scott, J. D. and T. W. Townsend. 1985. Characteristics of deer damage to commercial tree industries of Ohio. *Wildl Soc Bull* 13:135–143.

Seubert, J. L. 1984. Research on nonpredatory mammal damage control by the U.S. Fish and Wildlife Service. Pp. 553–571. In A. C. Dubbock (ed.). *Organization and Practice of Vertebrate Pest Control*. Imperial Chemical Industries PLC., Surrey, United Kingdom.

Terrill, C. E. 1988. Predator losses climb nationwide. *Nat Wool Grower* 78:32–34.

Tilghman, N. G. 1989. Impacts of white-tailed deer on forest regeneration in northwestern Pennsylvania. *J Wildl Manage* 53:524–532.

Turner, G. T. 1969. Responses of mountain grassland vegetation to gopher control, reduced grazing, and herbicide. *J Range Manage* 22:377–383.

Wade, D. A. and J. E. Bowns. 1982. Procedures for evaluating predation on livestock and wildlife. Bulletin B–1429. Texas Agricultural Extension Service, Texas A&M, College Station. 42 pp.

Related Web Sites

National Wildlife Control Operators Association. 2002. Homepage, 18 April, <http://www.nwcoa.com> (25 August 2002)

Wildlife Control Technology Magazine. 2002. Homepage, <http://www.wctech.com> (25 August 2002)

10 Companion Animal Ectoparasites, Associated Pathogens, and Diseases

Importance of Pets

Pet ownership, as well as the importance of pets in the lives of their owners, is increasing in the United States. To many, a pet completes the family unit. Among the benefits of ownership are companionship and enhanced safety for the elderly and those living alone, opportunities for children to practice responsibility and other social skills, and assistance for people with disabilities. But along with dogs and cats come their ectoparasites and the pathogens they transmit. Management of these ectoparasites relies primarily on insecticides and acaricides supplemented with mechanical, cultural, or biological control.

More than half of U.S. households have pets, totaling more than 60 million dogs and nearly 70 million cats (AVMA 2002). As the number of companion animals increases, their ectoparasites gain prominence, and expenditures for control of these pests and the diseases they transmit grow. Humans experience very close contact with companion animals, which they stroke, nuzzle, and often sleep next to. This physical contact also increases potential human exposure to ectoparasites, as well as to the chemicals used to control them. Pests such as fleas and ticks affect not only pet health, but also the human-pet bond; for instance, a dog scratching fleas may not be interested in playing. Insofar as pets share the human environment intimately, human exposure to both pests and pest control products is significant.

Pest Significance

Ectoparasites (insects and their eight-legged relatives, e.g., mites and ticks, occurring on the outside of the body) affect pet health and behavior and human health and comfort. Ectoparasites can affect animal health by causing blood loss or irritation and/or nervousness, or by transmitting pathogens. Their activity can lead to secondary infection. Likewise, ectoparasites can transfer from pets to their human companions, thereby causing irritation and transmitting causative agents of human disease.

The three most economically significant pet ectoparasites are fleas, ticks, and mosquitoes (Table 10.1). Fleas transmit the tapeworm *Dipylidium caninum* and cause anemia and flea-allergy dermatitis. Ticks transmit several pathogens, including those causing Lyme disease, Rocky Mountain spotted fever, ehrlichiosis, and babesiosis (Figure 10.1). The main health effect of mosquitoes is transmission of dog heartworm, *Dirofilaria immitis*.

Figure 10.1. Increasing size of blood-feeding tick. Photo courtesy of N. Hinkle, University of Georgia.

Nationwide, costs of flea treatment, combined with those for testing and treatment of ancillary flea problems such as flea-allergy dermatitis and tapeworm,

Table 10.1. **Major ectoparasite pests of dogs and cats**

Fleas	*Ctenocephalides felis* and *canis*, *P. irritans* and *simulans*, *Echidnophaga*
Lice	Chewing lice (Mallophaga)
	Sucking lice (Anoplura)
Ticks	*Ixodes, Dermacentor, Rhipicephalus, Amblyomma, Otobius*
Mites	*Cheyletiella, Sarcoptes, Notoedres, Chorioptes, Liponyssoides, Otodectes, Demodex*
Mosquitoes	Culicidae (in several genera)

amount to $3.8 billion/yr. The annual costs of testing for and treatment of tick-transmitted disease agents are estimated at $108 million, but because most products registered for tick control are marketed mainly for flea suppression, it is difficult to assess costs for tick control. Nationally, more than $600 million is spent on heartworm testing, prophylaxis, and treatment. Clearly, pet ectoparasites are significant in terms of their effects both on companion animals and on the economy.

Diseases Associated with Companion Animal Ectoparasites

Ectoparasite-associated diseases fall into two main categories: those caused by the arthropod itself and those caused by arthropod-vectored pathogens (Table 10.2) (Harwood and James 1979). Many of these diseases are *zoonoses* (i.e., disease agents capable of being transmitted to and affecting humans). Even parasites and pathogens such as dog heartworm, which is not considered to have great zoonotic potential, occasionally may produce human disease (Watson, Wetzel, and Burkhalter 1991).

Scabies and mange are caused by mites (Figure 10.2). Ear mites produce otoacariasis, which results in pain and itching in the ear. Scabies mites burrow in the skin, resulting in severe itching, hair loss, and secondary infection. Canine and feline lice can serve as intermediate hosts of tapeworms, but presence of lice constitutes a pathological condition in itself (Harwood and James 1979).

Similarly, certain animals experience severe allergic reaction to flea salivary secretions, a reaction termed *flea-allergy dermatitis* (FAD), which results in *pruritus* (i.e., itching), scratching, and hair loss (Dryden and Rust 1994). Flea-allergy dermatitis can be so debilitating to pets that owners may feel compelled to euthanize them. At times, pruritus can be alleviated by corticosteroids, but prolonged dependence on these drugs often precipitates other medical conditions and shortens an animal's life.

Certain ticks secrete salivary products causing a condition called *tick paralysis*, which can be lethal. Ticks also serve as vectors of several pathogens, including those causing babesiosis, ehrlichiosis, Rocky Mountain spotted fever, and Lyme disease (Hilton et al. 1999). Likewise, mosquitoes transmit the tiny nematodes causing heartworm disease in both dogs and cats (Hermesmeyer et al. 2000).

Control strategies are different for arthropods such as mites and lice, in which all stages are associated closely with the host, than for pests such as fleas and ticks, which undergo off-host developmental stages. Quite different means of control may be used against temporary parasites such as mosquitoes visiting the host only briefly to obtain a blood meal.

Table 10.2. Companion animal diseases and their vectors

Vector	Disease	Causative agent
Fleas	Tapeworm	*Dipyllidium caninum*
	Cat scratch disease	*Bartonella henselae*
Lice	Tapeworm	*Dipyllidium caninum*
Ticks	Lyme disease	*Borrelia burgdorferi*
	Rocky Mountain spotted fever	*Rickettsia rickettsii*
	Ehrlichiosis	*Ehrlichia* spp.
	Babesiosis	*Babesia* spp.
Mosquitoes	Dog heartworm (dirofilariasis)	*Dirofilaria immitis*

Fleas

Fleas are the most common companion animal ectoparasite, with the cat flea (*Ctenocephalides felis*) being the most frequently encountered flea on both dogs and cats (Figure 10.3). Veterinarians report that 70% of dogs and 55% of cats have fleas; 75% of flea-infested dogs and 40% of flea-infested cats experience FAD (Lyon 1997). Not surprisingly, flea control is a multibillion-dollar industry.

Fleas may cause not only irritating bites but also, as a result of salivary secretions, FAD in susceptible animals. Fleas serve as the obligate intermediate host

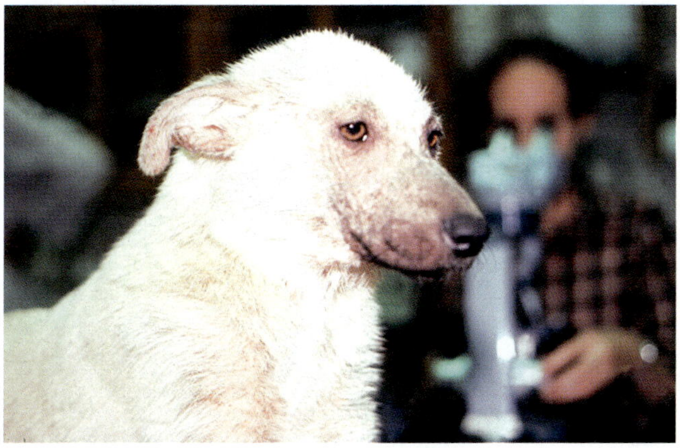

Figure 10.2. Dog with scabies. Photo by Gary Mullen, courtesy of N. Hinkle, University of Georgia.

of the most common canine and feline tapeworm and also may transmit the causative agent of cat scratch disease (Rust and Dryden 1997). Because fleas are a nuisance to humans, flea control often is initiated to alleviate human discomfort.

The most effective biological suppression for cat fleas is probably generalist predators that eat flea eggs and larvae. In nature, these indiscriminate predators account for substantial flea mortality (Hinkle, Rust, and Reierson 1997). The dominant mortality factor for immature fleas is desiccation, whereas host grooming or host absence are the most significant mortality factors for adult fleas. Adult fleas have no known parasitoids, no identified specific predators, and no significant pathogens. Pesticides therefore remain the basis of flea control (Rust and Dryden 1997).

The majority of pet ectoparasiticides are marketed primarily for flea control. These include products formulated as spot-ons, sprays, dusts, and shampoos.

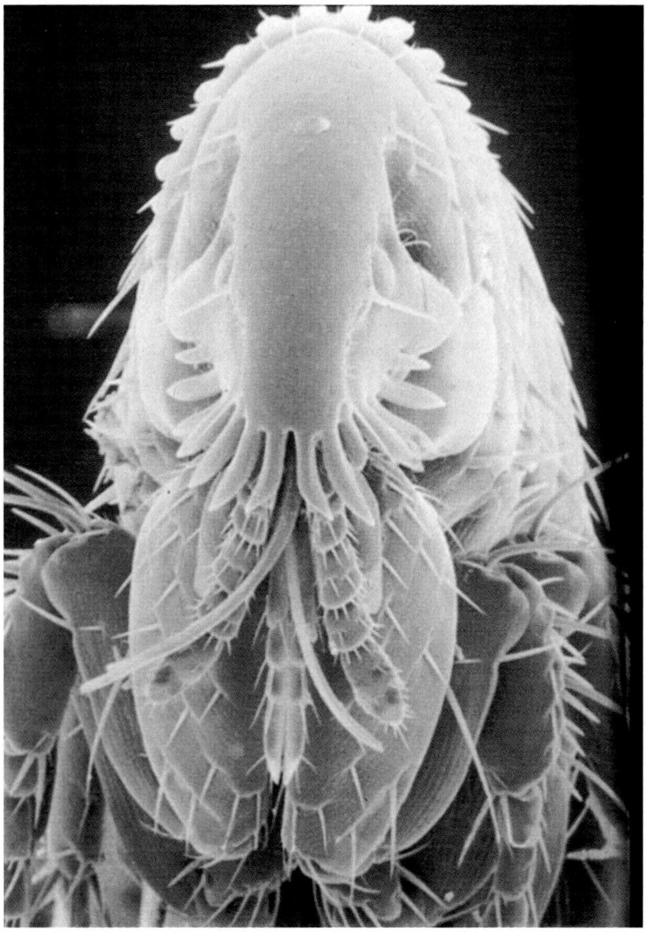

Figure 10.3. Face of cat flea. Photo courtesy of N. Hinkle, University of Georgia.

To suppress adult and immature stages simultaneously, adulticides have traditionally been combined with larvicides such as insect growth regulators (Dryden and Rust 1994). The advent of persistent on-host adulticides has decreased but not eliminated the need for environmental flea management (Rust and Dryden 1997).

Cat fleas are found on dozens of host species in addition to dogs and cats, so excluding alternative hosts (raccoons, opossums, foxes, skunks, etc.) from the premises is important to decrease reinfestation risk (Hinkle, Rust, and Reierson 1997). Sanitation also plays a role in flea suppression. Frequent vacuuming and regular washing of pet bedding decrease flea egg, larval, and pupal populations, thereby diminishing reinfestation potential.

Mites

Several types of mites, including ear mites and mange mites, infest dogs and cats. Mites that infest pets are not known to transmit disease agents, but the presence of large mite populations is irritating to the animal (Harwood and James 1979). All stages of the mite life cycle occur on the animal, so transmission occurs only from animal to animal. Mites are species-specific, meaning that canine mites are found only on dogs, and feline mites only on cats, etc. (Parish and Schwartzman 1993). Certain breeds and individuals are particularly susceptible to mite infestation.

Both cats and dogs are susceptible to ear mite infestations, which occur in the ear canal and cause ear canker accompanied by pain, itching, and discharge (Harwood and James 1979). Because ear mites can live on cats' bodies and will often survive topical treatment, using a systemic product such as an endectocide is generally more effective than using ear drops. In multicat households, the ability of mites to move from cat to cat makes infestations especially difficult to control.

Mange is caused by an infestation of scabies mites and typically is accompanied by a secondary bacterial infection (Harwood and James 1979). These tiny mites burrow just under the skin surface, thereby stimulating host allergic reactions to their feces and other by-products. Mangy animals typically have large bald patches accompanied by weepy sores produced by bacterial invasion.

Demodex mites are considered a ubiquitous component of normal canine dermal fauna. Demodicosis, or red mange, is an outbreak resulting in frank disease, including crusting and secondary infection, with hair loss occurring primarily in young animals (Mullen and Durden 2002). Without treatment, demodicosis is frequently fatal. In older animals, mange typically occurs secondary to underlying immunocompromisation, either inherent or induced (as by corticosteroid use).

Demodectic mange occurs in localized and generalized forms. Localized mange most often appears in dogs younger than one year. The first sign is a thinning of hair around eyelids, lips, corners of the mouth, and front legs. Most animals will self-cure as their immune systems mature.

Generalized demodectic mange can begin as a localized case or with sudden onset. Numerous patches appear on the head, legs, and trunk, then spread and coalesce. As hair follicles become congested with debris and mites, skin sores form, and crusting and infection follow. Treatment of dogs experiencing generalized demodectic mange can be quite prolonged. Response to treatment is slow and often requires frequent changes in medication. Generalized demodectic mange must be treated under veterinary supervision, and a cure is not always feasible. Older dogs developing demodectic mange, in either form, should be screened for underlying causative factors in immune system dysfunction (Caswell et al. 1997).

Sarcoptic mange, caused by *Sarcoptes scabiei*, results in irritation, itching, inflammation, and hair loss around the face. The skin wrinkles and thickens, and under certain conditions infestations may spread over the entire body. Left untreated, the condition may result in death (Kettle 1990).

Cats occasionally are infested with *Notoedres cati*, a mite attacking the head and ears and leading to crusting and baldness. Notoedric mange is highly contagious among cats and can prove fatal within a few months of onset (Kettle 1990).

Although conventional treatments such as sulfur dips and acaricides (e.g., the amidene products) can be effective against mite infestations, new endectocides are more efficacious and dependable and cause less mammalian toxicity (McKellar and Benchaoui 1996). No over-the-counter products are reliably effective against mites, and home remedies often entail health risks to both patients and applicators.

Mosquitoes

Female mosquitoes are intermittent feeders, taking several blood meals over a period of several days. Movement from one animal to another increases the opportunity for transmission of heartworm infectious stages. More than a dozen mosquito species in several genera are capable of transmitting heartworms (Kettle 1990).

Mosquito suppression often is undertaken on an areawide basis. Mosquitoes can travel great distances, so effective suppression must be accomplished on a large scale (Harwood and James 1979). Estimating the cost of heartworm suppression by vector control is nearly impossible, for much mosquito population management is undertaken for human health and comfort.

Effective heartworm prevention is based on prophylaxis using filaricides such as diethylcarbamazine or the macrocyclic lactones to maintain blood levels and to suppress filarial development (Knight and Lok 1998). Because adequate mosquito suppression is difficult to achieve, vector suppression is an impracticable means of controlling canine heartworm (Harwood and James 1979).

Lice

Lice are obligate ectoparasites, whose stages all occur on the host (Harwood and James 1979). Because transmission occurs from infested animals, exclusion and treatment of carriers interrupt the transmission cycle. Infestations can be eliminated with any insecticide registered for on-animal use.

Felicola subrostrata is the only louse occurring on cats and is found almost exclusively on elderly or sick animals. *Trichodectes canis*, the canine chewing louse, is found on dogs, with infestations more common on very young, very old, or sick animals (Kettle 1990). *Linognathus setosus*, the canine sucking louse, may be found all over the animal's body, but numbers are heaviest around the neck and shoulders, particularly on long-haired breeds (Kettle 1990). Large numbers may produce anemia, especially in puppies.

No biological controls have been developed against lice in companion animals although a *Bacillus thuringiensis* strain has been patented that is effective against lice. Host grooming and a strong immune system are the most effective suppressants of louse infestations.

The majority of products used to treat for lice are labeled for fleas also, so it is challenging to factor out the portion purchased for louse control. Louse infestations are relatively rare on dogs and cats, and their economic significance probably is minor (amounting to less than $10 million annually). Although an individual infestation may be costly, few cases are encountered by veterinarians.

Stable Flies

Stable flies produce scabbing and irritation around the periphery of the canine pinna. Protecting animals from stable flies involves keeping the flies away from dogs, and thus dogs need a protected area to which they can retreat and avoid flies. Stable fly feeding is decreased by the use of repellents such as diethyl toluamide (DEET) and permethrin (Thomas and Skoda 1992).

Stable fly breeding site elimination is the most effective means of eliminating stable fly problems. But because these flies have long flight ranges, suppression must be accomplished on an areawide basis (Thomas and Skoda 1992).

Ticks

Ticks carry a number of pathogens, some with zoonotic potential. Although both pets and humans can become infected with Lyme disease, ehrlichiosis, and Rocky Mountain spotted fever, these diseases are transmitted only by ticks; humans cannot become infected from pets (Daniels et al. 1998).

Ticks may be removed by hand picking, using a tick-pulling device or forceps (tweezers), which allows the owner to avoid contact with potentially pathogenic secretions. Contrary to popular belief, leaving a portion of the tick's mouthparts ("head") in the wound is not dangerous to the host (Faust, Russell, and Jung 1970).

Three effective acaricides are available for tick control: fipronil, permethrin, and amitraz (Dryden 2001). Fipronil and permethrin are formulated as topical spot-on products, and amitraz is available as a collar. Fipronil and permethrin are active against fleas, as well; amitraz, less so. Amitraz is toxic if ingested, so use of such collars is contraindicated in multiple-pet households or where the potential exists for other animals to be exposed. Permethrin is toxic to cats and so cannot be used on them, nor should it be used on dogs in close contact with cats. Products providing sustained activity without repeated applications are especially convenient, thereby increasing the likelihood of use and of pet protection (Dryden 2001).

Population Monitoring and Damage Assessment

Action thresholds depend on the pet owner's tolerance for ectoparasites and their effects on animal behavior, appearance, and health. Few nonsubjective measurement criteria exist for companion animal pests, unlike for agricultural pests, and there are no established economic injury levels for companion animal ectoparasites. In general, pet ectoparasites are not tolerated and action thresholds are low because human and animal health risks are perceived as high. Many pet owners consider the presence of *any* "vermin" objectionable, and for such owners the action threshold may be a single ectoparasite (Hinkle, Rust, and Reierson 1997).

Suppression of Companion Animal Ectoparasites

Pest management on pets is limited by owner acceptance, as well as by biological and operational constraints. Few specific parasites or predators have been identified for ectoparasites. Moreover, the level of control demanded by pet owners requires highly efficacious and dependable tactics, further limiting the range of viable strategies (Rust and Dryden 1997). Most pet owners have neither the time nor the inclination to implement and to monitor mechanical or biological control strategies, a fact limiting options further.

Current Status of Integrated Pest Management Implementation

Few effective alternatives exist to chemical control of ectoparasites or to pharmaceutical prophylaxis of the disease organisms they transmit. A number of compounds highly toxic to humans and to pets have been used to suppress ectoparasites, including such on-animal materials as nicotine, chlorinated hydrocarbons, organophosphates, carbamates, and pyrethroids (Hinkle 1997). With the advent of products displaying lower mammalian toxicities, such as insect growth regulators, chloronicotinyls (e.g., imidacloprid), and phenylpyrazoles (e.g., fipronil), on-animal pest suppression can be achieved with decreased risks to humans and pets (Rust and Dryden 1997).

Control of Companion Animal Pests

Biological Control

Biological control options are limited for most pet ectoparasites. Generalist predators can affect off-host stages of pests such as fleas and ticks. But no parasites or selective pathogens have been identified for pests such as mites and lice, whose life stages all occur on-host and so have limited exposure to parasites and predators (Hinkle, Rust, and Reierson 1997). Most ectoparasitic groups have no known parasitoids or host-specific predators. Although bacterial, viral, and fungal diseases may work against on-host ectoparasites, no such disease has been identified, and additional concerns arise regarding unanticipated effects on humans and pets.

One biocontrol agent developed for flea suppression is the parasitic nematode *Steinernema carpocapsae*, which was labeled for outdoor use against immature flea stages (Rust and Dryden 1997). Application requirements such as timing, site preparation, and conditions facilitating survival made these parasitic worms poor candidates for homeowner use. Biological control organisms do not have the shelf life of chemical insecticides, must be used in a rather narrowly proscribed manner, and require a relatively extensive knowledge base for effective use (Hinkle, Rust, and Reierson 1997).

Biocontrol options for ectoparasites are restricted further because pet owners usually are intolerant of any arthropods, even beneficial ones, being associated with their pets. Additionally, of the few parasites, predators, and pathogens identified, most produce only minimal pest mortality and thus fail to achieve the level of suppression required by pet owners. Because pet ectoparasites have such extremely low action thresholds and because pet owners will not tolerate the minimal pest population necessary to sustain beneficial parasitoids and predators, biological control options rarely are used. Furthermore, biological control almost never produces pest elimination, which is the goal of pet ectoparasite suppression.

Host Animal Resistance

Like plants, pets differ in terms of their susceptibilities to various pests; unlike plants, pets have not been selected for pest resistance. Put another way, ectoparasite refractoriness is not a prime objective in the breeding of cats or dogs. Characteristics such as susceptibility to fleas and ticks are not considered commonly in breed conformation, but the development of resistance may be possible by means of molecular genetic manipulations.

In addition to the use of classical breed-selection techniques and genetic manipulation, artificial stimulation of the animal's immune system can be accomplished to induce host animal resistance (Rust and Dryden 1997). These strategies of vaccinating against ectoparasites are being studied by research laboratories but are not yet available commercially.

Chemical Control

Insecticides from virtually every chemical class, including botanicals, chlorinated hydrocarbons, organophosphates, carbamates, pyrethroids, borates, and insect growth regulators, have been used against pet ectoparasites (Hinkle 1997).

Customers appreciate the convenience of products such as the injectable lufenuron formulation that provides seasonlong flea suppression, spot-on products claiming up to three months of effectiveness, and collars releasing acaricides for several months (Rust and Dryden 1997). Pet owners are interested not only in ease of use but also in products characterized by low mammalian toxicity and decreased risk to people, pets, and the environment.

Natural Products

Most "herbal" and "natural" products tested against ectoparasites have not been demonstrated to suppress pests effectively. When used according to specific procedures, alternatives such as limonene and linalool are capable of killing certain pests such as fleas in limited situations (Rust and Dryden 1997). Many natural products touted for ectoparasitic control are quite toxic in themselves, and plant-derived compounds have no intrinsic advantage over synthetic formulations (Hinkle 1995).

Newer biogenic endectocides, such as the macrocyclic lactones, demonstrate greater efficacy, broader activity spectrum, and lower mammalian toxicity than their predecessors did (McKellar and Benchaoui 1996). Their efficacy against both bloodsucking arthropods and internal parasites makes these microbially produced compounds valuable for combating a wide range of animal health problems.

Mechanical Control

Mechanical ectoparasite suppression is a useful adjunct to other control techniques. In particular, preventing pet exposure to infested animals or contaminated environments is essential to avoiding reinfestation. Abiotic environmental factors such as temperature and humidity significantly affect off-host pest stages, so environmental extremes can be used to make conditions inhospitable for development (Hinkle, Rust, and Reierson 1997).

Other types of mechanical control include sanitation and various forms of environmental manipulation. Mechanical pest suppression techniques typically require a great deal of care by the pet owner. For instance, flea combing or hand picking of ticks requires a tractable animal and a highly motivated owner (Figure 10.4). Devices such as light traps are useful in the interception of adult fleas before they find their way onto a host, as well as in flea population monitoring. Pet owner compliance depends to a great extent on ease of implementation of mechanical controls (Figure 10.5).

Figure 10.4. Flea comb. Photo courtesy of N. Hinkle, University of Georgia.

Monitoring and Action Thresholds

Instead of determining arthropod numbers, most pet owners assess the health status of their pets and use discomfort or illness as triggers for interventions. Whereas monitoring is possible using such implements as traps or flea combs, animal irritation and human discomfort are more commonly used means of monitoring.

Action thresholds depend on the region; where pests are common, pet owners generally tolerate a higher population before initiating therapy (Hinkle, Rust, and Reierson 1997). Response also depends on human discomfort. An owner who sees tapeworm segments exiting his or her pet's anus is more likely to schedule a vet visit than the owner of an animal with heartworm is, even though the latter condition is life threatening and the former is not.

Future Developments

The future of integrated pest management for ectoparasites will include the use of innate animal physiology to suppress ectoparasites and to minimize their deleterious effects on pet and owner health. Insecticides and acaricides will continue to play necessary roles. Regardless of the tactic used against companion animal pests, consumers expect both maximum efficacy in prevention and suppression and minimal impact on themselves and their pets. This is a high-value market segment, for pet owners view ectoparasite control as an investment in their animals' health and comfort as well as in the health and comfort of human members of their households. As consumers become increasingly concerned about the health effects of ectoparasites, both on themselves and on their pets, willingness to invest in ectoparasite suppression will increase.

Figure 10.5. Pet owner compliance depends to a great extent on ease of control implementation. Photo courtesy of N. Hinkle, University of Georgia.

Appendix A. Abbreviations and Acronyms

DEET	diethyl toluamide
FAD	flea-allergy dermatitis

Appendix B. Glossary

Ectoparasites. Insects and their eight-legged relatives, e.g., mites and ticks, occurring on the outside of the body.

Flea-allergy dermatitis. Severe allergic reaction to flea salivary secretions.

Mange. An infestation of mites accompanied by skin inflammation and hair loss.

Pruritus. Itching.

Tick paralysis. A potentially lethal condition caused by the salivary products secreted by ticks.

Zoonoses. Disease agents capable of being transmitted to and affecting humans.

Literature Cited

American Veterinary Medical Association (AVMA). 2002. *U.S. Pet Ownership and Demographics Sourcebook, 2002*. American Veterinary Medical Association, Statistical Research Group, Schaumburg, Illinois.

Caswell, J. L., J. A. Yager, W. M. Parker, and P. F. Moore. 1997. A prospective study of the immunophenotype and temporal changes in the histologic lesions of canine demodicosis. *Vet Pathol* 34:279–287.

Daniels, T. J., T. M. Boccia, S. Varde, J. Marcus, J. Le, D. J. Bucher, R. C. Falco, and I. Schwartz. 1998. Geographic risk for Lyme disease and human granulocytic ehrlichiosis in southern New York state. *Appl Environ Microbiol* 64:4663–4669.

Dryden, M. W. 2001. Internal and external parasites of the racing Greyhound. In *Proceedings, 17th Annual International Canine Sports Medicine Symposium*, January 2001, Orlando, Florida. Pp 16–23.

Dryden, M. W. and M. K. Rust. 1994. The cat flea: Biology, ecology and control. *Vet Parasitol* 52:1–19.

Faust, E. C., P. F. Russell, and R. C. Jung. 1970. *Craig and Faust's Clinical Parasitology*. Lea & Febiger, Philadelphia, Pennsylvania.

Harwood, R. F. and M. T. James. 1979. *Entomology in Human and Animal Health*. 7th ed. Macmillan Publishing, New York.

Hermesmeyer, M., R. K. Limberg-Child, A. J. Murphy, and L. S. Mansfield. 2000. Prevalence of *Dirofilaria immitis* infection among shelter cats. *J Am Vet Med Assoc* 217:211–212.

Hilton, E., J. DeVoti, J. L. Benach, M. L. Halluska, D. J. White, H. Paxton, and J. S. Dumler. 1999. Seroprevalence and seroconversion for tick-borne diseases in a high-risk population in the northeast United States. *Am J Med* 106:404–409.

Hinkle, N. C. 1995. Natural born killers. *Pest Control Technol* 23:54, 56, 116.

Hinkle, N. C. 1997. The history of flea control. *Pest Control Technol* 25:34, 36, 82.

Hinkle, N. C., M. K. Rust, and D. A. Reierson. 1997. Biorational approaches to flea (Siphonaptera: Pulicidae) suppression: Present and future. *J Agric Entomol* 14:309–321.

Kettle, D. S. 1990. *Medical and Veterinary Entomology*. CAB International, Oxon, United Kingdom.

Knight, D. H. and J. B. Lok. 1998. Seasonality of heartworm infection and implication for chemoprophylaxis. *Clin Tech Small Anim Pract* 13:77–82.

Lyon, W. F. 1997. Pet pest management. Bulletin 586. The Ohio State University, Columbus.

McKellar, Q. A. and H. A. Benchaoui. 1996. Avermectins and milbemycins. *J Vet Pharmacol* 19:331–351.

Mullen, G. R. and L. Durden. 2002. *Medical and Veterinary Entomology*. Academic Press, San Diego, California.

Parish, L. C. and R. M. Schwartzman. 1993. Zoonoses of dermatological interest. *Semin Dermatol* 12:57–64.

Rust, M. K. and M. W. Dryden. 1997. The biology, ecology, and management of the cat flea. *Annu Rev Entomol* 42:451–473.

Thomas, G. D. and S. R. Skoda. 1992. The stable fly: A pest of humans and domestic animals. Institute of Agriculture and Natural Resources, Agricultural Research Division, University of Nebraska–Lincoln.

Watson, J., W. J. Wetzel, and J. Burkhalter. 1991. Human disease caused by dog heartworm. *J Miss State Med Assoc* 32:399–401.

11 Urban Integrated Pest Management

Integrated pest management (IPM) has its roots in contemporary agriculture, in which questions of economics—cost of treatments versus the value of benefits—are a driving force. Recently, IPM techniques and philosophies with slightly different emphases have been applied to urban pest management, which concerns the direct interactions between people and pests, whether in the innermost parts of a city, in a suburban area, or in a comparatively rural location. Like agricultural IPM, urban IPM applies pest management decisions on the basis of determined need, instead of on the basis of preventive, or prophylactic, philosophy. Calling for a multidisciplinary approach, urban IPM uses various nonchemical and chemical techniques in an overall management strategy.

Unlike agricultural IPM decisions, however, urban IPM decisions cannot be made on an entirely economic basis (Gibb 1999), for often in urban environments no commodity is produced, harvested, or sold. Urban IPM balances the cost of treatments with a human comfort value. Compared with economic gains in agricultural IPM, urban IPM quantifies human comfort value less readily and becomes more subjective when human tolerance of pests and quality of life are considered. Thus, a management decision based solely on the economic return of pest control inputs is unrealistic in an urban environment. Rather, urban IPM is based on the premise that, even in the absence of monetary value, damage can occur and pest management can be justified.

Urban IPM includes a human factor rather than an economic factor in the pest management equation. For example, qualitative aspects such as aesthetics, comfort, health, and peace of mind substitute in many ways for the quantitative economics involved in agricultural IPM decision making. Because many of the qualifying aspects for urban IPM are subjective, they cannot be measured economically. This subjectivity represents a major challenge in urban IPM and can cause it to become very complex and subject to individual interpretation. But the unifying and distinguishing characteristic of all urban IPM approaches is the human factor. Whether a pest is a pathogen, an insect, a weed, or a rodent, an organism that damages homes, structures, clothing, food, or landscape plantings or harms, annoys, or otherwise interferes with people and their activities is considered an *urban pest.*

In urban IPM, social concerns replace economic gains or losses as the driving force behind decision making. Among these concerns are public attitudes, perceptions, and prejudices regarding pests and pesticides and their effects on human activity and the environment. Within the last few years, environmental and human health issues clearly have become prime considerations of regulatory agency personnel. A massive push to decrease human exposure to toxicants of all kinds is under way. Concurrently, a substantial increase in litigation relating directly to pest management and pesticide application is occurring. Rights to be informed of pesticide use, residues, and toxicity levels; rights to pollution-free public environments in which to work, visit, or study; and rights to be free from harmful parasites, disease, and nuisance pests are being legislated. Attaining acceptable levels of pest management without exposing people and the environment to excessive risks from pesticides is the major objective of urban IPM.

The definition of *urban IPM* has evolved over the years as various environmental and social pressures have been brought to bear. Olkowski, Daar, and Olkowski (1991), Owens (1986), and others have had insights into IPM concepts. Their insights and discussions have been considered in the development of the following definition:

> Urban integrated pest management is a process that employs physical, mechanical, cultural, chemical, biological, regulatory, and educational techniques to prevent pests from establishing populations capable of causing unacceptable economic, social, medical, or aesthetic damage. Urban IPM utilizes regular inspections to determine if and when a treatment is needed. Treatments are chosen based on proven records of success and become part of the overall IPM strat-

> egy. Chemicals, when required, are target-specific, low-impact and are made only when and where monitoring has indicated that the pest will cause intolerable damage or annoyance. The IPM program must, as a result, be environmentally, socially, and economically compatible to meet public expectations. (Gibb 1999)

Five main components of urban IPM, whether inside buildings or out-of-doors, should be considered essential. These include the need for (1) inspection, (2) monitoring, (3) situation-specific decision making, (4) pest management technique application, and (5) evaluation and record keeping.

Inspection

In its strictest sense, IPM is simply a decision making process that begins with a thorough inspection to identify the pest and the habitat in which it exists and the nature of its interactions with humans. Identification of the pest and its demographics is key to determining the seriousness of the problem and the management steps that must be taken.

Identification is a step sometimes taken for granted. Nevertheless, accurate identification of the pest must always be confirmed. Although this process is one of the most basic elements of pest management, mistakes in identification still are made commonly when many urban pests are being studied, especially pests that are similar in appearance or behavior. Accurately identifying the pest, recognizing which life stages are present, and understanding the life history of the pest and how it interacts with people help a pest manager exploit the weak links in pest biology.

An equally important component of an inspection is that of assessing potential human/pest interactions. How the building or landscape is constructed and used, how it is accessed and maintained, how its use and construction affect the local and general environments, and whether any additional regulatory influences are at work all are important pieces of information that can be provided by an inspection. Assessing the building and landscape habitat allows for an evaluation as to relative pest-conducive situations and, therefore, allows prophylactic or preventative solutions to be implemented. Also implicit in the pest equation is a specific evaluation of the human factors involved—personal and company biases, fears, concerns, and pest tolerance levels. Together, these interactions inform decisions regarding appropriate pest management techniques for each situation.

Monitoring

The primary objective of monitoring is to obtain an estimate of the pest population size and distribution. Monitoring begins during the initial inspection and should become an ongoing component of the IPM program. Data obtained from monitoring, coupled with knowledge of pest tolerance levels, allow the manager to assess a problem as severe, moderate, or minor. This process also confirms whether pest damage is evident, increasing, or decreasing. It is not uncommon to find pest treatments applied inappropriately in situations because pests have become inactive or have reached a life stage during which treatments no longer are effective.

Information obtained from monitoring plays an important role in determining the type, intensity, and duration of management techniques required. Only through monitoring can a decision be made regarding whether an action is justified. Monitoring provides demonstrative proof that pesticides or other management tactics need to be applied. Often such justification is necessary to reassure the public. Monitoring helps a pest manager determine a time both to begin a pest control action and to end it.

Perhaps monitoring's most important advantage is that it can be used as an evaluation tool to assess the effectiveness of any pest control strategy over time. Monitoring should be considered one of the basic tenets of effectively designed urban IPM.

Establishing Pest Tolerance Levels

It is unrealistic to expect to eradicate all pests either outside or inside buildings. Notwithstanding the broad-spectrum, long-residual pesticides used in the 1960s and 1970s, it has become evident that eradication of pests is impossible, even if society were willing to accept the negative environmental consequences of the use of these products and regimens. Rather, in most situations, low pest-populations, monitored and managed carefully over time so that they do not increase beyond a certain tolerance threshold, are acceptable. When compared with the potential negative human and environmental health effects of total reliance on chemical pesticides, a low, well-managed pest population is, in fact, preferable.

Education is key to helping establish realistic tolerance levels. Educating clientele about whether, when, how, and where pests harm humans is the first step. Urban-pest managers must recognize that tolerance levels are situation specific. Very sensitive situations may dictate that the presence of even a single pest such as a brown recluse spider exceeds the highest tolerance level; in other situations, even great numbers of pests such as ants may not be a serious threat, and thus a "no action" policy may be the correct response. Tolerance levels also depend on specific sites, clientele groups, locations, and, even, time. For example, pest tolerance levels may be lower in hospitals, daycare centers, or nursing homes because the sick, the young, or the elderly are more sensitive to the pest. Established pest tolerance levels may decrease when large gatherings are expected either in a building or in an outdoor setting, only to rise again after the event ends.

The decision regarding whether to use pest management tactics pivots on tolerance level. Because such decisions in urban IPM are based on more than economics, they can be difficult to make.

Tolerance Levels in Urban Landscapes

In plantings such as ornamentals and turfgrasses, whose values cannot be translated easily into dollars, the IPM issue becomes complicated. Assessment comes to hinge on the relative worth of the plant in its surroundings or on its aesthetic worth. In landscape pest management, even the relative value of a planting can depend on such factors as its location in the landscape. For example, in a residential setting, limited pest damage might be tolerated in a backyard, where relatively few people might see it; the same amount of damage in the front yard would tend to be unacceptable, however. In golf courses, limited pest damage and invasion are tolerable in the rough areas because playability of the course is unaffected. Much lower levels of the same damage are acceptable, however, in intensively maintained fairways. Even less pest damage is tolerated on the tees and greens, and especially on those closest to the clubhouse, where they are played and seen by more people more often.

A decision to manage a pest population depends on two major interacting factors: the size of the pest population and the associated tolerance level. Together, pest population level, species, and host susceptibility determine potential damage to landscapes. Damage usually occurs as a result of the pest's feeding on or otherwise injuring a plant. Tolerance levels must, therefore, also include assessments of potential damage.

Tolerance Levels in Urban Buildings and Structures

Perhaps even more subjective than the process of determining treatment thresholds in landscapes is that of determining pest tolerance levels in urban buildings. As in landscape IPM, tolerance levels are determined according to the pest's potential to cause harm or annoyance and by the residents' personal tolerance levels, factors both influenced by the site/environment. The issue may be confounded further when personal health or safety issues are at stake. For example, the number of ants that personnel in an office building might tolerate before implementing a costly management practice is greater than that of brown recluse spiders, simply because of the potential negative consequences associated with each pest. Even if all buildings and locations were identical, tolerance levels would differ because each person has a unique zone of comfort. For instance, even if an individual became convinced that he or she would not be bitten, stung, or contaminated in any way by a specific pest, comfort levels might remain low insofar as the individual was uncomfortable "just knowing" that a pest was present in home or office.

Situation-Specific Decision Making

By definition, IPM in urban environments (buildings or surrounding landscapes) must be site specific. Because of the multiplicity of possible site-specific variations, a list of universally acceptable thresholds for pest tolerance cannot be established. Significant differences occur depending, for instance, on whether the landscape is a home lawn or an athletic field. Parks, cemeteries, and rights-of-way all have significantly different uses. Even within a landscape, pest tolerances depend on several factors known only to those managing the grounds. Factors such as upcoming events, expected traffic use differences, site-specific weather conditions, and irrigation stresses, soil types, and seasonal changes all play roles in IPM decision making.

On landscape plantings, factors such as pest life stage and plant variety and development, health, and susceptibility to the pest must be known. Environmental factors such as irrigation, soil compaction, plant susceptibility to the pest, as well as the interaction of myriad other possible stresses potentially affecting the plant also must be accounted for in a formula predicting when a pest population should be controlled. In certain instances, specific numbers have been generated to give a general idea of pest thresholds and acceptable damage tolerances (Schumann et al. 1997). It must be remembered, however, that these are general guidelines only and should be adapted to the specific circumstances of the site and to the pest in question.

Likewise, certain urban structural environments such as hospitals or food production plants are especially sensitive to pests. Health and food contamination concerns drive pest tolerances in these structures to levels lower than those in parking garages or in nonfood warehouses. Even within a single residential or public structure, localized areas such as kitchens and bedrooms are more sensitive than basements, attics, and patios, and a unique pest threshold exists for each environment.

As a result of these variables, no published table of pest tolerance levels could apply universally: pest tolerances must be established on a case-by-case, situation-by-situation basis. Before attempts to establish a tolerance level are made, it is crucial to thoroughly understand the pest, its habitat, and the people it affects. Prior understanding of these issues helps a manager avoid the pitfall of basing treatments on impulses.

Potential advantages gained by applying pest management tactics must be weighed against application costs. This comparison is always at the heart of urban IPM. Although direct treatment-costs such as personnel time and resources, golf course down time, or inconvenient use restrictions of an area in a building while pesticides are being applied are measured with relative ease, the less-direct costs, such as strained public relations and the potentials for negative environmental side effects and for associated lawsuits, often are more important.

Urban pests must be managed so that people can function and live in the way they have become accustomed to. Urban IPM encourages pest management tactical choices based on a comparison of the advantages gained with the possible costs and disadvantages accompanying a decision to treat. Site-specific judgments must be made in each instance and must take into account potential perils to humans and pets (e.g., direct exposure to chemicals). Preventing environmental contamination by considering the fate of the pesticide through time and space, as it is affected by factors such as proximity of surface- or groundwater, type of soil, and density of plants also is required. Experience and sound judgment, together with a conscientious consideration of each of the aforementioned factors, are at the heart of site-specific determinations.

Human-Health Concerns

Urban pests may damage plantings, structures, and furnishings; cause annoyance; or create serious human-health concerns. Wasps can sting, fleas can bite, and flies can spread germs and other disease agents. In addition to harming people directly through biting or stinging, pests also can cause serious health threats simply by being present. For example, fecal material, dander, the discarded skins of insects during growth, as well as the expired bodies of past generations, break down over time. During decomposition and over time, fine particles can become airborne and, when inhaled by a person already predisposed to asthma, can cause severe complications. Cockroaches, dust mites, and, recently, house mice and lady beetles all have been implicated directly in human allergies and asthma (Rosenstreich and Eggleston 1997; Sporik et al. 1990).

It also should be recognized that many individuals are intolerant of pesticide residues in the environment. Some people may oppose pesticides on the basis of fear or prejudice whereas others are sickened physically by pesticides, even at low concentrations. A condition called *environmental sickness*, or *multiple chemical sensitivity*, is linked to exposures to chemicals in the environment (Ziem 1992). A theory called *toxicant-induced loss of tolerance* might explain why overexposure to a single environmental pollutant (not always a pesticide) leads to heightened sensitivity to several other unrelated substances, including pesticides, in the environment (Miller 1997).

Concerns have arisen and debate has ensued about the issue of children's exposure to pesticides (NRC 1993) and whether there is a correlation between exposure and acute or chronic health disorders. Although this issue still is being debated, neither side questions that human health can be affected negatively when pesticides are misused. What should be recognized by all is that pesticides used properly are part of a safe and effective IPM program. In decisions regarding whether to use pest management tactics, *trade-offs* (advantages due to decreases in pest populations, and potential disadvantages due to management applications) must be considered.

Over the years, society has come to depend on chemical pesticides as the primary tool with which to maintain pest-free landscapes and buildings. Strict reliance on and overuse of chemicals have, at times, created a pesticide treadmill, by which more and more chemicals have been required to achieve consistent results. Because of the constant and intimate human presence in urban environments, perceived health risks due to pesticide applications in these areas have increased. The potentials for direct human exposure to pesticides as well as for indirect exposure through environmental pollution have become important and volatile social issues. The dilemma faced by urban IPM is that such concerns have not lessened public expectations for pest-free buildings or for high-quality, damage-free plantings (Sadof and Raup 1997). Even without the aid of many of the chemical management tools of the past, pest managers still are expected to provide pest eradication in homes and buildings as well as pest- and damage-free plantings in the landscape. Fortunately, new pest management technologies are helping make this possible in the urban environment.

Unrealistic expectations regarding pest-free environments often confound IPM efforts. Realistic IPM expectations recognize that a degree of pest presence may be acceptable, provided the long-term health of individuals or plants is unaffected. Urban IPM requires extensive public education. Changes in public perceptions and tolerances of pest presence and damage must be effected before the full range of IPM benefits can be achieved.

Integrated pest management in the urban environment remains, in many ways, similar to that in the agricultural environment. Protecting human interests from damage—and not necessarily killing pests—is the basis of urban landscape IPM. Management strategies can be chosen from among chemical, cultural, mechanical, biological, and/or regulatory options. Decisions should be made on a site- and situation-specific basis and be consistent with an overall pest management plan.

Application of Pest Management Techniques

The terms *pest control* and *pest management* often are interpreted as synonymous. There is a fundamental, though subtle, difference between them, however. *Pest control* traditionally has involved a one-dimensional emphasis on pesticidal remedies for pest problems. *Pest management* is not a reaction but a deliberate process of treating a pest with a management tool such as a pesticide. Pest management requires an understanding of pest population levels and of possible applications of a number of different control tactics in a pest management framework. In pest management, pest thresholds are established and used as decision making guides to clarify whether action against a certain pest is desirable. In addition, pest management, instead of stressing reliance on a pesticide, stresses integration of at least two approaches to managing complex or severe pest problems.

In any IPM program, accurate determination of the need for intervention depends on many situation-specific and use-specific factors. Therefore, IPM in urban environments has come to focus on (1) preventing buildings and landscapes from being invaded by pests and (2) offering an array of potential management techniques when infestations do occur.

In IPM, the word *integrated* implies a multidisciplinary approach whereby several management options can be brought to bear on a single problem. Such an approach is considered the most environmentally healthy and viable long-term strategy available. Choosing from a variety of possible management strategies ensures that the best management/site fit is achieved. The better the fit, the less the chance of undesirable consequences.

Once it has been determined that a management practice should be implemented, a decision should be made regarding appropriate tactics. Arriving at such a decision seldom is a straightforward process. It is made simpler, however, when the pest manager has a thorough understanding of the environment, the biology, the life cycle, and the ecology of the infesting pest and of the available management options. Often the best method for choosing a tactic is to compare all evident advantages with all possible limitations. Factors such as efficacy, application ease, environmental impact, and a host of other on-the-job experiences will be of value to decision makers.

Pest Prevention by Exclusion

Preventing pest entry or spread is a fundamental IPM objective. On a national and a statewide basis, regulatory agencies help prevent the spread of pests by enforcing requirements on shipping food commodities, sod, and other landscape materials from one location to the next. A great number of diseases and insect pests are held in check by such regulations. Understanding how pests move and are introduced into a new location allows circumvention of pest distribution. Routinely inspecting, cleaning, and disinfecting equipment, transportation vehicles, and other materials in which pests might be introduced from one site into another is also an important practice requiring knowledge of pest presence and biology.

Additionally, efforts to eradicate a small pest-population before it becomes established are desirable. Governmental agencies actively monitor for the invasion of suspected pests and stand ready to perform localized, intensive eradication procedures in an effort to stave off permanent establishment of new pests. On a local basis, urban-pest managers using similar tactics can help prevent movement and establishment of pests from an infected area into a new site. The first step is routine monitoring.

A significant part of pest management is done well before a building or a landscape is constructed. Designing with pest management in mind is one of the most effective urban IPM techniques available. Both structures and landscapes can be designed that will (1) make it difficult for a new pest to enter and to live and (2) make subsequent management tactics much more efficient.

Landscape Design

Urban landscape IPM begins in the designing phase, with a goal of minimizing potential pest problems. Early decisions as to variety selection and placement, soil and seedbed preparation, planting, irrigation, fertilization management, use, and aesthetics are aspects considered by effective IPM planners. Continuously maintaining plant vigor and decreasing environmental stress through appropriate cultural management are key to warding off pests and allowing plants to recover from infestation rapidly.

Landscape pest problems often can be avoided simply by selecting plant varieties that are well suited to the site and to the purpose for which the landscape is being used. Choosing plant species that are certified tolerant or resistant to pests is part of that design. Laboratory and field screening of plant varieties and cultivars resistant to pests has been underway for several decades. Methods involve choosing plant genotypes relatively tolerant of, resistant to, or undesired by pests and, subsequently, combining these genetically resistant traits with other desirable traits. Lists of resistant or pest tolerant plants should be studied, and plant selections made as part of a predetermined landscape design. Even minimal efforts in the initial design of landscapes and in the selection of plants can pay huge dividends later if pests invade.

Well before the first seedling is planted, many more pest problems can be prevented by planning. Site selection and seedbed preparation are two often-overlooked areas. Type and grade of soil, movement of water and air, slope of the land, and proximity of other landscapes, buildings, and water sources all will influence future pest potentials.

Seeding properly prepared beds with the best match of plant variety and ensuring that optimal irrigation and fertilization practices can continue also are important preplant decisions that a pest manager should help make. These factors should be considered carefully and, when necessary, modified to facilitate decreased pest problems.

Building Design

Like landscape design, building architecture and layout are potentially important factors to consider in pest management (Figure 11.1). "Building pests out" of a structure simply means "making the invasion and the persistence of pests there difficult." Decreasing a pest's initial attraction to a building by modifying lights, eliminating food and water sources, and restricting harborage sites greatly diminishes the susceptibility of a building to pest invasion. Easy entrance to the building can be denied pests by means of open door and window policies, screens, and door sweeps, and through the use of sealants in foundation cracks and in plumbing and electrical conduits. Ensuring that air intake and air handling systems are working properly and are closed off adequately is important. Controlling temperature and humidity and restricting or eliminating food sources through sanitation and food handling procedures can prevent pest proliferation inside a building and are essential elements in structural IPM. Choosing and modifying mechanical equipment such that it is easy to inspect, clean, and treat for pests makes IPM possible in many urban structures.

Figure 11.1. Like landscape design, building architecture and layout are potentially important factors to consider in pest management. Photo courtesy of T. Gibb, Purdue University.

Cultural Controls

Once established, pests become part of the urban environment. They affect their surroundings, but also are affected—directly or indirectly—by every practice occurring there. Many cultural practices can be manipulated to the detriment of pests. Making the pest environment less conducive to reproduction is the goal of cultural pest control. In urban buildings, sanitation often is the most important cultural control. Denying pests access to food, water, and harborage through increased sanitation makes it difficult for pests to enter or to persist once they are present.

In landscape management, it is accepted generally that healthy, vigorously growing plants can withstand more pressure from disease, weed, and inspect pests than stressed plants can. Doubtless, cultural practices such as mowing, fertilizing, pruning, mulching, and irrigating all affect pest populations indirectly.

Inattention to these management practices can create stresses that encourage the development of pests. Proper identification and alleviation of these stress factors through cultural management changes are some of the longest-term and most environmentally conscious methods of pest control in the landscape. For example, understanding that high populations of weeds in a turfgrass stand are good indicators of stressed turfgrass and subsequently resolving such stress problems, be they irrigation, traffic, soil compaction, or other, can diminish the need for herbicides. Likewise, trees under stress due to any cause are relatively attractive and susceptible to wood borers and other insect pests. Landscape managers must understand the interactions that irrigating, fertilizing, mowing and pruning, soil compacting, and other human activities have on potential plant pest problems.

Biological Control

Using living organisms to control pests is not a new science, but this approach is beginning to see greater acceptance in urban pest control. Biological control lends itself more readily to landscape situations but has implications to indoor pest populations, as well.

Evidence of biological control efforts exists from ancient times, and significant control was achieved through the introduction of biological controls in the early part of the twentieth century. With the introduction of modern synthetic chemical insecticides, especially during the latter half of the twentieth century, however, biological control became a largely forgotten science. New chemical insecticides were cheap and seemed to be "cure-alls." But, as has been discussed, this perception was unfounded. As the disadvantages of synthetic chemicals became evident and public demands for more environmentally friendly methods of pest control grew, biological controls again were sought. Biological controls offer many advantages over conventional chemicals. Chief among these advantages are the facts that biological controls can be safer both to people and to the environment and, once introduced, can continue to be effective without human intervention. Certain instances of introduced biological controls have succeeded whereas other types of controls have been disappointing. Of course, care must be taken to avoid introducing biologicals into the urban environment only to realize later that they have become pests themselves.

Figure 11.2. In nature, many different parasites or predators can be found, each with a unique ability to persist, to kill, and to seek out potential hosts. Photo by Scott Bauer, Agricultural Research Service, U.S. Department of Agriculture, Beltsville, Maryland.

Predatory or parasitic organisms feeding directly on landscape pests have shown promise in their abilities to control landscape pests. In nature, many different parasites or predators can be found, each with a unique ability to persist, to kill, and to seek out potential hosts (Figure 11.2). Pest-specific diseases also have been used with different degrees of success in the management of pest populations. Successful organisms, mostly fungi and bacteria, have been propagated in the laboratory and now are available commercially.

Naturally occurring biological control agents probably are more important than one might imagine (Zenger and Gibb 2001). Insect pests usually are controlled naturally by beneficial arthropods and pathogens that can moderate or often prevent outbreaks of pest populations. It is telling that landscapes not managed intensively seldom are prone to serious pest outbreaks. Society has learned, through hard-won experience, that when the natural balance is upset by chemical or cultural interference, pests readily move in. Thus, careful consideration must be given to the effect that each management input can have on a system's beneficial organisms.

Identifying and conserving naturally existing floral and faunal pest-control organisms are logical steps in IPM implementation. Maintaining an equilibrium between pests and their natural controls will diminish the need for pesticides, save urban dwellers money, and benefit the urban ecosystem greatly.

Overuse of nonselective chemicals interferes directly with the potential of naturally occurring beneficial organisms. Spot treatments and appropriate monitoring of pest distributions allow urban-pest managers to avoid the overuse of chemicals and, thus, the negative effects on nontarget beneficials. Releasing a pest population from its natural biological checks and balances may set the stage for a dramatic and often devastating resurgence of a pest population.

In summary, urban-pest managers need not wait for the discovery of new or better biological controls but can develop their own through conservation and encouragement of naturally occurring predators and parasites.

Alternative Pesticides

Toxins such as azadiractin, rotenone, abadilla, and pyrethrin all are derived from plants. Although these are not, technically, biological pesticides, they are materials with a botanical origin that offer some of the same pest management advantages that true biological controls do.

As discussed under "Biological Controls," fungal, bacterial, or viral diseases can be used to control certain pests. Biological microbes (or microbe derivatives) also are effective tools in urban IPM. Several such materials have been developed for use in landscape as well as in structural pest control.

Discovering new and better alternative control agents is an active area of research promising to affect the future of urban-pest control significantly. To date, these agents have been effective as bait or as broadcast or foliarly applied sprays. New technology and biochemistry also have aided the proliferation of urban IPM. Certain pesticides have been developed that are engineered to be taken up and transported systemically through a plant, thereby improving efficacy and decreasing nontarget toxicity. Biotechnology now aims to incorporate the genes responsible for producing toxins through insertion into the plant. Significant progress in the field of molecular genetics has been made on a number of plants.

Chemical Pesticides

Use of synthetic chemical pesticides also is an important aspect of most IPM strategies. Chemical treatments are especially effective because they can be applied with relative ease, are quick acting, and can bring a high pest-population down to an acceptable level rapidly. When used in this manner and for this purpose, chemicals can be an integral part of an IPM program. Only when pesticides, to the exclusion of other techniques, are relied on continually to maintain pest populations at low levels do problems arise.

Recently, great strides have been made in the development of low-impact chemistries. Environmentally friendly, narrow-spectrum, selectively less-toxic pesticides have been developed by the chemical industry. As these have been registered by the Environmental Protection Agency, older, higher impact, less environmentally compatible pesticides have been removed from the market.

Targeting in Time and Space

The solution to a certain pest problem does not always depend on a new or better pesticide. Often the difference between success and failure in decreasing a pest population lies in knowing where, when, and how to apply the selected control technique. Such knowledge usually is based on a study of pest habits and biologies.

Application timing of many landscape management techniques can be improved through the use of pathogen, weed, and insect phenological models that track the development of pests by using temperature as the dependent variable. The use of temperature-based models or of tracking phenological indicator plants can suggest to pest managers an appropriate timing for applications. Timing specific treatment applications according to these models has proved much more advantageous than timing them according to a calendar spray schedule.

Targeting pesticide applications to those sites where monitoring has determined a need for control (*spot treating*) decreases the amount of pesticide applied and conserves natural controls already in place (Figure 11.3). Integrated pest management dictates that spot treatments replace blanket treatments wherever possible.

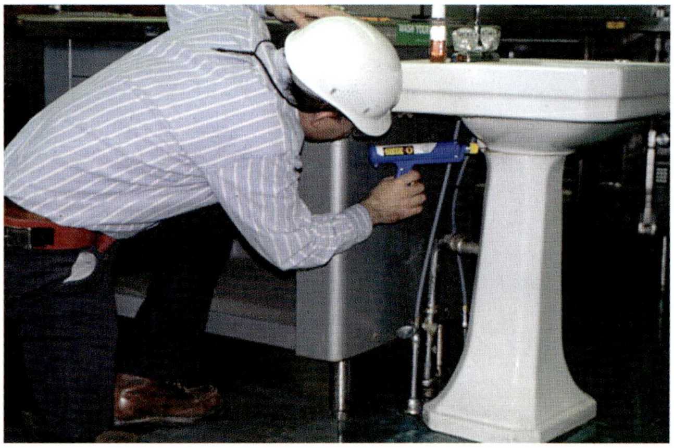

Figure 11.3. Targeting pesticide applications to those areas in which monitoring has determined a need for control (spot treating) decreases the amount of pesticide applied and conserves natural controls already in place. Photo courtesy of T. Gibb, Purdue University.

Using new technologies to deliver pesticides directly into locations where pests are present (that is, in the *target zone*) diminishes the probability of exposure to nontarget organisms, decreases the quantity of pesticides needed for treatment, and increases pesticide efficacy. Replacing baseboard with crack and crevice applications is one way of targeting pesticides in buildings and structures. Injection techniques may be used to deliver ever-smaller amounts of pesticides into soils or trees in the urban landscape. Select pesticides also may be taken up systemically and diffused throughout the plant, ultimately killing only those insects feeding directly on it.

Professional urban IPM continually incorporates new procedures and technologies into pest management, thereby decreasing both the populations of pests and the negative effects of pesticide applications. Development of pesticide-laced baits has improved greatly the ability of structural-pest managers to control rodent and insect pests such as ants, cockroaches, and termites. Baiting techniques significantly decrease the amount of toxic materials applied in the urban environment and yet achieve pest population control equal to or better than that of traditional methods.

Evaluation and Record Keeping

Not only is methodical record keeping an important tool providing historical data, but it can help the manager evaluate control techniques over time. After pest management practices are incorporated into urban areas, their efficacy must be evaluated. Decisions to continue, to increase, or to suspend a pest management practice should be made only in light of the effects of its previous use. Evaluations can be educational in the long run and are an especially crucial feature of IPM.

Site-specific factors such as pest-infected area size, exact location, population estimates, damage amounts, symptoms, dates, and, where appropriate, weather conditions leading up to infestation are types of information that should be recorded. When examined in relation to the effect of current IPM practices, such information allows pest managers to make informed decisions. As a rule, thorough record keeping provides a database from which future pest management decisions can be made.

In conclusion, urban IPM programs are based on the same basic philosophies that agricultural IPM programs are. Minor differences occur because urban IPM involves pest/human interactions that may occur in buildings or landscapes in which health and social concerns are crucial factors. Inspection and monitoring tools dictate the application of various management tactics through a decision-making process that considers potential injury by pests as well as potential injury from pest control tactics. Urban IPM strategies are, ultimately, a compilation of many common sense decisions based on a sound understanding of the pest, the environment, and the social implications of one or more control tactics.

Appendix A. Glossary

Environmental sickness. An illness linked to exposures to chemicals in the environment.

Integrated. The quality of being an outcome of several management options brought to bear on a single problem.

Multiple chemical sensitivity. See environmental sickness.

Pest control. A one-dimensional emphasis on pesticidal remedies for pest problems.

Pest management. A deliberate process of treating a pest with a management tool such as a pesticide.

Spot treating. Targeting pesticide applications to those areas in which monitoring has determined a need for control.

Target zone. Locations in which pests are present.

Toxicant-induced loss of tolerance. A theory that may explain why overexposure to a single environmental pollutant (not always a pesticide) leads to heightened sensitivity to several other unrelated substances, including pesticides, in the environment.

Trade-off. Potential advantages due to decreases in pest populations, and potential disadvantages due to management applications.

Urban integrated pest management. A process with five main components, whether inside or outside of buildings: (1) inspection, (2) monitoring, (3) situation-specific decision making, (4) pest management technique application, and (5) evaluation and record keeping.

Urban pest. An organism that damages homes, structures, clothing, food, or landscape plantings or harms, annoys, or otherwise interferes with humans and their activities.

Literature Cited

Gibb, T. 1999. *Turfgrass Pest Management: A Correspondence Course*. Center for Urban and Industrial Pest Management, Purdue University, West Lafayette, Indiana. 452 pp.

Miller, C. S. 1997. Toxicant-induced loss of tolerance: An emerging theory of disease. *Environ Health Perspect* 105:445–453.

National Research Council. 1993. *Pesticides in the Diets of Infants and Children*. National Academy Press, Washington, D.C.

Olkowski, W., S. Daar, and H. Olkowski. 1991. *Common Sense Pest Control*. Taunton Press, Newtown, Connecticut. 716 pp.

Owens, J. M. 1986. Urban pest management: Concept and context. P. 112. In G. Bennett and J. Owens (eds.). *Advances in Urban Pest Management*. Van Nostrand Reinhold, New York. 399 pp.

Rosenstreich, D. L. and P. Eggleston. 1997. The role of cockroach allergy and exposure to cockroach allergen in causing morbidity among inner-city children with asthma. *N Engl J Med* 336:1356–1363.

Sadof, C. S. and M. J. Raup. 1997. Aesthetic thresholds and their development. Pp. 203–226. In L. G. Higley and L. P. Pedigo (eds.). *Economic Threshold for Integrated Pest Management*. University of Nebraska Press, Lincoln. 327 pp.

Schumann, G. L., P. J. Bittum, M. L. Elliott, and P. P. Cobb. 1997. *IPM Handbook for Golf Courses*. Ann Arbor Press, Chelsea, Michigan. 264 pp.

Sporik, R., S. T. Holgate, T. A. E. Platts-Mills, and J. J. Cogswell. 1990. Exposure to house-dust mite allergen (Der p 1) and the development of asthma in childhood: A prospective study. *N Engl J Med* 323:502–507.

Zenger, J. T. and T. J. Gibb. 2001. Identification and impact of egg predators of *Cyclocephala lurida* and *Popillia japonica* (Coleoptera: Scarabaeidae) in turfgrass. *Environ Entomol* 30:425–430.

Ziem, G. E. 1992. Multiple chemical sensitivity: Treatment and follow-up with avoidance and control of chemical exposures. Advancing the understanding of multiple chemical sensitivity. *Toxicol Ind Health* 8:181–202.

Related Web Sites

Sustainable Urban Landscapes Information Series. 2002. Homepage, <http://www.sustland.umn.edu/ index.html> (9 September 2002)

Technical Resource Center for IPM in Public Schools. 2002. Homepage, <http://www.entm.purdue.edu/schoolipm>

12 Economics in the Design, Assessment, Adoption, and Policy Analysis of Integrated Pest Management

How Economics Relates to Integrated Pest Management

Why do farmers sometimes fail to adopt integrated pest management (IPM) practices that have succeeded on experiment stations? Would crop insurance encourage IPM adoption? How much should the government invest in promoting IPM? What is the value of IPM research?

These questions center not on pests and control methods, but on farmers and society—specifically, on what motivates human behavior and on how the social value of IPM products and services is or should be measured. Answers to these and similar questions are central to motivating individual decisions about IPM and to evaluating public IPM programs. The economics section of the last Council for Agricultural Science and Technology report on IPM focused exclusively on profitability analysis (CAST 1982), but in the intervening two decades the explosion of IPM economic analyses has extended over a much more diverse terrain.

Perhaps the most commonly asked question is whether IPM is worthwhile from the perspectives of producer, consumer, and/or society at large. *Benefit-cost analysis* (BCA) is the tool used most commonly to answer this question. Although straightforward in theory, BCA can become quite complex when costs or benefits are measured with difficulty, as is the case for many IPM applications. Benefit-cost analysis can be *financial*, when it includes only cash costs and benefits, or *economic*, when it includes the cost of alternatives not pursued and the external effects of a pest management practice on other parts of society. Because both society and individuals often care about attributes beyond average profitability to farmers, BCA sometimes calls for estimating the value of the seemingly priceless, such as clean water, biodiversity, or more-stable crop yields. Benefit-cost analysis is used not only in assessing projects but also in assessing potential adoptability and in designing computerized decision support tools that incorporate pest damage thresholds.

Understanding of producer objectives and constraints can aid in the design of readily adoptable IPM methods. Integrated pest management research has identified traits associated with adopters. Such information can guide extension education, new IPM technology development, and IPM-friendly public policy design.

Finally, aggregate effects of IPM adoption are of interest to both public- and private-sector decision makers. The value of changes brought about by IPM is important to government officials attempting to determine whether a program is worthwhile; the same information may help the owner of a private firm decide how much to invest in research and development of IPM-related goods and services. Where social benefits are substantially greater than private ones realized by growers, it may make sense to create public policies encouraging IPM adoption. Policy research aids in the determination of which tools are effective and how justified the government is in investing in them.

This chapter will review economic analysis as used in the following endeavors:

1. designing IPM decision systems;
2. predicting private, producer-level profitability of IPM strategies;
3. weighing IPM effects on broader producer objectives such as decreasing risks to crop yield, human health, and environmental quality;
4. assessing IPM adoption patterns;
5. evaluating public IPM programs; and
6. designing public policy related to IPM.

Benefit-Cost Analysis and Pest Damage Thresholds

"Is it worth it?" is the question at the heart of any assessment of technology. For IPM, the question is useful at three levels: (1) design of pest damage thresholds, (2) potential adoptability for individual producers, and (3) public assessment of IPM projects and programs.

The original notion of *economic threshold* (Stern et al. 1959) was based on the insight that sometimes the value of yield saved is less than the cost of pesticide applied. At the heart of the original IPM concept, pest damage thresholds exemplify a class of ex ante BCA; that is, they predict likely future value rather than measure actual value after the fact.

The simplest analyses of benefits and costs use partial budgets. These assume typical conditions, no carryover effects, predictable prices, predictable yield effects from pests, and decision-maker focus on profitability. Partial budgets evaluate whether benefits (due to increased revenue and decreased costs) outweigh burdens (due to decreased revenue and increased costs). Partial budgets are the basis for the simple economic or action thresholds for pesticide spraying (Cousens 1987; Pedigo, Hutchins, and Higley 1986; Stern et al. 1959). The *net gain function* illustrates the idea of gain in gross margin (total revenues minus costs that vary) in relation to pest density (Auld, Menz, and Tisdell 1987). Figure 12.1 illustrates that when pest density is very low, *pest control costs* outweigh benefits, but as pest density increases above threshold density, or Ds*, benefits from yield protection begin to exceed control costs. This is the kind of threshold most often used in the first generation of bioeconomic IPM decision support software described in Textbox 12.1.

Many IPM practices have effects over more than one season. Killing a pest today not only protects against damage the pest would have done but also may prevent the pest from reproducing and thereby protect against damage its offspring would have done.

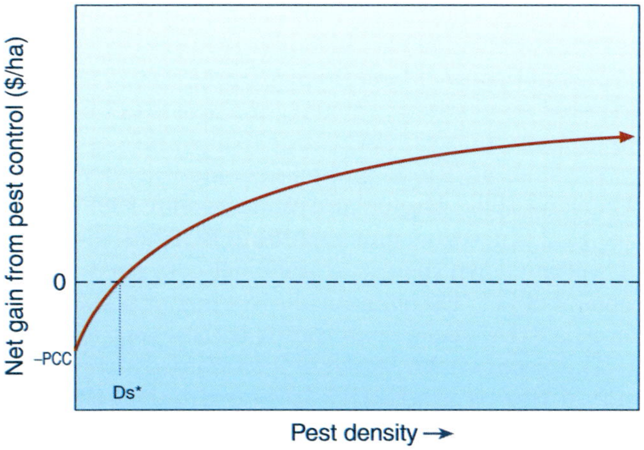

Figure 12.1. Static net gain function from pest control.

This observation has led to the consideration of *dynamic thresholds (Dd*)*, which take into account future effects, typically by predicting pest population dynamics, cropping patterns, and crop values. Dynamic thresholds often use net-present value methods to discount the value of future income (Auld, Menz, and Tisdell 1987; Cousens 1987; Pedigo, Hutchins, and Higley 1986; Swinton and King 1994; Taylor and Burt 1984). Dynamic thresholds for pesticide-based control tend to increase pesticide use because they factor in future as well as present benefits from pest control. The shift in the net gain function is illustrated in Figure 12.2. Apart from providing dynamic thresholds for chemical control of pests, multiperiod BCAs also are useful for predicting the value of investment in biocontrol methods such as the release of parasitoids to control insect pests.

Textbox 12.1. Economics and decision support systems

In the design of decision support systems (DSSs), economics has contributed directly to the development of IPM technologies. Decision support systems are computerized tools for assisting managers in making complex decisions (King et al. 1993). When they involve changing prices, multiple pest-species, nonlinear yield reductions, multiyear effects, or environmental costs, IPM thresholds can become very complex. All IPM DSSs that the authors of this report are aware of have been designed to implement threshold decision rules for pesticide application.

Although IPM thresholds first were developed for insects, weed management has led in the development of DSSs for IPM. Key reasons for this leadership are that weeds are stationary, and species differ in their susceptibilities to specific herbicides and their yield-decreasing effects on crops. Weed management DSSs first were developed as herbicide selection models that identify the herbicides most effectively killing a given mixture of weeds while not harming the growing crop (Mortensen and Coble 1991). Most such DSSs were developed for widely planted field crops such as corn and soybean (e.g., Kells and Black 1991; Kidder, Posner, and Miller 1989; Renner and Black 1991). The second generation of weed management DSSs were the so-called *bioeconomic models*, which predicted yield effects from mixed-weed populations and identified which treatment, including the no-control option, would maximize expected net gains from weed control (Lybecker, Schweizer, and Westra 1994; Swinton and King 1994; Wiles et al. 1996; Wilkerson, Modena, and Coble 1991).

Producer objectives in addition to profit maximization sometimes can be converted into monetary values to fit into a benefit-cost framework. Such attributes as aesthetic appeal (of a weedless field) or environmental costs (ECs) (due to harmful pesticides) give rise to aesthetic and environmental thresholds (Cousens 1987; Higley and Wintersteen 1992). Figure 12.2 illustrates how including ECs has the effect of shifting down the entire static net-gain curve by the amount of EC. This shift results in a higher pest-density threshold (De*) before pest control becomes optimal.

The recognition that environmental thresholds may differ dramatically from ordinary economic thresholds has prompted a surge of attempts to measure producer willingness to pay for decreased pesticide risks (Beach and Carlson 1993; Higley and Wintersteen 1992; Mullen, Norton, and Reaves 1997; Swinton, Owens, and van Ravenswaay 1999). Not only have producers expressed willingness to pay for this benefit, but studies of pesticide-related sickness and death have found that by decreasing farmers' pesticide exposure, IPM may decrease the cost of medical treatment and lost work days (Antle and Pingali 1994; Crissman, Antle, Capalbo 1998).

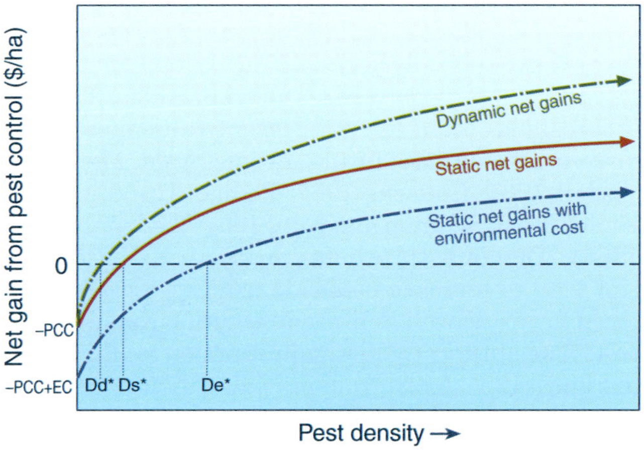

Figure 12.2. Dynamic and environmental net gain functions.

A new frontier for IPM thresholds is the inclusion of spatial variability because spatial information allows improved targeting of IPM thresholds. Sensing and mapping technologies allow for the focusing of pesticides on areas where pests are present (Johnson, Cardina, and Mortensen 1997; Weisz, Fleischer, and Smilowitz 1995). Given that many pest populations follow highly skewed spatial-density distributions (Johnson, Cardina, and Mortensen 1997; Wiles et al. 1992), significant areas may go unsprayed when pest control is targeted only at those locations where a threshold is exceeded. Although these spatial technologies have not reached the farm level yet, preliminary economic analyses for weeds have shown that, under certain circumstances, spatial pest management technologies may be profitable (Bennett and Pannell 1998; Oriade et al. 1996).

The threshold-based IPM methods just described rely chiefly on chemical controls once the threshold is reached. But purely chemical controls are only one end of the IPM continuum (Benbrook et al. 1996). The burgeoning field of biological pest control (Landis and Orr 2002) has yet to benefit from economic analysis. Yet the U.S. federal government has invested significantly in biocontrols as well as in areawide insect eradication programs irrelevant to threshold-based analyses. These economic analyses of biological controls will call for new research that involves dynamic modeling of component pest and predator populations and that is linked to measuring changes in the economic value of yield saved as a result of using these instead of alternative methods.

Crops genetically modified for pest resistance or herbicide tolerance have, since about 1995, represented a new approach to pest control. Economic assessments of these crop varieties began to appear from 1998 to 1999. Of special interest has been discovering whether farm profits actually increase after seed technology fees charged for patented seeds are accounted for. National survey results from 1997 found that planting herbicide-resistant corn or soybean did not affect the profit level of U.S. farmers. A 1998 survey of Iowa farmers found similar results for genetically modified soybean and for *Bacillus thuringiensis* (*Bt*) corn (Duffy 1999). United States cotton growers did, however, achieve increased profitability by using both herbicide-resistant and *Bt* varieties (Fernandez-Cornejo and McBride 2000).

Integrated Pest Management and Agricultural Income Risk

Pest attacks constitute one of the most important sources of risk to crop yield. Not only can pests decrease yields, they can decrease quality, thereby exposing producers to price risk. Prophylactic use of pesticides can act as a form of insurance against pest attack (Feinerman, Herriges, Holtkamp 1992; Smith and Goodwin 1996): that is, risk-averse producers may choose to use more pesticide than strictly necessary because by doing so they expect to decrease the likelihood of crop damage. Although IPM does not necessarily increase yield variability (Lamp, Nielsen, and Dively 1991; Napit et al. 1988), it may do so (Szmedra, Wetzstein, and McClendon 1990). More important, many growers *perceive* IPM as exacerbating yield risk.

Yield risks from threshold-based IPM strategies arise at two levels: (1) pest infestation prediction and (2) control strategies. Pest density thresholds are based on damage predictions. These, in turn, depend on accurate and timely pest demographic predictions. Such predictions can be faulty because of poor scouting and/or pest population assessment, weather conditions unexpectedly favoring pest populations, or poor predictive models. Even if the pest action threshold is predicted properly, poor weather or competing tasks may prevent the grower from timely treatment. Certain growers view calendar spraying as less risky on these accounts.

Economic research has proposed IPM insurance to insulate farmers from the income risk associated with adopting threshold-based IPM practices (Feinerman, Herriges, and Holtkamp 1992). Given the public benefits from decreased pesticide use that have the potential to result from more extensive adoption of IPM, public cost-share programs have been introduced to decrease the cost of adopting IPM practices. One example can be found in the Environmental Quality Incentives Program (EQIP) of the 1996 Federal Agricultural Improvement and Reform Act. In selected U.S. counties, the EQIP program compensates participating farmers for certain IPM related costs such as scouting. These cost-share programs do not, however, address yield and income risk associated with the performance of threshold-based IPM methods. Such risks have been addressed directly in the design of new crop-insurance products that, as a result of collaboration between the U.S. Department of Agriculture's (USDA) Risk Management Agency and the Agricultural Conservation Innovation Center (ACIC) (See Textbox 12.2), have been developed for IPM users (ACIC 2001).

Textbox 12.2. Two integrated pest management insurance policies

Corn Root Worm Integrated Pest Management Policy

Insurance soon will be available for farmers using corn root worm IPM systems. This policy will permit a farmer to rely on the advice of an expert who uses approved scientific scouting procedures. If the farmer decides to follow the "don't treat" recommendation and does not trust the scouting procedure fully, he can purchase an insurance policy. The policy will cost about $5 acre (a.)/compared with the $12 to 15/a. cost of the root worm application. The policy will work as follows:

Step 1: A certified crop advisor scouts for corn root worm beetle in July and August and makes "treat" or "don't treat" recommendation for the subsequent corn crop.

Step 2: The grower applies for insurance and follows the "don't treat" recommendation during the subsequent spring.

Step 3: A root rating analysis is performed to determine if a loss occurred. An insurance claim is made if the root rating is 3.5 or higher (based on the Iowa State University scale of 1 to 6).

Step 4: If the rating is 3.5 or more, the policy is adjusted (an indemnity is paid).

Step 5: At crop maturity, if the insured determines that harvest will be slowed significantly due to lodging of the insured acres, an insurance claim may be filed.

Step 6: If the insured crop is adjusted for having increased harvest expenses, an indemnity will be paid to cover the increased expenses. Maximum additional harvest expenses are equal to the average custom-harvesting rate for the local region where the insured acres are located.

Potato Late Blight Policy

This policy permits potato farmers to follow the "wait until fungus conditions exist" announcement made by extension in certain states. By spraying after this recommendation is made, the farmer could possibly avoid 1 to 3 fungicide sprays/season.

Maine, Wisconsin, and New York have the weather systems to make this policy work. Under this program, a grower purchases the insurance policy and waits to spray until the recommendation is made by the local extension forecasting system. If late blight is detected before or within ten days of recommendation to spray, an indemnity is paid. The indemnity covers the value of the lost crop, cost of destroying infected plants, and recovery treatment.

Excerpted from: Agricultural Conservation Innovation Center. 2001. Promoting conservation innovation in agriculture through crop insurance, May 1, <http://www.agconserv.com/promote.html> (27 September 2002) NOTE: site was valid 1999–2002. Find current related information at <http://www.aftresearch.org/researchresource/taking.htm>)

Adoption of Integrated Pest Management: Why Do Growers Adopt?

Public benefits from IPM—notably from decreased pesticide risks—have attracted government interest in fostering adoption by farmers. Whereas the use of BCA for IPM thresholds focuses on the individual producer, IPM adoption research tends to focus on the aggregate producer population. This focus on the producer means not only measuring the effects of a set of IPM practices on a single producer, but also measuring which factors affect producer adoption and how many producers have adopted (or will adopt) these IPM practices.

The first step in adoption studies is to describe what proportion of the producer population has adopted the IPM practices in question. The next step is to determine which factors encourage adoption. Adoption studies typically are cross-sectional surveys targeted at clarifying why certain farmers take up a given IPM practice whereas others do not.

Numerous cross-sectional studies of IPM adoption have taken place in the United States (Caswell and Shoemaker 1993; Ferguson and Yee 1995; Fernandez-Cornejo, Beach, and Huang 1994; Fernandez-Cornejo, Jans, and Smith 1998; Harper et al. 1990; McNamara, Wetzstein, and Douce 1991; Napit et al. 1988; Vandeman et al. 1994). Characteristics influencing adoption can be divided roughly into four categories based on the technology, the farmer, the farm physical environment, and the farm institutional environment (Feder, Just, and Zilberman 1985). In general, adopters of IPM practices are younger and more educated than the average grower (Drost 1996). Integrated pest management adopters also tend to have less farming experience and to be more likely to use a computer (Leslie and Cuperus 1993, as cited in Chapter 9).

Public Program Assessment

The economic methods discussed so far address the questions "Would an IPM practice be profitable if adopted?" "Would it be risky?" and "What are the characteristics of adopters?" Public program evalua-

Table 12.1. The impact of IPM on pesticide use, yields, and profits—summary of empirical results (Fernandez-Cornejo et al. 1998)

			Pesticide use			
Commodity	IPM techniques	Total number of studies	Most common effect	Range (percentage)	Yield	Profit (net returns per acre)
Cotton	Scouting only	10	Increase	–64 to +92	Increase[1]	Increase[1]
Cotton	Scouting/others[2]	9	Decrease	–98 to +34	Increase[1]	Increase[3]
Soybean	Scouting only	5	Decrease	–21 to +83	Increase[4]	Increase[4]
Soybean	Scouting/others[2]	2	Decrease	–100 to –85	NA	Increase
Corn	Scouting	1	Increase	+15 to +47	Increase	Increase
Corn	Scouting/others[2]	2	Decrease	–50 to +67	Increase[5]	NA
Peanut	Scouting only	5	Decrease	–81 to +177	Increase[6]	Increase[5]
Fruits/nuts	Scouting only	6	Decrease	–43 to +24	Increase[7]	Increase[7]
Fruits/nuts	Scouting/others[8]	4	Decrease	–41 to –12	Same[5]	Same[5]
Vegetables	Scouting/others[8]	7	Decrease	–67 to +13	Same	Increase[5]

NA = not available or not applicable

[1]Only six studies reported results.

[2]Scouting plus other techniques or other techniques alone.

[3]Only eight studies reported results.

[4]Only four studies reported results.

[5]Only one study reported results.

[6]Only three studies reported results.

[7]Only two studies reported results.

[8]All studies but one considered insect IPM only.

tion asks a broader question: "What is the net effect on social welfare of this IPM practice (or program)?" Answering this last question calls for aggregating the individual-level profitability analysis according to IPM adopter numbers and adoption timing. Because social welfare is about more than agricultural producers, a public program assessment also must integrate effects on consumers and nature. Economic evaluations of IPM programs are sufficiently widespread to spawn published literature reviews (Fernandez-Cornejo, Jans, and Smith 1998; Norton and Mullen 1994).

The most comprehensive summary of private, producer-level economic evaluations of IPM programs to date was developed by Fernandez-Cornejo, Jans, and Smith (1998) and updated the work of Norton and Mullen. Table 12.1 demonstrates that although most of the 51 IPM programs studied increased profits, increased yields, and decreased pesticide use, these effects did not occur universally. For no commodity group did IPM decrease pesticide use across the board. In fact, IPM in cotton increased pesticide use more often than not.

Measuring Cumulative Adoption of Integrated Pest Management

To facilitate prediction of long-term program effects, future technology adoption trends must be projected. The diffusion of a new technology tends, over time, to follow a sigmoid curve (Rogers 1983), as shown in Figure 12.3. At first, only the daring, experimental few adopt. But as the new technique becomes recognized as attractive, the adoption rate accelerates. Adoption pace tapers off as only a few laggards remain among those who might find the technology worthwhile. Most empirical attempts at estimating adoption curves have followed the lead of Griliches (1957), who fitted a logistic function to time-series data on the adoption of hybrid corn in the United States. Recently, Fernandez-Cornejo, Alexander, and Goodhue (2002) extended this line of research with a model of agricultural technology disadoption that explains how adoption levels can decline, rather than only increase.

Fernandez-Cornejo and Castaldo (1998) statistically estimated logistical adoption curves for a variety of IPM practices in the major fruits produced in the United States. Their work identified the target date for 75% adoption of each technique in each crop. They also studied factors affecting adoption rate. Stock of public and private research turned out to be the most important determinants of scouting adoption, and public research was the single most significant determinant of decreased pesticide use.

Measuring the Value of Health and Environmental Effects

As has been mentioned, BCA for public IPM programs differs from individually oriented programs in terms not only of aggregating adopters, but of measuring effects on other individuals. Specifically, BCA for public IPM programs factors in the unintended effects that economists call *externalities* by virtue of their being external to the immediate interests of the decision maker. When a cotton farmer burns crop residues to destroy overwintering boll weevil eggs and the smoke triggers an asthmatic attack in a neighbor's child, the asthmatic attack is an externality not considered by the farmer. Public-policy analyses aiming to evaluate net social benefits attempt to estimate the value of changes in the level of such external effects (Carlson 1989).

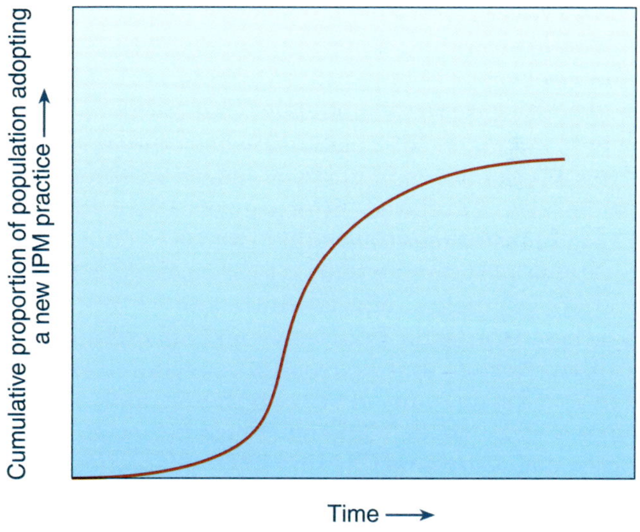

Figure 12.3. Classic curve illustrating increasing cumulative technology adoption over time.

Attempts to measure the value of IPM methods in efforts to decrease environmental and health risks have given rise to three major types of research. The first type aims to measure the effects of pesticides on human health. Summarizing the large and growing literature on the medical epidemiology of pesticide exposure is, however, beyond the scope of this chapter. Generally, the epidemiological studies are costly analyses of large human-population samples. These studies attempt to relate pesticide exposure to changes in probability of death or of illness from various causes.

The very costliness of these methods has triggered the second type of research, which aims at developing low-cost indicators of both human health and environmental risk. The growing need for sound indicators of these risks spawned at least two major workshops in 1998 (Day 1998; Waibel et al. 1999). The challenge is to strike a reasonable compromise between, on the one hand, the formidable expense of comprehensively measuring environmental effects and, on the other, the questionable reliability of measuring health risk in terms of facile impact-indicators such as weight of pesticide active ingredient/hectare, a measure that ignores both toxicity and exposure likelihood. No single indicator is used widely at present to measure health and environmental risks; alternative measures being debated include risk-ratios, scoring tables or rankings, and fuzzy expert systems.

The third type of research related to externalities of pest management aims to develop economic measures of (1) environmental and health effects of pesticides and (2) related influences of IPM programs. These measurement attempts can be situated on a spectrum placing monetary value on human health and environmental effects (Antle and Pingali 1994; Beach and Carlson 1993; Harper and Zilberman 1989; Mullen, Norton, and Reaves 1997; Swinton, Owens, and van Ravenswaay 1999) and those simply identifying a risk-benefit trade-off (Bouzaher et al. 1992; Crissman, Antle, and Capalbo 1998). Valuation analyses have substantiated that both the public and the pesticide user are willing to pay to decrease pesticide risks, a goal that IPM can be used to achieve. But the specific numerical values emerging from valuation studies are controversial, both ethically—for suggesting that it is possible to place a value on what many consider priceless (i.e., human health)—and methodologically—for omitting certain types of risk, which most studies do. Trade-off analyses can be useful for decision making purposes but do not contribute usefully to measurements of aggregate program benefits.

Overall Returns to Research and Outreach in Integrated Pest Management

The difficulties in measuring and valuing IPM effects may account for the scarcity of studies estimating economic returns to public IPM research and outreach activities. Because forms of IPM involve information use or subtle changes in the rationale for management practices, adoption is made much more difficult to quantify than for discrete commodity-specific technologies such as crop varietal introductions (Alston, Norton, and Pardey 1998; Waibel et al. 1999). In fact, most thoroughly documented cases of returns to IPM research are based on plant breeding for genetic pest resistance, as in the instance of resistance to soybean cyst nematode (Bradley and Duffy 1982).

The only comprehensive IPM program assessments available that the authors know of are Napit and colleagues' 1988 evaluation of extension IPM effects across nine commodity-state combinations in the United States, and Waibel's 1999 attempt to evaluate the returns to IPM at international agricultural research centers. Napit and colleagues found that across a diverse set of commodities and U.S. states, IPM mainly generated less variable and higher net returns to growers, and greater economic gains to consumers, than traditional production methods did. Unexpectedly, pesticide costs rose with IPM use in several states. The researchers conducted economic surplus analyses in only the two instances where they found statistically significant differences in net returns in cotton, both between nonusers and low users of IPM and between low and high users of IPM. As a result, very high annual internal rates of return to IPM extension programs were calculated (452% for Texas cotton and 300% for Mississippi cotton). Although internal rates of return for other IPM practices were not published, it can be inferred that they were lower, for they would be calculated from smaller differences in annual net returns between non-, low, and high IPM user groups. Unlike Napit et al. (1988), Waibel (1999) found it impossible to conduct a quantitative economic-surplus analysis of returns to research in IPM. Instead, he surveyed the scientists involved, reviewed publication productivity, and illustrated his research with economic case studies.

Public Policy: Intended and Unintended Effects on Integrated Pest Management

United States government policies have affected IPM adoption and research through various channels, directly and indirectly. Direct efforts to foster IPM adoption include cost sharing for selected adopters of IPM practices under the EQIP program, and federal subsidies for IPM extension and research under the USDA's regional IPM programs and regional research projects. As has been noted, the USDA's Risk Management Agency and ACIC are conducting a pilot IPM insurance program for corn root worm management in the Midwest.

But the policies with *indirect* effects on IPM adoption probably have greater effects. For years, U.S. federal price supports and deficiency payments for wheat and feed grains discouraged IPM by raising crop prices, which implicitly decreased the threshold for pest control (Reichelderfer and Hinkle 1989). Because exchange rate misalignment distorted the relation between chemical inputs (often imported) and crop products (often exported) (Norton, G. W. 2000. Personal communication), similar effects have occurred in other nations.

Environmental policy, notably federal pesticide policy, has had mixed effects on IPM adoption. Pesticide policy has been a bastion of rigid command and control rules during a period when much federal environmental policy has been becoming more flexible (Ogg 1999). Under the Federal Insecticide, Fungicide and Rodenticide Act and its successor, the Food Quality Protection Act of 1996, the Environmental Protection Agency (EPA) is charged with registering pesticides for specified uses. But in banning certain uses, the EPA applies a very blunt tool that removes those pesticides from the arsenal available to IPM practitioners (Swinton and Batie 2001; Zilberman and Millock 1997). Although one might expect them to encourage IPM adoption, EPA policies have not necessarily done so. For instance, because EPA inspectors encounter difficulty in monitoring surface- and groundwater quality, serious attempts to curtail nonpoint source water pollution—the very type of pollution that IPM has the greatest potential to alleviate—are discouraged.

Private Sector Initiatives

Private sector and nongovernmental organizations recently have begun using market methods to promote IPM. In response to surveys revealing consumer willingness to pay for foods with decreased pesticide residues, or otherwise produced in an environmentally friendly manner, "ecolabels" have been developed to certify the quality of the production process (van Ravenswaay and Blend 1999). In Europe and the United States, a small number of food retailers have begun using these labels. Among these are IPM certification labels, such as those used on canned vegetables sold by Wegman's food stores in western New York State.

Such labeling practices can have two effects. If processors require IPM of their growers, then IPM is mandated by the market. If IPM is a voluntary activity fetching a higher price, then its adoption is compensated. Either by regulation or inducement, there exists an incentive for producers to adopt practices necessary to achieve certification. So far, the first instance seems to predominate in the United States; that is, IPM is becoming a prerequisite for growers to obtain access to vegetable and fruit production contracts.

Meanwhile, nongovernmental organizations and producer commodity organizations are collaborating on IPM certification programs. Such programs certify that growers are using best pest management practices. A current example is a joint project between the World Wildlife Fund, the Wisconsin Potato and Vegetable Growers Association, and the University of Wisconsin (See Textbox 12.3). The collaboration has developed an IPM certification program that is gov-

Textbox 12.3. Public-Private partnerships: World Wildlife Fund and Wisconsin Potato and Vegetable Growers Association

In 1996, the World Wildlife Fund (WWF) and the Wisconsin Potato and Vegetable Growers Association (WPVGA), an environmental organization and an agricultural commodity association respectively, established a precedent setting partnership to work toward more ecologically sound agricultural practices. The WPVGA represents about 175 farmers who raise about 85,000 a. of potatoes/yr. The goal of this unique collaboration is to promote development and wider use of economically viable farming systems that are safer for farm families, consumers, and the environment.

The WPVGA's proactive approach shows that adoption of biointensive IPM can diminish substantially growers' reliance on high-risk pesticides. The impressive first-year results of the collaboration—a 25% decrease in pesticide toxicity in 1997 compared with the 1995 baseline—testify to the effectiveness of these efforts. The key components of the project, namely, setting ambitious IPM adoption and pesticide risk reduction goals, promoting research and extension on IPM practices, and agreeing on risk-reduction indicators, provide a promising model for other agricultural groups to follow when addressing their own pest management challenges.

Wisconsin's experience shows that committed growers, backed up by a proactive organized trade association and a strong university research team, can innovate around pest and pesticide regulatory problems (see Chapter 2, textbox 2.2).

For more information on the WWF program, see the World Wide Web sites <http://www.worldwildlife.org/toxics/progareas/ap/alternatives.htm> and <http://www.ipmalmanac.com/solutions/200104/panda.asp>. For more information on the WPGVA, see <http://wisconsinpotatoes.com/about.htm>.

erned by the nonprofit organization Protected Harvest.

In conclusion, economics during the last 20 years (yr) has played a key role in IPM technology assessment and policy analysis. Economic analysis has been applied to evaluate expected profitability, ex ante and ex post adoption, social welfare effects, returns to research, and policies affecting pest management generally. In specific instances, economic analysis has been significant in the development of threshold-based IPM decision support software.

For all that has been accomplished, important unfinished business remains in at least two areas. First, the economic assessment of biological pest management requires more thorough development. This will require (1) dynamic modeling of interactions between pest and predator or parasitoid populations, (2) estimating changes in pest effects on valued commodities, (3) comparing biological pest management with nonbiological benchmark pest control methods, and (4) assessing effects on profitability, human health, and environmental quality. Pest resistance to pesticides, too, will need to be considered.

The second area needing more economic input is the measurement of returns to research. The daunting problems with defining and measuring IPM continue to hamper attempts at comprehensive assessment of IPM research, and no major IPM extension assessment has been completed in the United States since the mid-1980s. This failure to measure benefits comprehensively is likely to deprive IPM programs of the public support that the available evidence, scanty though it may be, suggests they deserve.

Appendix A. Abbreviations and Acronyms

a.	acre
ACIC	Agricultural Conservation Innovation Center
BCA	benefit-cost analysis
Bt	*Bacillus thuringiensis*
DSS	decision support system
EC	environmental costs
EPA	U.S. Environmental Protection Agency
EQIP	Environmental Quality Incentives Program
IPM	integrated pest management
USDA	U.S. Department of Agriculture
yr	year

Appendix B. Glossary

Dynamic thresholds. Cut-off levels for decisions taking into account future effects, typically by predicting pest population dynamics, cropping patterns, and crop values.

Economic benefit-cost analysis. Tool used to detrmine whether IPM is worthwhile from the perspectives of producer, consumer, and/or society at large; including the opportunity cost of sacrificing current income and the external effects of a pest management practice on other parts of society.

Externalities. Unintended economic effects experienced by someone other than the decision maker.

Financial benefit-cost analysis. Tool used to detrmine whether IPM is worthwhile from the perspectives of producer, consumer, and/or society at large; including only cash costs and benefits.

Gross margin. Total revenues minus costs that vary.

Net gain function. Gain in gross margin over pest control costs as pest population increases.

Literature Cited

Agricultural Conservation Innovation Center (ACIC). 2001. Promoting conservation innovation in agriculture through crop insurance, May 1, <http://www.agconserv.com/promote.html> (27 September 2002)

Alston, J. M., G. W. Norton, and P. G. Pardey. 1998. *Science under Scarcity: Principles and Practice for Agricultural Research Evaluation and Priority Setting*. CAB International, New York.

Antle, J. M. and P. L. Pingali. 1994. Pesticides, productivity, and farmer health: A Philippino case study. *Am J Agric Econ* 76:418–430.

Auld, B., K. M. Menz, and C. A. Tisdell. 1987. *Weed Control Economics*. Academic Press, London.

Beach, E. D. and G. A. Carlson. 1993. Hedonic analysis of herbicides: Do user safety and water quality matter? *Am J Agric Econ* 75:612–623.

Benbrook, C. M., E. Groth, III, J. M. Halloran, M. K. Hansen, and S. Marquardt. 1996. *Pest Management at the Crossroads*. Consumers Union, Yonkers, New York.

Bennett, A. and D. J. Pannell. 1998. Economic evaluation of a weed-activated sprayer for herbicide application to patchy weed populations. *Aust J Agric Econ* 42:389–408.

Bouzaher, A., D. Archer, R. Cabe, A. Carriquiry, and J. F. Shogren. 1992. Effects of environmental policy on trade-offs in agri-chemical management. *J Environ Manage* 36:69–80.

Bradley, E. B. and M. Duffy. 1982. The value of plant resistance to soybean cyst nematodes: A case study of forest soybeans. Staff Report Number AGES 820929. Natural Resource Economic Divison, Economic Research Service, U. S. Department of Agriculture, Washington, D.C.

Carlson, G. A. 1989. Externalities and research priorities in agricultural pest control. *Am J Agric Econ* 71:453–457.

Caswell, M. F. and R. A. Shoemaker. 1993. Adoption of pest management strategies under varying environmental conditions.

Technical Bulletin Number 1827. Resources and Technology Division, Economic Research Service, U.S. Department of Agriculture, Washington, D.C.

Council for Agricultural Science and Technology (CAST). 1982. *Integrated Pest Management*. Report Number 93. Council for Agricultural Science and Technology, Ames, Iowa.

Cousens, R. 1987. Theory and reality of weed control thresholds. *Plant Protect Quart* 2:13–20.

Crissman, C., J. M. Antle, and S. M. Capalbo. 1998. *Economic, Environmental, and Health Tradeoffs in Agriculture: Pesticides and the Sustainability of Andean Potato Production*. Kluwer Academic Publishers, Dordrecht, The Netherlands.

Day, E. 1998. *Integrated Pest Management Measurement Systems Workshop,* Chicago, Illinois, June 12–13, 1998. Working Paper CAE/SP 981. Center for Agriculture in the Environment, American Farmland Trust, Dekalb, Illinois.

Drost, D. 1996. Barriers to adopting sustainable agricultural practices. In *J Extension* 34, <http://www.joe.org/1996december/index.html> (27 September 2002)

Duffy, M. 1999. Does planting GMO seed boost farmers' profits? *Leopold Letter* 11:1, 4–5.

Feder, G., R. E. Just, and D. Zilberman. 1985. Adoption of agricultural innovations in developing countries: A survey. *Econ Develop Cult Change* 33:255–298.

Feinerman, E., J. A. Herriges, and D. Holtkamp. 1992. Crop insurance as a mechanism for reducing pesticides usage: A representative farm analysis. *Rev Agric Econ* 14:169–186.

Ferguson, W. L. and J. Yee. 1995. A logit model of cotton producer participation in professional scout programs. *J Sustain Agric* 5:87–96.

Fernandez-Cornejo, J. and C. Castaldo. 1998. The diffusion of IPM techniques among fruit growers in the USDA. *J Prod Agric* 11:497–506.

Fernandez-Cornejo, J. and W. D. McBride. 2000. Genetically engineered crops for pest management in U.S. Agriculture: Farm-level effects. Agricultural Economics Report 786. Resource Economics Division, Economic Research Service, U. S. Department of Agriculture, Washington, D.C.

Fernandez-Cornejo, J., E. D. Beach, and W.-Y. Huang. 1994. The adoption of integrated pest management technologies by vegetable growers in Florida, Michigan, and Texas. *J Agric Appl Econ* 26:158–172.

Fernandez-Cornejo, J., S. Jans, and M. Smith. 1998. Issues in economics of pesticide use in agriculture: A review of empirical evidence. *Rev Agric Econ* 20:462–488.

Fernandez-Cornejo, J., C. Alexander, and R. E. Goodhue. 2002. Dynamic diffusion with disadoption: The case of crop biotechnology in the USA. *Agric Resource Econ Rev* 31:112–126.

Griliches, Z. 1957. Hybrid corn: An exploration in the economics of technological change. *Econometrica* 25:501–522.

Harper, C. R. and D. Zilberman. 1989. Pest externalities from agriculture. *Am J Agric Econ* 721:692–702.

Harper J. K., M. E. Rister, J. W. Mjelde, B. M. Drees, and M. O. Way. 1990. Factors influencing the adoption of insect management technology. *Am J Agric Econ* 72:997–1005.

Higley, L. G. and W. K. Wintersteen. 1992. A novel approach to environmental risk assessment of pesticides as a basis for incorporating environmental costs into economic injury level. *Am Entomol* 38:34–39.

Johnson, G. A., J. Cardina, and D. A. Mortensen. 1997. Site specific weed management: Current and Future Directions. Pp. 131–147. In F. J. Pierce and E. J. Sadler (eds.). *The State of Site-Specific Management for Agriculture*. Agronomy Society of America, Crop Science Society of America, and Soil Science Society of America, Madison, Wisconsin.

Kells, J. J. and J. R. Black. 1991. CORNHERB: Herbicide Options Program for Weed Control in Corn: An Integrated Decision Support Computer Program. Version 2.0. Agricultural Experiment Station, Michigan State University, East Lansing.

Kidder, D., B. Posner, and D. Miller. 1989. WEEDIR: Weed Control Directory. AG–CS–2163. Version 3.0. Minnesota Extension Service, University of Minnesota, St. Paul.

King, R. P., D. W. Lybecker, A. Regmi, and S. M. Swinton. 1993. Bioeconomic models of crop production systems: Design, development and use. *Rev Agric Econ* 15:393–405.

Lamp, W. O., G. R. Nielsen, and G. P. Dively. 1991. Insect pest-induced losses in alfalfa: Pattern in Maryland and implications for management. *J Econ Entomol* 84: 610–618.

Landis, D. A. and D. B. Orr. 2002. Biological control: Approaches and application. In E. B. Radcliffe and W. D. Hutchinson (eds.). *Radcliffe's IPM World Textbook*, July 25, <http://ipmworld.umn.edu> (27 September 2002)

Leslie, A. R. and G. W. Cuperus (eds.). 1993. IPM and growers: An evolution in thinking. Chap. 9. *Successful Implementation of Integrated Pest Management for Agricultural Crops*. Lewis Publishers, Boca Raton, Florida.

Lybecker, D. W., E. E. Schweizer, and P. Westra. 1994. WEED-CAM Manual. Department of Agricultural Economics, Colorado State University, Ft. Collins.

McNamara, K. T., M. E. Wetzstein, and G. K. Douce. 1991. Factors affecting peanut producer adoption of integrated pest management. *Rev Agric Econ* 13:129–139.

Mortensen, D. A. and H. D. Coble. 1991. Two approaches to weed control decision-aid software. *Weed Tech* 5:445–451.

Mullen, J. D., G. W. Norton, and D. W. Reaves. 1997. Economic analysis of environmental benefits of integrated pest management. *J Agric Appl Econ* 29:243–253.

Napit, K. B., G. W. Norton, R. F. Kazmieczak, Jr., and E. G. Rajotte. 1988. Economic impacts of extension integrated pest management programs in several states. *J Econ Entomol* 81:251–256.

Norton, G. W. and J. Mullen. 1994. *Economic Evaluation of Integrated Pest Management Programs: A Literature Review*. Virginia Cooperative Extension Publication 488–120. Virginia Polytechnic Institute and State University, Blacksburg.

Ogg, A. C. W. 1999. Evolution of EPA programs and policies that impact agriculture. Pp. 27–42. In F. Casey, A. Schmitz, S. Swinton, and D. Zilberman (eds.). *Flexible Incentives for the Adoption of Environmental Technologies in Agriculture*. Kluwer Academic Publishers, Boston.

Oriade, C. A., R. P. King, F. Forcella, and J. L. Gunsolus. 1996. A bioeconomic analysis of site-specific management for weed control. *Rev Agric Econ* 18:523–535.

Pedigo, L. P., S. H. Hutchins, and L. G. Higley. 1986. Economic injury levels in theory and practice. *Annu Rev Entomol* 1:341–361.

Reichelderfer, K. and M. K. Hinkle. 1989. The evolution of pesticide policy: Environmental interests and agriculture. Pp. 147–177. In C. Kreamer (ed.). *Political Economy of U.S. Agriculture: Challenges for the 1990s*. National Center for Food and Agricultural Policy, Resources for the Future, Washington D.C.

Renner, K. A. and J. R. Black. 1991. SOYHERB: A computer program for soybean herbicide decision making. *Agron J* 83:921–925.

Rogers, E. M. 1983. *Diffusion of Innovations*. 3rd ed. Free Press, New York.

Smith, V. H. and B. K. Goodwin. 1996. Crop insurance, moral hazard, and agricultural chemical use. *Am J Agric Econ* 78:428–438.

Stern, V. M., R. F. Smith, R. van Den Bosch, and K. S. Hagen. 1959. The integrated control concept. *Hilgardia* 29:81–101.

Swinton, S. M. and R. P. King. 1994. A bioeconomic model for weed management in corn and soybean. *Agric Syst* 44:313–335.

Swinton, S. M. and S. S. Batie. 2001. FQPA: Pouring out (in?) the risk cup. *Choices* 16:14–17.

Swinton, S. M., N. N. Owens, and E. O. van Ravenswaay. 1999. Health risk information to reduce water pollution. Pp. 263–271. In F. Casey, A. Schmitz, S. Swinton, and D. Zilberman (eds.). *Flexible Incentives for the Adoption of Environmental Technologies in Agriculture*. Kluwer Academic Publishers, Boston.

Szmedra, P. I., M. E. Wetzstein, and R. W. McClendon. 1990. Economic threshold under risk: A case study of soybean production. *J Econ Entomol* 83:641–646.

Taylor, C. R. and O. R. Burt. 1984. Near-optimal management strategies for controlling whorled oats in spring wheat. *Am J Agric Econ* 66:50–60.

Vandeman, A., J. Fernandez-Cornejo, S. Jans, and B.-H. Lin. 1994. Adoption of integrated pest management in U.S. agriculture. Agriculture Information Bulletin Number 707. U.S. Resources and Technology Division, Economic Research Service, U.S. Department of Agriculture, Washington, D.C.

van Ravenswaay, E. O. and J. Blend. 1999. Using ecolabeling to encourage adoption of innovative environmental technologies in agriculture. Pp. 119–138 In F. Casey, A. Schmitz, S. Swinton, and D. Zilberman (eds.). *Flexible Incentives for the Adoption of Environmental Technologies in Agriculture*. Kluwer Academic Publishers, Boston.

Waibel, H. 1999. Evaluation of the impact of integrated pest management at the IARCs. Document Number ICW/99/080C. Consultative Group on International Agricultural Research, Washington, D.C.

Waibel, H., G. Fleischer, P. E. Kenmore, and F. Feder. 1999. Evaluation of IPM programs and methodologies. Pesticide Policy Project Publication Series Number 8. Papers presented at the First Workshop on Evaluation of IPM Programs, University of Hannover, Germany, March 16–18, 1998.

Weisz, R., S. Fleischer, and Z. Smilowitz. 1995. Site-specific integrated pest management for high value crops: Sample units for map generation using the Colorado potato beetle (Coloptera: Chrysomelidae) as a model system. *J Econ Entomol* 88:1069–1080.

Wiles, L. J., G. W. Oliver, A. C. York, H. J. Gold, and G. G. Wilkerson. 1992. The spatial distribution of broadleaf weeds in North Carolina soybean (*Glycine max*) fields. *Weed Sci* 40:554–557.

Wiles, L. J., R. P. King E. E. Schweizer, D. W. Lybecker, and S. M. Swinton. 1996. GWM: General Weed Management Model. *Agric Syst* 50:355–376.

Wilkerson, G. G., S. A. Modena, and H. D. Coble. 1991. HERB: Decision model for postemergence weed control in soybean. *Agron J* 83:413–417.

Zilberman, D. and K. Millock. 1997. Financial incentives and pesticide use. *Food Policy* 22:133–144.

Related Web Sites

American Farmland Trust. 1998. Integrated pest management measurement systems workshop, October 14, <http://www.aftresearch.org/researchresource/wp/sp98-1/index/htm> (13 February 2003)

U.S. Department of Agriculture–Environmental Research Service. 1998. Pest management. In *Adoption of Agricultural Production Processes*, <http://www.ers.usda.gov /publications/aer792/aer792.pdf> (27 September 2002)

U.S. Department of Agriculture–Environmental Research Service. 2001. Pest management in U.S. agriculture, July 24, <http://www.ers.usda.gov/publications/ah717> (27 September 2002)

13 Integrated Pest Management Education and Delivery Systems

The definition of *integrated pest management* (IPM) has evolved continually since Stern and colleagues introduced the concept of "integrated control" in 1959. This concept, which at first included the most basic idea of insect management and eventually came to include the ideas of integrated biological and chemical, and ultimately cultural, biological, and physical control of insects, weeds, and crop pathogens, now refers to a very complex integrated system of management. Conceptualizations of IPM education, adoption, and evaluation have undergone parallel evolutions.

As Swinton and Williams (1998) suggest, *IPM* can be defined broadly as either input- or output-oriented. Both categories have evolved to reflect how IPM is taught, delivered, adopted, and evaluated. Input-oriented definitions focus on pest management practices, and specifically on biological, cultural, and chemical practices. These definitions have gained popularity in programs promoting IPM adoption or using certification and labels to determine whether growers are following prescribed practices. Output-oriented definitions focus on IPM-use outcomes and primarily on decreased environmental risk and increased profit. Measurable in many different ways, profitability has had a major influence on IPM adoption from the start. Impacts on profitability are easier to measure than those on environment and health (Swinton and Williams 1998). Lately, definitions of *IPM* have addressed more than the two latter concerns.

Education, Past and Present

Pest management personnel must be trained in a wide range of disciplines including ecology, toxicology, biology, entomology, nematology, weed science, and plant pathology. As IPM implementation has grown more and more widespread, researchers and educators have seen the need to prepare students to take responsibility for the development, teaching, and application of IPM concepts. As early as 1975, IPM professionals recognized the necessity of crop-protection workers' becoming familiar with the intricacies of IPM, and of growers' recognizing the growing importance of IPM (Glass 1975). Early on, the importance of the connections among educators, crop protection personnel, and growers was understood clearly.

A Brief Chronology: The 1970s

In early 1972, after discussions with personnel in the U.S. Department of Agriculture (USDA) and in other federal agencies about the need for possible training programs for crop protection specialists, the Resident Instruction Committee on Organization and Policy established the Committee on Plant Protection.[1] The curriculum proposed differed from then-current curricula. It was to take into account all plant pests and ideally would include internships in crop management. Students in the proposed program would gain an understanding of all parts of the agroecosystem, through the application of biological, chemical, and integrated systems that can contribute to a sustainable environment (Browning 1972).

Degree Training

In late 1972, presumably the first formal workshop on IPM curriculum development and training needs was held at the East-West Center in Hawaii (Obein and Motooka 1972). Workshop participants from Asia, the Pacific, and the United States developed curricula and training programs. At the time, there was a common need in the United States and the Asia-Pacific region for more personnel to work with problems caused by pests, as interrelated instead of as distinct issues.

By the late 1970s, however, IPM began to infiltrate and ultimately to permeate much of the curriculum, virtually unnoticed and certainly undocumented. Table 13.1 provides an overview of IPM education during that decade.

[1]The RICOP is an organization of deans and directors of U.S. universities and colleges of agriculture.

Table 13.1. IPM Education in the 1970s

University	Curriculum	Purpose
Ohio State University, 1972	B.S. in plant protection	Integrate weed, insect, and plant disease control that would "prepare a person much more adequately for farming, pesticide application, and advising agricultural producers in all facets of plant protection. Opportunities for employment by industry in sales and technical services will [be] enhanced" (Bendixen 1972).
Cornell University, 1972	Ph.D. in pest management B.S and M.S. planned for the future	Ph.D. focus on training pest management researchers and teachers. B.S. and M.S. for pest control consultants with less emphasis on research planned for the future (Glass 1972). Cornell realized that there was a need for nondegree training in IPM as well, especially for corn and apple production (even though IPM had not reached implementation in other areas of production). Pilot demonstration and training programs were planned for the future.
University of California, Berkeley	B.S., M.S., and Ph.D. in pest management	Positions should develop at universities to teach pest managers and for research. Should Extension expand, Ph.D. and M.S. people would be required in supervisory or research positions. Industry should replace the "pure salesperson" with trained pest management personnel at the undergraduate level (Huffaker 1972).
University of California, Davis	B.S. in crop protection	The basic need for trained IPM people in California was due to the state requirement that pest control operators be licensed, as well as to the lack of trained pest management personnel at the sales level.
California State Polytechnic College	B.S.	Pest Management (Thomas 1972).
California State University, Fresno	B.S.	Agricultural Sciences Option I: Plant Protection (Thomas 1972).
University of California, Riverside	M.S.	Pest Management (Thomas 1972).
Texas A&M University	Interdisciplinary M.S. in IPM (by 1978)	Insect IPM and interdisciplinary host plant resistance classes and weed ecology and insect biocontrol courses added later, combined with toxicology and field-oriented disease courses offered through the departments of entomology, plant pathology, and agronomy. This curriculum for the master's degree in IPM remained essentially the same until now (Harris 1999. Personal communication).

Table 13.2. IPM nondegree training

People	Training
Administrators (directly or indirectly involved in pest management)	Series of informal short courses and workshops around the country
Researchers	Short courses and workshops—multidisciplinary and pest problem-specific
Teachers and Extension personnel	Short courses and workshops to infuse teaching with IPM
Field survey personnel	On-the-job training, no formal training required other than interest in agriculture
Pest control applicators and consultants	Special nondegree programs
Growers	Use of pilot projects to provide understanding of IPM benefits

Nondegree Training

By the mid-1970s, nondegree training was recognized as extremely important. Crop protection personnel needed to understand IPM concepts, philosophies, and goals (Glass 1975). Table 13.2 indicates the types of individuals and training involved at that time.

Moving Toward More Current Integrated Pest Management Education: The 1980s and the 1990s

During the 1980s, universities became more and more involved in IPM training. For example, in 1980, the University of California (UC) initiated the Statewide Integrated Pest Management Project, which is still in place. From the start, the project recognized that synthetic pesticides provided benefits such as improved productivity but could also lead to pest resistance and to environmental concerns. Programs were designed to educate growers to limit their reliance on pesticides and to consider a wide range of IPM tactics. By the end of 1990, the project had funded 222 IPM projects in 35 crops. Today, the UC publishes pest management guidelines for more than 30 commodities and sells a series of IPM manuals combining the expertise of UC researchers and Extension specialists, designed for professional pest control advisors and others involved directly in pest management. In addition to these publications, a web site provides much useful information, and Cooperative Extension and the university hold extensive grower education meetings throughout the year.

Also in 1980, researchers from the Washington State University, the University of Idaho, and Oregon State University established the Solution to Environmental and Economic Problems (STEEP) research program. Funded by the USDA, STEEP, a multidisciplinary program, focused on limiting soil erosion from agriculture by implementing sustainable systems. The realization that tillage systems also affect soilborne pests led to expansion of the program's scope by way of incorporating a comparatively integrated approach taking into account both weed and disease control. Research by STEEP, with its hands-on field experiments, linked producers, agronomists, and suppliers in the Pacific Northwest.[2]

[2]Although IPM education continued to evolve in the 1980s, information about its evolution in that decade is difficult to obtain. A study of IPM education throughout the decades would be interesting but is beyond the scope of this chapter.

Integrated pest management has become a widely studied subject. Table 13.3 lists degree programs and major areas of study related to IPM in the Northeast. Although very few universities offer IPM degrees as such, most degrees have an IPM component within the traditional departments of entomology, weed science, and plant pathology.

There are, however, clearly defined IPM degrees. To compare curricula in the 1970s with those in the 1990s for a B.S. in IPM, see Table 13.4. Although the titles of most courses have not changed since the 1970s, course content most likely is quite different, for much has been learned in the fields of genetics, disease, and insect and weed management.

Integrated pest management now reaches more students as well, despite changes in the availability of IPM degrees. For example, all agronomy undergraduates at North Carolina State University take an IPM course even though the university eliminated the IPM degree program in 1994. North Carolina's farmers, agents, and agribusiness professionals essentially come from this pool (Linker, M. 1999. Personal communication). A similar situation exists at Purdue University (Ontman, E. 1999. Personal communication). As IPM has become more accepted and used, IPM course materials have gone into broader use at universities.

Delivery Systems

Much of the information about IPM is delivered by papers, books, and magazines. But computers and the Internet are becoming more and more valuable for IPM growers. Almost every land-grant university has an IPM web site with many links to other topics, organizations, and government agencies. A recent survey of IPM growers in several regions showed, however, that neither the computer nor the Internet is used as widely as many had hoped it would be. In a small survey of IPM producers in several regions, 40% used computers, but only 12.7% used decision-making software for pest control, whereas 31.1% used the Internet to obtain information about the latest IPM technologies (Sorensen, Day, and Stewart 2000).

The Internet represents a powerful new information tool, but its potential is increasing rapidly. Sites are established and abandoned at remarkable rates. The value of the Internet to IPM as both an information system and a decision aid is growing, and its future usefulness will be limited primarily by web site developers' imaginations (See Textbox 13.1 for a sampling of IPM web sites).

Table 13.3. Degree programs and major areas of study related to IPM in the northeastern United States

University	Department	B.S. (major)	M.S.	Ph.D.
U. of Connecticut	Plant science	Crop science, turf science, horticulture	Agronomy, horticulture	Agronomy, horticulture
U. of Delaware	Entomology and applied ecology	General entomology, plant protection	Entomology and applied ecology	Entomology and applied ecology
U. of Maine	Biological sciences Applied ecology and environmental science	 Sustainable agriculture	Botany and plant pathology, ecology and environmental science, entomology Various areas of emphasis	Botany and plant pathology, ecology and environmental science, entomology Various areas of emphasis
U. of Maryland	Biological resource engineering Agronomy Entomology	Natural resource management Agronomy	 Crop science Entomology	 Crop science Entomology
U. of Massachusetts	Entomology Microbiology Plant and soil science	 Microbiology Floriculture, general studies, fruit production, turf management, ornamental horticulture, vegetable crops, sustainable agriculture	Entomology Research fields: virology, pathogenic bacteriology, plant pathology Plant and soil science	Entomology Research fields: virology, pathogenic bacteriology, plant pathology Plant and soil science
U. of New Hampshire	Plant biology Natural resources Microbiology	B.S. Plant biology, environmental horticulture, B.A. plant biology Wildlife management	Research fields: breeding and genetics, crop management, genetic engineering, pathology Environmental conservation Microbiology	Research fields: breeding and genetics, crop management, genetic engineering, pathology Natural resources Microbiology
Rutgers University, New Brunswick	 Entomology	Botany/plant biology, plant science Entomology	 Entomology	 Entomology
Cornell University	Entomology Plant breeding and biometry Plant pathology Soil, crop and atmospheric science Fruit and vegetable science Plant science	Entomology Plant breeding Plant pathology Weed science Plant breeding, plant pathology	Entomology Plant breeding Plant pathology Weed science Vegetable crops	Entomology Plant breeding Plant pathology Weed science Vegetable crops
Penn State U.	College-wide Agronomy Entomology Plant pathology	Ag sciences, agronomy, environmental and renewable resources economics, environmental resource management, forest science, horticulture, landscape contracting, turfgrass science Agronomy	 Agronomy Entomology Plant pathology	 Agronomy Entomology Plant pathology
U. of Rhode Island	Plant science Natural resources	Plant science, golf and sports turf management Environmental science and management, wildlife biology and management		

Adoption of Integrated Pest Management

Rate and degree of IPM adoption have been focal points since the mid 1970s, when implementation began. Barriers to and factors affecting adoption are well documented (See Textbox 13.2).

Profile and Diffusion

According to diffusion theory, adopters of IPM practices go through certain stages, characterized as awareness, interest, evaluation, trial, and adoption (Nowak 1996). In the earlier stages, producers become aware that a specific IPM technique exists. According to Nowak (1996), this awareness comes from two distinct sources. Either the producer has a problem and seeks a solution, or a third party calls attention to an unrecognized problem. The producer then seeks information about the solution and evaluates cost and use of this new technology, practice, or set of practices. The technical information can come from scientists, educators, neighboring producers, dealers, or other sources. According to a national evaluation of the USDA Extension IPM programs in 1986, IPM technology adopters share a number of characteristics (Leslie and Cuperus 1993):

1. A greater proportion of IPM adopters than nonadopters have at least a college education.
2. Integrated pest management producers are likely to be younger than 50 years (yr) of age, which is below the average age—55 yr—of U.S. farmers.
3. Generally, adopters tend to have less farming experience than nonadopters do; and in most states a larger proportion of adopters than nonadopters have less than 30 yr experience.
4. Adopters tend to use Extension technical assistance and publications on IPM technologies, one-on-one meetings with Extension personnel, production meetings, field demonstrations, and neighbors as their sources of information. Adopters tend to use the computer.
5. The most persuasive arguments for IPM are improved pest control, increased crop quality and yield, increased returns to management, improved protection of individual and public health, and decreased environmental damage.

Adoption Barriers

To manage agricultural pests, IPM builds primarily on biological and ecological interventions. Cost

Table 13.4. Curricula, past and present

Ohio State University—1972 B.S. in Plant Protection (Core Courses only)	Mississippi State University—1990s B.S. in IPM (Core Courses only)
Biology	Plant biology
Chemistry	General chemistry
Botany	Experimental chemistry
Organic chemistry	Elementary organic chemistry
Agricultural economics	Introduction to agricultural economics
Agricultural economics or accounting or computer science	Survey of agriculture
Math	College algebra
General zoology	Principles of zoology
Entomology	Introduction to statistical inference
Agronomy	General entomology
Weed control	Soils
Plant pathology	Soils lab
Genetics	Introduction to weed science
Ecology	Introduction to plant pathology
Microbiology	Genetics I
Insect control	Weed biology and ecology
Plant physiology	Taxonomy of spermatophytes
Pesticide regulation	Field crop insects
Environmental concerns	Genetic plant physiology
Animal science	Herbicide technology
Disease control	Genetic plant ecology
Application equipment	Diseases in crops
Food science and nutrition	Plant disease management
Cereal, forage, vegetable and tree fruit production	Principles in insect pest management
Engineering in agriculture	

Textbox 13.1. Institutions offering IPM degrees/training/information on the Internet (Source: E. Day, American Farmland Trust 2000)

University of California Statewide Integrated Pest Management Project: http://www.ipm.ucdavis.edu/

The New York State Integrated Pest Management (IPM) Program at **Cornell University**: http://www.nysaes.cornell.edu:80/ipmnet/ny/; Global Crop Pest Identification and Information Services in Integrated Pest Management (IPM) on the World Wide Web. http://www.nysaes.cornell.edu/ent/hortcrops/

University of Illinois, Masters in Integrated Pest Management. http://w3.aces.uiuc.edu/CropSci/grad/Options/ipm/faculty.html

Iowa State University, Master of Science in Agronomy includes one summer course in Integrated Pest Management. http://www.ipm.iastate.edu/ipm/icm/indices/1996/pesticideeducation.html
(http://www.ipm.iastate.edu/ipm/)
The Corn Rootworm Home Page: http://www.ent.iastate.edu/pest/rootworm/

Michigan State University: http://www.msue.msu.edu/ipm/about.htm, http://www.canr.msu.edu/ipm/index.htm, http://www.cips.msu.edu/

University of Minnesota: http://www.ent.agri.umn.edu/academics/classes/ipm/chapters/macrae.html, http://ipmworld.umn.edu/

Mississippi State University: Agricultural Pest Management and Integrated Crop Management. http://www.msstate.edu/dept/pss/public_html/intcropmgmt.html, http://WWW.msstate.edu/Dept/APM/info.html

Montana State University – Bozeman, Department of Entomology, graduate degrees in Insect Pest Management/Economic Entomology: http://scarab.msu.montana.edu/, and **University of Florida** IPM: urban IPM classes through Montana University: http://www.ifas.ufl.edu/~schoolipm/montana.htm

Oregon State University, Pesticide Applicator Training (PAT)Program. http://ippc.orst.edu/pat/

Purdue Cooperative Extension Service, Purdue Pesticide Programs: Commercial Pesticide Applicator Training. http://www.btny.purdue.edu/PPP/PPP_CPAT.html

Purdue University, Department of Botany and Plant Pathology, BS is Crop Protection: http://www.btny.purdue.edu/, National Pest Management Materials Database: http://www.entm.purdue.edu/ipmdb.html

Texas A&M University, Department of Rangeland Ecology and Management. BS in Integrated Resource Management: http://cnrit.tamu.edu/rlem/integrated.html

University of Wisconsin, Madison, College of Agricultural and Life Sciences, Department of Agronomy, course offered in Integrated Weed Management: http://agronomy.wisc.edu/pages/undergrad/undergrad.html

Washington State University, Pesticide Information Center On-Line: http://picol.cahe.wsu.edu/. Pesticide and Environmental Stewardship: http://pep.wsu.edu/

Kansas State University, Department of Agronomy, BS offered – Crop Consultant Option: http://www.ksu.edu/agronomy/ACADEMIC/cnsltopt.htm, Integrated Management of Arthropod Pests of Livestock and Poultry – Pesticide Guide for livestock pests, extension information by state. http://www.oznet.ksu.edu/pr_LP-pests/welcome.

North Carolina State University, The Industry/University Center for Integrated Pest Management: Newsletter for IPM: http://ipmwww.ncsu.edu/cipm/newsletter.html, Center for IPM: http://ipmwww.ncsu.edu/cipm/cipm.html

University of Arizona, College of Agriculture, Pesticide Information & Training Office. Pesticide and Training Information. http://Ag.Arizona.Edu/pito/apptrain/current.html

Miner Institute, Educational Programs. http://www.whminer.com/EDU.htm

Clemson University, Pesticide Applicator Training (PAT). http://entweb.clemson.edu/pesticid/program/PAT.htm.

USDA, Cooperative State Research, Education, and Extension Service. Pesticide Applicator Training. Programs in every state. http://www.reeusda.gov/pas/programs/pat/pest.htm; On-line: http://pwd.reeusda.gov/pwd/view.asp

Auburn University, Alabama, Alabama Pesticide Information: http://www.aces.edu/department/ipm/

University of Tennessee Extension, Entomology and Plant Pathology Section, Biological Control and IPM Sites (under construction). http://funnelweb.utcc.utk.edu/~extepp/eppipm.htm

Clark Consulting International, agLINKS Index. Subject specific Internet directory which includes modeling, software, agronomy and crop specific information. http://www.agpr.com/consulting/aglinks.html

Association of Applied Insect Ecologists, http://www.aaie.com/index.html

Textbox 13.1 (continued).

Gempler, IPM Solutions Newsletter. A newsletter of IPM products and policy news for the IPM Professional. http://www.agriculture.com/contents/gempler/ipm/1997/v2i5/index.html

Database of IPM Resources is an information retrieval/referral system and a compendium of customized directories of worldwide IPM information resources accessible through the Internet. http://www.ippc.orst.edu/cicp/

The Natural Resources Institute, The Pest Management Department. IPMForum is a member of the international IPM Information Partnership: http://www.nri.org/IPMForum/main.htm

American Crop Protection Association (ACPA): http://www.acpa.org/

The Certified Crop Adviser (CCA) program is a membership service of the American Society of Agronomy (ASA): http://www.agronomy.org/cca/

Pennsylvania State University, PA IPM Program. Includes training of IPM, newsletters, working groups, research and extension. http://paipm.cas.psu.edu/index.html

Michigan Department of Agriculture, Integrated Pest Management Training Offered for Workers in Public Buildings: http://www.mda.state.mi.us/news/archive/1997/0897/081197.html

Maryland Cooperative Extension Entomology Agricultural IPM. Agricultural Integrated Pest Management. Extension entomology faculty offer a variety of courses, workshops, and meetings: http://pest.umd.edu/training.html

NIPMN. State Contacts in every state for Integrated Pest Management, Pesticide Impact Assessment and Pesticide Applicator Training: Illinois, Indiana. http://www.ent.iastate.edu/ipm/nipmn/statecontacts.html

National IPM Network - Colorado State University. Agricultural Integrated Pest Management and Crop Production: http://www.colostate.edu/Depts/IPM/csuipm.html

CTN Educational Services, Inc. Continuing education classes for re-certification of pest control license holders under the Texas Structural Pest Control Board and the Texas Department of Agriculture: http://pestnetwork.com/ctn/about_ctn/about_ctn.html

PestNetwork.com, provides an IPM pest control training video catalogue: http://ctnedu.com/video/videocatalog.html

American Association of Pesticide Safety Educators. http://aapse.ext.vt.edu/aapse.html

effectiveness and sustainability of IPM systems depend on their capacities to maintain pest populations below economic thresholds across different and usually unpredictable situations. Systems must withstand a wide range of weather, pest pressure, and agronomic conditions and also must perform well over time by continually undermining the ability of pests to adapt to management practices.

Ecological theory suggests and field experience confirms that IPM systems based on a variety of control tactics are especially resilient because redundancy is built into the system. Where a single tactic is relied on for control, growers are vulnerable when pests adapt. Regardless of tactic used, however, certain combinations of weather, agronomic conditions, and pest adaptations are likely to lead to problems. The best way to ensure cost effective management across conditions is to build redundant mechanisms of control into production systems—in other words, to build bio-intensive, prevention-based IPM systems (Coble, H. 1998). Producers choose not to adopt a technology for two basic reasons: inability or unwillingness. As Nowak (1992) points out, these reasons are not mutually exclusive. Producers may be able but unwilling, willing but unable, or unwilling *and* unable. If unable, there generally is an obstacle preventing growers from adopting (Nowak 1992):

1. Information may be lacking or scarce.
2. Cost of obtaining information may be prohibitive.
3. Complexity of the technology may be too great—complexity of the technology is related inversely to rate and degree of adoption. In recent years, IPM technologies have become quite complex.
4. Technology may be too expensive—investment costs and returns on investment are primary concerns of producers.
5. Labor costs may be prohibitive.
6. Planning horizon for the operation may be too small compared with the times associated with learning costs and with recuperation of initial investment costs.
7. Availability and accessibility of support may be limited (e.g., other producers may share successes and failures, and chemical dealers may assume parts of the risk).

Textbox 13.2. Case studies: Adoption barriers

A study conducted with Georgia peanut producers (McNamara et al. 1991) suggests that there are four primary categories considered in decisions regarding adoption: producer component, management practices, farm structure, and institutional factors. Variables from three of the four categories, with the exception of farm structure, were significant. None of the variables in the farm structure category were associated with whether a producer was going to adopt IPM technologies, although it was hypothesized that farm assets and farm debt should have a positive (assets) and negative (debt) relation to adoption. If a producer had many assets, he/she was relatively likely to take on real or perceived risks from adopting IPM.

Thomas, Ladewig, and McIntosh (1990) conducted a study among Texas cotton growers to determine the characteristics of IPM adopters. These researchers also defined specific categories to test their hypotheses that certain variables do or do not impede producer adoption decisions. Earlier studies had determined that a major factor inhibiting adoption of IPM was efficacy—for example, the value of foregoing early crop planting to preserve beneficial insects. In addition, many producers have seemed unwilling to change the traditional plow-plant-spray-harvest routine and therefore have depended heavily on pesticides. For this study, farm and grower characteristics, information sources, technology beliefs, and adopted technologies were taken into account. Producers, in fact, differed in their use of diffusion agents, i.e., in the way they used human resources, information, etc., to facilitate IPM adoption. The percentage of irrigated land, the level of education, and the amount of gross farm sales were the most important variables studied. Gross farm sales, group meetings, private consultants, and IPM philosophy all had a positive influence on adoption. Additionally, producers who used pheromone traps generally had high levels of education, were young, and valued information provided by individual contacts and group meetings. Unlike other studies, this study found that growers who used less irrigation on their cotton fields adopted more IPM practices than other producers did. Furthermore, IPM belief measures (economic threshold and perceived benefits of IPM) positively affected the number of IPM practices, whereas producer age, farm size, and most information sources had no effect.

Kovach and Tette (1988) conducted a survey of the use of IPM by New York apple growers, IPM users, and nonusers. More than 80% of apple growers used IPM techniques. Generally, IPM adopters were younger and had more education and less experience than nonadopters. Cornell Cooperative Extension played a vital role in the provision of information regarding pesticide use and pest management. Producers who had adopted IPM used less pesticides than producers who had not.

Fernandez-Cornejo, Beach, and Huang (1994) examined the adoption of IPM techniques by vegetable growers in Florida, Michigan, and Texas, using survey data from individual vegetable growers. At the time of the study, the distribution of adopters varied widely, from 31% in Florida to 60% in Michigan. The results support the notion that adopters tend to be less risk averse than nonadopters. Farm size is significant, confirming that owners of larger farms are more likely to adopt than owners of smaller farms. Furthermore, farm operator and unpaid family labor variables were significant, showing that the quality as well as the quantity of labor affect adoption decision and that managerial time is essential to the decision-making process.

8. Managerial skills may be inadequate—IPM systems sometimes are designed for the above-average manager.
9. Control over adoption may not exist when decisions require a third party's (a landlord's, a bank's, or a partner's) approval.

Producers also may be unwilling to adopt an IPM technology or practice because of one of the following reasons:

1. Information is difficult to access or irrelevant—information about other states, crops, or even regions may be perceived as unimportant.
2. Conflict exists between current production goals and new IPM technology. Too often, new techniques do not fit into production systems, especially when the technique is relatively inflexible. Conflict with IPM strategies also can arise when the goal is to produce a perfect, unblemished product (Fernandez-Cornejo and Castaldo 1998; Kovach and Tette 1988).
3. Producers lack the opportunity to learn about the technology.
4. The IPM practice or technology is inappropriate for the operation's production system.
5. Perceived or real risk of negative outcome such as yield loss is heightened.
6. Traditional production methods have credibility.

Finally, if IPM advice does not take into account the biological variability within specific production units and instead advocates treating those units as homogeneous units for decision making, farmers may conclude that IPM is a "one-size-fits-all" approach (Nowak, P. 1999. Personal communication).

Experts clearly need to continue building on what they have learned about IPM adoption. As production becomes more technologically sophisticated and information intensive, specialists will have to train producers in increasingly complex technologies.

Measurement Systems

Consumer concern about pesticide residues in food is widespread and has been in the media spotlight for some time (Byrne et al 1991; van Ravenswaay and Wohl 1995). Although research regarding residues is ongoing, negative responses are prevalent in Europe, the United States, and elsewhere; those responses typify concerns about pesticide use that have resulted in worldwide efforts to decrease the amount applied to crops. In response to these concerns, researchers have begun to investigate the health and environmental effects of pesticides, especially in light of the fact that pesticides have been detected in surface- and groundwater, in air and soil, and residually in food (NAS 1993). Because economic thresholds and potentially decreased profitability are easier to measure, however, environmental impact models are a fairly recent phenomenon (Higley and Peterson 1996).

Starting with the 1992 National Integrated Pest Management Forum (Sorensen 1994) and extending beyond the Third National IPM Symposium in 1996 (Lynch, Greene, and Kramer-LeBlanc 1996), the need to assess the effects of IPM programs has been a dominant theme. Although consensus has been reached regarding the need for improved documentation, none has been reached regarding appropriate assessment methods. But given the federal 75% IPM goal, the concomitant goal of decreasing reliance on high-risk pesticides (as outlined in the Food Quality Protection Act) and the demand for greater accountability for public expenditures (the Government Performance and Results Act), measurement systems are here to stay. Indeed, the USDA IPM Initiative and National IPM Implementation Plan require integration of assessment activities in future IPM funding proposals (Benbrook et al. 1996; Jacobsen 1996).

Environmental Indicator Models

Sixteen environmental indicator model, or measurement, systems have been developed by both U.S. and international researchers (Table 13.5). In different ways, the indicators calculate potential risk to the environment. Some calculate a risk ratio, others a scoring table or ranks; yet others use a "fuzzy" expert system. The indicators have different purposes and are designed for different levels of geography. Because they incorporate information on individual pesticide applications, they are used as decision aids for farmers; for very fine-tuning producer risk management; for providing information on pesticide use in a specific crop for the purpose of green labeling; and, increasingly, for aiding the decisions of policymakers. The latter use is especially evident in the European Union.

Decision makers and farmers in the field need information and tools that will allow them to choose the "best" pest control practices with the fewest negative effects on the environment and human health. Additionally, farmers want to avoid pesticides that harm beneficial parasites and predators of the target pest. Environmental measurement system methodologies include the following:

1. simulation of environmental effects;
2. sampling, monitoring, and tracking changes in biological indicators (e.g., biodiversity, soil respiration rate, and pesticide level);
3. surveys and qualitative research methods, including observations and interviews regarding tabulation results; and
4. indexing or ranking of the extent and severity of chemical effects on one or more environmental indicators.

Most environmental indicators were developed (and still are in the process of being refined) in Europe. These efforts recently were accelerated by an appeal from the European Commission to all European member states to develop, in light of the General Agreement on Tariffs and Trade, the Common Agricultural Policy, and the European Agenda 2000 (Reus et al. 1999), environmental indicators as possible decision-making tools. A key indicator system developed by the Wisconsin potato IPM collaboration is used in tracking potato pesticide use and risks associated with potato production in that state (Benbrook et al. 2002). Table 13.5 summarizes the environmental indicator models currently available for testing with field data.

These indicators primarily take into account the number of applications, label information, and kilograms of active ingredient. A complete set of environmental and social parameters in the assessment of consequences of pesticide use should, ideally, include

Table 13.5. Environmental indicator models

Name of indicator	Organization	Intended use	Measurements	Result
Environmental Impact Quotient (EIQ)	Kovach et al., U.S.	Tool for farmers	• Groundwater and aquatic effects • Terrestrial effects	Rating system
Stemilt Growers Integrated Fruit Production Responsible Choice Point Summary	Fruit packing company in Washington State, U.S.	Tool for farmers	• Consumer and farm worker safety • Pesticide persistence • Surface- and groundwater • Agricultural sustainability • Economic feasibility	Combines ratings (target pest specific)
PestDecide®	New South Wales, Australia	Tool for farmers	• Application activity • Timing, site of application • Persistence, efficacy, cost	Weight-based rating system (based on EIQ)
Environmental Yardstick for Pesticides (EYP)	Centre for Agriculture and Environment, The Netherlands	Tool for farmers	• Risk to water organisms • Risk of groundwater contamination • Risk to soil organisms	Risk-based ratio between concentration and toxicity
Consumers Union's Agricultural Pesticide Risk Index	Consumers Union, U.S.	Advice to policymakers	• Acute mammalian toxicity • Chronic mammalian toxicity • Ecotoxicity • Impacts on biointensive IPM systems	Weight-based risk index
The Hasse Diagram (HD)	Danish Institute of Agricultural Sciences	Advice to policymakers	Soil, surface, and groundwater Ecotoxicological effects	Relative and total ranking
UC Berkeley Environmental Health Policy Program Ranking System	University of California, U.S.	Advice to policymakers	• 12 indicators, three categories of impact on human health, ecosystems, natural resources	Weight-based ranking system
USDA ERS Chronic and Acute Risk Indicators of Pesticide Use	USDA, Economic Research Service, U.S.	Advice to policymakers	• Acute and chronic toxicity based on "toxicity/persistence units."	Weight-based potential risk indicator
SYNOPS_2	Institute of Technology Assessment in Plant Protection (BBA), Germany	Advice to policymakers	• Surface water, soil • Time • Ecotoxicological effects on aquatic and soil organisms	Ratio between concentration and toxicity and translated into relative "risk graphs"
Environmental Performance Indicator for Pesticides (p-EMA)	Univ. Hertfordshire Dept. of Environmental Sciences, UK	Tool for farmers	• Potential hazard for humans, wildlife, bees, and aquatic organisms; potential emission to air; surface and groundwater; and bioaccumulation	Software-based environmental performance assessment
Pesticide Environmental Impact Indicator (Ipest)	National Institute of Agricultural Research (INRA), France	Tool for farmers	• Groundwater • Surface water • Air • Dose rate	Fuzzy expert system
Environmental Potential Risk Indicator for Pesticides (EPRIP)	Università Cattolica del Sacro Cuore, Inst. of Environ. and Ag. Chemistry, Italy	Tool for farmers	• Groundwater • Surface water • Soil • Air	Ratio between concentration and toxicity, translated into score and total score
System for Predicting the Environmental Impact of Pesticides (SyPEP)	Vet. and Agrochemical Research Centre, Belgium	Advice to policymakers	• Groundwater • Surface water	Individual and total score
CHEMS-1: Chemical Hazard Evaluation for Management Strategies	University of Tennessee	Advice to policymakers	• Seven toxicological (and ecotoxicological) endpoints and four indicators of exposure potential	Scoring system
Pesticide Environmental Risk Indicator (PERI)	Swedish University of Agricultural Sciences, Sweden	Tool for farmers	• Mobility data (DT50, Koc, etc.) toxicity data (soil organisms, water organisms, bees, bioaccumulation)	Scoring system, individual scores, and total score
Pesticide Risk Assessment Tool	Collaboration of Wisconsin Potato and Vegetable Growers Assoc., Univ. of Wisc., and World Wildlife Fund	Tool for farmers and advice to policymakers	• Multiattribute toxicity factors calculated to reflect each pesticide's acute and chronic toxicity, and compatibility with biointensive IPM programs	Pesticide risk assessment system

these variables:

1. health of farm workers, consumers, and the general public (including vulnerable subpopulations such as the elderly and children);
2. lethal and sublethal effects on other nontarget biota;
3. direct and indirect effects on natural- and agroecosystems, including effects on habitat and food resources;
4. parameters to calculate pollution of air, soil, and water; and
5. costs and benefits to producers and society for decreasing pesticide use.

Most environmental indicator models have not yet been tested with field data. Pesticide application, timing, location, and method are the main components of these models, and unless producers are willing to cooperate, researchers will have a difficult time working with actual data to test the proposed systems and to determine how accurately they estimate effects on the environment.

In conclusion, with IPM evolving toward more complex integrated systems approaches, changes will continue to occur in the education of IPM practitioners, the delivery of information to producers, and the evaluation of effects. Because of these complexities, additional barriers to adoption are likely to arise. The speed at which technologies are being introduced also has a bearing on the agricultural community's ability to provide necessary training and information to incorporate new tools into existing IPM systems. Because IPM is first and foremost a philosophy of pest management, however, the concept certainly is sufficiently resilient to survive these challenges.

Appendix A. Abbreviations and Acronyms

IPM	integrated pest management
STEEP	Solution to Environmental and Economic Problems
UC	University of California
USDA	U.S. Department of Agriculture
yr	year

Literature Cited

Benbrook, C. M., E. Groth, J. Halloran, M. Hansen, and S. Marquardt. 1996. *Pest Management at the Crossroads.* Consumers Union, Yonkers, New York. 272 pp.

Benbrook, C. M., D. L. Sexson, J. A. Wyman, W. R. Stevenson, S. Lynch, J. Wallendal, S. Diercks, R. Van Haren, and C. A. Granadino. 2002. Developing a pesticide risk assessment tool to monitor progress in reducing reliance on high-risk pesticides. *Amer J Potato Research* 79:183–200.

Bendixen, L. E. 1972. A curriculum for plant protection at the Ohio State University. *Workshop on Pest Management. Curriculum Development and Training Needs*, the East-West Center and the College of Tropical Agriculture, University of Hawaii, Honolulu, December 4 to 7, 1972.

Browning, C. B. 1972. Pest management and plant production. *Workshop on Pest Management.* Curriculum Development and Training Needs, the East-West Center and the College of Tropical Agriculture, University of Hawaii, Honolulu, December 4 to 7, 1972.

Byrne, P., J. Gempsaw, M. Conrado, II, and U. Toensmeyer. 1991. An evaluation of consumer pesticide residue concerns and risk information sources. *South J Agric Econ* 23:164–174.

Coble, H. 1998. A new tool for measuring the resilience of IPM systems: The PAMS Diversity Index. *IPM Measurement Systems Workshop*, Chicago, Illinois, June 12 to 13, 1998.

Fernandez-Cornejo, J., E. D. Beach, and W.-Y. Huang. 1994. The adoption of IPM techniques by vegetable growers. *J Agric Appl Econ* 26:158–172.

Fernandez-Cornejo, J. and C. Castaldo. 1998. The diffusion of IPM techniques among fruit growers in the USDA. *J Prod Agric* 11:497–506.

Glass, E. H. 1975. *Integrated Pest Management: Rationale, Potential, Needs and Implementation.* Special Publication. Entomological Society of America, Lanham, Maryland. 141 pp.

Higley, L. G. and R. K. D. Peterson. 1996. Environmental risk and pest management. In E. B. Radcliffe and W. D. Hutchison (eds.). *Radcliffe's IPM World Textbook.* Regents of the University of Minnesota, St. Paul.

Jacobsen, B. 1996. USDA Integrated Pest Management Initiative. In E. B. Radcliffe and W. D. Hutchison (eds.). *Radcliffe's IPM World Textbook.* Regents of the University of Minnesota, St. Paul.

Kovach, J. and J. P. Tette. 1988. A survey of the use of IPM by New York apple producers. *Agric Ecosyst Environ* 20:101–108.

Leslie, A. R. and G. W. Cuperus (eds.). 1993. IPM and growers: An evolution in thinking. *Successful Implementation of Integrated Pest Management for Agricultural Crops.* Lewis Publishers, Boca Raton, Florida.

Lynch, S., C. Greene, and C. Kramer-LeBlanc (eds.). 1996. *Proceedings of the Third National IPM Symposium / Workshop: Broadening Support for Twenty-First Century IPM.* Miscellaneous Publication Number 1542. Natural Resources and Environment Division, Economic Research Service, U.S. Department of Agriculture, Washington, D.C.

McNamara, K. T., M. E. Wetzstein, and G. K. Douce. 1991. Factors affecting peanut producer adoption of integrated pest management. *Rev Agric Econ* 13:129–139.

National Academy of Sciences (NAS). 1993. *Pesticides in the Diets of Infants and Children.* Committee on Pesticides in the Diets of Infants and Children. National Academy Press, National Academy of Sciences, National Research Council, Washington, D.C.

Nowak, P. 1992. Why farmers adopt production technology. *J Soil Water Conserv* 47:14–16.

Nowak, P. 1996. Practical considerations in assessing barriers to IPM adoption. Pp. 93–115. In S. Lynch, C. Green, and C. Kramer-LeBlanc (eds.). *Proceedings of the Third National IPM Sym-*

posium / Workshop: Broadening Support for Twenty-First Century IPM. Miscellaneous Publication Number 1542. Economic Research Service, U.S. Department of Agriculture, Washington, D.C.

Obein, S. R. and P. S. Motooka (eds.). 1972. *Workshop on Pest Management: Curriculum Development and Training Needs.* The East-West Center and University of Hawaii, Honolulu, December 4 to 7, 1972.

Reus, J., P. Leendertse, C. Bookstaller, I. Fromsgaard, V. Gutsche, K. Lewis, C. Nielsson, L. Pussemier, T. Trevisan, H. ven der Werf, F. Alfarroba, S. Blümel, J. Isart, D. McGrath, and T. Seppälä (eds.). 1999. Concerted Action for Pesticide Environmental Risk Indicator Project, Centre for Agriculture and Environment, Utrecht, The Netherlands, March 18 to 20, 1999.

Sorensen, A. A. 1994. *Proceedings of the National Integrated Pest Management Forum.* Center for Agriculture in the Environment, American Farmland Trust, DeKalb, Illinois.

Sorensen, A. A., E. Day, and P. Stewart. 2000. Factors affecting the adoption of new technologies. Pp. 12–31. In G. G. Kennedy and T. B. Sutton (eds.). *Proceedings of the International Conference on Emerging Technologies in IPM: Concepts, Research, and Implementation.* APS Press, American Phytopathological Society, St. Paul, Minnesota.

Stern, V. M., R. F. Smith, R. van Den Bosch, and K. S. Hagen. 1959. The integrated control concept. *Hilgardia* 29:81–101.

Swinton, S. M. and M. B. Williams. 1998. Assessing the economic impacts of integrated pest management: Lessons from the past, directions for the future. Staff Paper Number 92–12. Department of Agricultural Economics, Michigan State University, East Lansing.

Thomas, J. K., H. Ladewig, and W. A. McIntosh. 1990. The adoption of integrated pest management practices among Texas cotton growers. *Rural Soc* 55:395–410.

van Ravenswaay, E. O. and J. Wohl. 1995. Using contingent valuation methods to value the health risks from pesticide residues when risks are ambiguous. Pp. 219–256. In J. A. Caswell (ed.). *Valuing Food Safety and Nutrition.* Westview Press Incorporated, Boulder, Colorado.

References for Labeling and Regulation

Barham, E. 1997. What's in a name? Eco-Labeling in the global food system. Joint Meetings of the Agriculture, Food, and Human Values Society and the Association for the Study of Food and Society, Madison, Wisconsin, June 5 to 8, 1997.

Browner, C. M., R. Rominger, and D. L. Kessler. September 22, 1993. Testimony before the Subcommittee on Department Operations and Nutrition and the Committee on Agriculture. U. S. House of Representatives, Washington, D.C.

Bruno, K. 1992. *The Greenpeace Book of Greenwash.* Greenpeace International, Amsterdam, The Netherlands.

California Clean Growers. 1998. Homepage, <http://www.californiaclean.com> (30 September 2002)

Core Values. 2002. Homepage, <http://www.corevalues.org> (30 September 2002)

Fernandez-Cornejo, J. and S. Jans. 1999. Pest management in U.S. agriculture, August, <http://www.ers. usda.gov/publications/ah717/ah717.pdf> (30 September 2002)

The Food Alliance. 2001. The Food Alliance's seal of approval, <http://www.thefoodalliance.org/ about.html> (30 September 2002)

New York State Integrated Pest Management Program. 2002. Homepage, September 25, <http://www. nysipm.cornell.edu> (30 September 2002)

Rainforest Alliance. 2002. Conservation programs, <http://www.rainforest-alliance.org/programs/cap/ program-description2.html> (30 September 2002)

Stemilt Growers. 2002. Homepage, <http://www.stemilt.com/ index.php> (30 September 2002)

14 Future Challenges and Directions

This report by the Council for Agricultural Science and Technology (CAST) addresses advances in integrated pest management (IPM) that have occurred since publication of the first CAST IPM report (1982), which focused on pesticide reduction. Recent advances include more-efficient pest management tools and strategies, new technologies, innovative pest management systems, more sustainable crop/animal production systems, and an expanded public appreciation of the utility of IPM (Cavigelli et al. 2000; Kennedy and Sutton 2000; Madden 1992). Although certain IPM tactics have been adopted widely (that is, on more than 70% of crop acres), the use of biointensive IPM remains limited (on 18% of crop acres) (GAO 2001). Changing resources, increasing pest diversity, evolving government policy, and the ongoing need for more widespread adoption of comprehensive IPM all affect IPM development and implementation while offering many exciting but often significant challenges.

Extensive discussions of the goals of IPM and the work that remains unfinished have been published in recent years. Key agents of change in the ever-evolving agricultural production systems include economic factors, market opportunities, government polices and regulations, technological advances (including those related to IPM), exotic-pest introductions, and pest control crises (GAO 2001; Kennedy 2000; Sorensen, Day, and Stewart 2000). In the United States, a dramatic example of change is the increasing number of acres of biotechnology crops grown, which amounted to more than 75 million acres (a.) in 2000 as compared with 4 million a. in 1996 (CAST 2001). Integration of transgenic crops with cost-effective and logistically compatible IPM practices in sustainable cropping systems poses many challenges. Incorporation of transgenic crops often results in increases in secondary pests and a narrowing of germplasm use, thus decreasing the robustness of an IPM program in cropping systems (Jacobsen 2000). This trade-off highlights the need for increased systems research in evolving farming systems (Merrigan 2000). Such research also should address the need for multiple tactics and strategies for managing an increasingly complex and changing array of pests that is usually site specific. Insofar as history has shown that there is no "silver bullet" for crop or IPM management (Kogan 1998), management options must be deployed with great care (Kennedy 2000). In 1996, the National Research Council (NRC) published a call for the development of ecologically based pest management; this, however, has been a long-term IPM goal for decades (Kogan 1998; Rabb and Guthrie 1970). In a recent book, Liebman, Mohler, and Staver (2001) similarly emphasized the potential for ecology-based weed management.

As farming and industrial organizations grow, so does the need for interdisciplinary and interorganizational collaboration. Especially vexing are the increasing diversity of pest species, including introduced taxa and new strains, and the complexity of IPM and crop production systems. These systems are becoming more technology and information intensive, and the importance of information transfer and communication is increasing. Private organizations, with their breadth of resources—pesticides, crop varieties, grower information—are assuming considerable responsibility for information transfer and development of crop solutions (whole production packages) and IPM/integrated crop management (ICM) systems (Carroll 2000). How will these developments affect traditional university/Extension information transfer and training for IPM practitioners and others? Will a more clearly defined collaboration between private and public IPM/ICM sectors evolve? Recently developed long-distance teaching/learning technologies also are coming to the forefront in IPM and in most information transfer systems.

General constraints on IPM adoption have been categorized as technical, educational, institutional, and, especially, interorganizational, interpersonal, interdisciplinary, and financial (GAO 2001; Goodell and Zalom 2000). Examples of specific challenges faced by IPM practitioners include decreased funding; unfavorable public perceptions of IPM and pesticides; the negative impact of the Food Quality Protection Act (FQPA) and other government polices on IPM; the limited commitment to IPM by environmental and consumer groups; the increasing need for coordination among federal, state, and private sectors in IPM research planning and implementation; and the growing role of marketplace issues such as IPM risk insurance (Goodell and Zalom 2000; Merrigan 2000). Specific research and Extension IPM needs for various crops and animals also have been enumerated (Geden and Hogsette 2001; Kennedy and Sutton 2000). Government policy and regulation continue to influence IPM by means as varied as local and state governments to presidential executive orders and national farm bills. While the national and international phase-out of methyl bromide is ongoing, the FQPA will eliminate many other pesticides currently in use. Crop insurance for IPM and IPM labeling may be on the horizon. Finally, the 2002 Farm Bill presents additional opportunities and challenges for IPM. That bill provided the National Resources Conservation Services with funds for use in IPM implementation. The source of these funds is the Environmental Quality Incentives Program (EQIP); decisions regarding their use will be made at the state level (Coble, H. 2002. Personal communication; Kelley 2002; USDA–RMA 2003).

Key Issues for the Future

The aforementioned IPM-related issues, and many others, must be considered if recent and ongoing advances in technology are to influence the general direction of IPM and to enhance its implementation. Because detailed discussions of various facets of IPM advances were treated in earlier chapters, the remainder of this concluding chapter focuses on *seven primary areas*: gene technology constraints; genetic diversity and pest adaptability; ecologically based IPM; systems approaches; evolving pools of trainers and speed of technology transfer; government policy and regulations; and assessments of IPM.

Gene Technology Constraints

As described in this report, genetic resistance to pests has been and will continue to be an important feature of IPM. Pest-resistant crops developed by means of conventional breeding and genetic engineering have helped decrease pesticide use. Future opportunities are great, and crops may be modified for increased compatibility with existing and future IPM systems. Nevertheless, key unresolved issues include the extent to which genetically engineered crops will be used in production systems, the rate at which they will be adopted, their compatibility with IPM systems, their acceptance or lack of acceptance by the public, and their ultimate beneficiaries.

The NRC (2000) has stated that deployment of *pest-protected plants* (i.e., plants resistant to pests) is not without problems. The council specified toxicity, allergenicity, effects of gene flow, development of resistant pests, and effects on nontarget species as concerns for both conventional and transgenic pest-protected plants. And the council cited as another area of concern *pleiotropic* effects, or unanticipated effects on plant physiology. The NRC has provided recommendations for these major areas of concern, which are beyond the scope of this CAST report. But even with such constraints in place, the importance of host resistance to pests will continue to be central for IPM on most major crops.

Three key steps in developing pest resistance in crops are (1) identification of a source of resistance, (2) transfer from donor source to recipient plant, and (3) evaluation of the resulting plant. At each of these steps, constraints exist, regardless of breeding method. Identification of useful donors can pose unique difficulties. The utility of a donor gene is judged only after genetic transfer succeeds. With genetic engineering, the genes responsible for resistance must be identified, isolated, and inserted into a plant transformation deoxyribonucleic acid (DNA) vector that may contain genes providing a marker for their presence and allow efficient integration of alien genes. With traditional methods, techniques for identifying resistant phenotypes must be developed and used to evaluate large populations of hybrids. Understanding of the mechanisms or causes of resistance, the manner of genetic translation into resistance phenotypes, the features of multifaceted interactions between pests and hosts, and the complexity of resistance phenotypes might alleviate many concerns. Additional basic problems include potential narrowing of the germplasm base used in crop production, for such narrowing may enhance vulnerability to new pests, as well as unexpected effects on nontarget organisms, pest resistance development/management, and grower/consumer acceptability.

Theoretically, any gene present in any organism, or the molecular alteration or creation of any gene, is a potential resource of genetic variability for the genetic engineer. In contrast, breeders using only conventional methods are limited to those genes present in the crop species and in relatives closely related to the crop. The availability of a wide array of genetic material raises exciting possibilities of novel crops. Simultaneously, this array of genetic possibilities raises fears of crops with unanticipated consequences such as the devastation of ecosystems, or of the detrimental effects of long-term production or consumption (University of Guelph 2002). The challenge to the IPM community is to evaluate critically the use of pest-protected plants and to envision creative solutions achievable through their development and deployment.

The legal and ethical environments in which genetically engineered crops are created differ from those in which conventional varieties are created. For example, the intellectual property underlying a genetically engineered crop has been defined precisely and therefore is protected and defended legally with greater facility than the intellectual property underlying the improvement of conventionally bred crops. The intellectual property of the genetic engineer is perceived as more interchangeable and, therefore, as more attractive for commercialization by the private sector. For conventionally bred varieties, resistance characterization, gene pool enrichment and, often, variety development have occurred in the public sector, with the primary goal of improving the profitability and sustainability of farmers. Because the IPM community places a high priority on improving farmer sustainability through biological or ecological means, the differences in value systems between the public sector and the private sector need to be bridged so that pest-protected plants can be more suitable for IPM systems. Moreover, any genetic advances must address resistance management (RM), avoid further narrowing of the gene pool, and convince consumers of the desirability of the concept.

Constraints to genetically engineered crop varieties range from the high cost of developing and gaining regulatory approval for crops with biotechnology traits to new management practices aimed at minimizing their potentially negative effects on the environment and on nontarget organisms. The cost of product development and regulatory approval can limit the number of resistant crops commercialized. On the other hand, engineered pest resistance has several advantages. In biotechnology crops, a small amount of DNA containing the gene or genes responsible for the desirable trait is introduced and acts as single dominant traits segregating in a Mendelian fashion (cf NRC 2000). Because of the ease of identifying and tracking transgenic progeny in a breeding program, an engineered trait can be introduced into several elite varieties of crops fairly quickly.

Resistance traits that are quantitative, inherited, or controlled by more than a single gene present additional constraints. Regarding transgenic approaches, the ability to transfer quantitative traits remains in the early stages of development. Quantitative resistance traits can be difficult to use with traditional breeding approaches, especially if they are present only in wild relatives or in unadapted donor species. Quantitative traits usually are more difficult to evaluate or to score, and they require re-evaluation methods of greater precision than those for qualitative traits. The greater the number of alien genes transferred, the more likely the desirable characteristics will diminish performance; thus, additional breeding improvement steps are required. At times, as a result of undesirable linkages and other factors, economic traits do not perform as expected when recovered in combination with resistance.

Implicit in IPM practice is the goal of minimizing the potentially undesirable consequences of deploying a pest control tactic. Decreased predator populations and enhanced pesticide resistance in target pests are just two examples of the potentially undesirable consequences of pesticide use that IPM practitioners seek to avoid. At this time, however, scientists are unable to predict with reliability the consequences of deployment. A report by the NRC (2000) confirms that genetic traits conferring pest resistance on plants created either by conventional breeding or by transgenic methods can be toxic to humans, cause allergic reaction, and affect nontarget species. All these constraints must be dealt with appropriately.

Pest populations also may be selected or may develop resistance to the genes protecting crops from pests. These potential undesirable consequences of resistance deployment must be considered and minimized or eliminated in the development and the deployment of pest-resistant plants. Such concerns are important and should be considered before recommendation of management practices for biotechnology crops with enhanced pest resistance traits (Fitt and Wilson 2000).

If transgenes are to succeed, several special requirements must be addressed. A biointensive IPM approach—including industry support, monitoring, rotations, carefully planned pesticide applications, and a truly integrated production/management system—is essential for long-term sustainability of these new tools (Ferro 2000; Kennedy and Sutton 2000). There is evidence for the importance of integrated tactics in the cotton system. For example, the use of *Bacillus thuringiensis* (*Bt*) cotton has resulted at times in decreased scouting; as a result, control of secondary pests has suffered. The problems with transgenic single-tactic approaches to pest management are no different from those within a single-pesticide system. Multiple-tactic systems must be retained if sustainability is to be achieved. Management systems also should encompass biodiversity and richness (of species/genetic variation) and strategies/tactics to address potential new pest problems.

Although conventional and molecular crop genetics have led to much progress in the production of major crops and the control of related pests, minor crops continue to receive limited attention. It is to be hoped that future genetic products will bring benefits, also. Regardless of crop, grower and societal acceptance are crucial to the success of new crop varieties and gene products—conventional as well as genetically engineered. Challenges and difficulties encountered with *Bt* corn (Starlink) illuminated the magnitude of the problems that can surface as a result of hastily implemented new technologies. *Bacillus thuringiensis* corn received federal approval for use as animal feed only but was found in human foods, resulting in huge economic and product losses and, perhaps more important, diminished public trust in regulatory safeguards. Whereas the debate regarding the merits of biotechnology continues, an extensive recent case study offered very encouraging results. For example, varieties of six crops (soybean, corn, cotton, papaya, squash, and canola) developed via biotechnology yielded some four billion pounds of additional food and fiber on the same acreage, resulted in increased farm income of $1.5 billion, and decreased pesticide volume by 46 million pounds (NCFAP 2002).

Genetic Diversity and Pest Adaptability

Animal and crop pests are endowed with a vast genetic diversity and capacity to adapt to almost limitless habitats. Their ecological elasticity often is extended by (1) their ability to reproduce vegetatively and by means of crossfertilization and (2) their active or passive capacity to travel over very great distances. Thus, typical pest population dynamics involve great change over time and space and frequently pose challenges to IPM scientists and practitioners. Major IPM problems result in the adaptation to given niches, through mutation and/or natural selection, of pest population dynamics. These adaptations may allow pests to attack specially bred resistant crop cultivars or to overcome the effects of pesticides by becoming tolerant of or resistant to given compounds (See Chapter 2). The capacity of pests to survive harsh conditions, including long-distance transport, has allowed them to invade agricultural and natural ecosystems. This phenomenon is being fostered by increasing high-speed international trade and may be accentuated by global warming.

The genetic diversity within most crop pests continues to limit the utility of plant varieties developed with resistance to one or more pests. Thus, the durability of such varieties can be maximized only by a combination of management practices, including rotations with nonhosts and susceptible varieties (See Chapter 4). Certain pests have alternative hosts or may move with air currents, both factors that must be considered in developing management strategies to protect the usefulness of a pest resistance gene. Although genetically engineered resistant varieties offer new IPM tools, they, too, often will be overcome by pest adaptation, especially when used improperly.

Many pesticides, like resistant crop varieties, are rendered ineffective by the target pest's ability to adapt genetically. As discussed in detail in Chapter 2, hundreds of species of insects, pathogens, and weeds have developed resistance to pesticides. At times, pests may develop *cross-resistance*, whereby the pest becomes resistant to two or more pesticides without having been exposed to all chemicals involved. Somewhat similar to IPM programs using and managing host resistance, pesticide RM programs are necessary to ensure the durability of many pesticides. Resistance management strategies/tactics for this purpose range from rotating pesticides with different modes of action to developing models to assist in designing RM programs. Because of the needs to customize RM programs and to evaluate pest populations, extensive information is required and must be modified continually to fit the current pest situation. Crucial for this purpose are data regarding how best to minimize selection pressure for resistance within the pest while maximizing long-term benefits of available pesticides and genetically modified organisms (GMOs).

Highly adaptable invasive pests are moving into all habitats, including bodies of water, parks, forests, pastures, rangelands, croplands, and other natural areas (See Chapter 2). Thousands of nonnative species of weeds, insects, and pathogens from many countries have been introduced into the United States. The annual cost of damage due to pest species is estimated at more than $100 billion. For example, the imported fire ant (*Solenopsis invicta*) is now a permanent resident of several states and causes damage to livestock, wildlife, and public health—damage amounting to more than $2 billion/year (yr). The Invasive Species Council established in 1999 by a presidential executive order to address this enormous problem will require significant resources.

Ecologically Based Integrated Pest Management

As has been discussed, in 1996 the National Academy of Sciences (NAS) published the book *Ecologically Based Pest Management—New Solutions for a New Century* (NRC 1996). Prepared at the request of the U.S. Department of Agriculture (USDA), the U.S. Environmental Protection Agency (EPA), and the Board on Agriculture, this book focused on the need for more sustainable pest management strategies and emphasized that they should be based on ecological principles. The clear thesis of this publication and of other recent publications (e.g., Benbrook et al. 1996; Liebman, Mohler, and Staver 2001) was to shift the existing IPM paradigm from a focus on pest management relying primarily on pesticide management to a systems approach relying primarily on biological knowledge of pests and their interactions with crops. In biointensive IPM systems, knowledge of the managed ecosystem and its natural processes that suppress pest populations is a primary tactic, and intervention with pesticides or physical or biological supplementation a secondary tactic. Although ecological principles long have been central to IPM (Rabb and Guthrie 1970), the recent calls for a new ecologically based pest management paradigm also have involved greater investment in research to develop the necessary biological and ecological knowledge base.

Although the NAS and the Consumers Union (CU) each call for a new IPM paradigm, there are vast differences in proposed paths to the new paradigm. The CU, in *Pest Management at the Crossroads*, calls for accelerated movement along the pest management continuum toward more biologically based management, by implementation of existing IPM knowledge and tools. The organization also sets specific goals (e.g., 75% of crop acreage under medium- to high-level IPM by 2010) (Benbrook et al. 1996) that are, given current estimates, in large part attainable with modest investments (Fernandez-Cornejo and Jans 1999). On the other hand, to attain economic best management practices (EBMP) as described by the NAS, massive investments will be required to build the ecological knowledge base needed for the multitude of cropping systems, environments, and pest complexes that comprise U.S. agriculture. Currently, this knowledge base is incomplete for most pests in all pest disciplines, and the development of interdisciplinary pest complex-based and agroecologically based system knowledge has hardly begun. The goals of EBMP should be viewed as long-term strategic goals that can be achieved only incrementally by members of the pest disciplines working with farmers and encouraging their movement along the IPM continuum toward a greater degree of EBMP. Only then will the necessary public/private financial and regulatory support be developed. Lack of success by the Clinton Administration in achieving the relatively modest funding increases for the USDA/EPA IPM initiative reflected the challenges facing proposals for increased federal funding for research and Extension education. In addition to developing increased funding, members of the pest disciplines, economists, ecologists, soil and crop scientists, sociologists, public interest groups, and, most important, farmers will have to develop new dialogues and blueprints for interdisciplinary funding. As farmers move along the IPM continuum, they develop confidence in IPM tactics and techniques and become leaders in calling for new research and technology transfer programs. Clear examples are found with apple, potato, and cotton growers, considered by many the most advanced producers in terms of implementing IPM.

The term *IPM* is familiar to many growers who identify it specifically with tactics such as rotating, scouting, monitoring temperature/humidity or other environmental features, using predictive models for fungicide or bactericides applications, using certified seeds, and/or disease-resistant varieties, improving sanitation, controlling alternative or overwintering hosts, using biological controls, maintaining plant health via fertility and water management, using high-quality seed, and planting into favorable conditions for rapid germination and seedling emergence. To achieve EBMP or biointensive IPM, involved scientists and others must see these tactics as elements in an IPM continuum that growers should implement based on information from a multitude of disciplines. The interdisciplinary research and Extension education needed to achieve this goal will require focusing on the implementation needs of the grower, not just on high-quality disciplinary science or Extension education. To be successful in implementing IPM, the IPM team clearly will need to involve more than the pest disciplines alone. Implementation will require pest discipline partnerships with economists, sociologists, ecologists, horticulturists, agronomists, agricultural engineers, geographic information specialists, soil scientists, food processors, crop consultants, pesticide applicators, regulatory agencies, computer scientists, consumers, public policy interest groups, and, again, most important, farmers.

Systems Approaches

As the aforementioned goals are addressed, another related, major long-term goal for maximizing the benefits of IPM and many related cropping systems should focus on significantly increased understanding of the synchrony and interactions of microflora/fauna in natural ecosystems. Further, through systems research, information from natural harmonic biological microhabitats and from advanced sustainable agriculture and organic production systems should have great potential in the development of large-scale, biointensive cropping systems. To bring such systems to fruition, knowledge of the unending ecological interactions among beneficial bacteria, fungi, insects, nematodes, protozoa, and other microflora/fauna with crop plants and associated pests, as well as knowledge of soil factors (organic matter, etc.) and the general environment and production practices will be essential (Magdoff and van Es 2000). Central to achieving this goal will be effective collaborative research and Extension programs, including appropriate interfacing with funding/support agencies, as have been described. Equally important, a systems approach (Merrigan 2000) can be adapted to facilitate the management/direction of the biological symphony composed of thousands of beneficial-crop-growth promoting, pest-suppressing organisms. As in most natural ecosystems, a harmonic balance among beneficial organisms can minimize pest problems. Although this approach is not a cure-all for pests, its potential warrants increased research on numerous field and vegetable crops.

Evolving Pool of Trainers and Speed of Technology Transfer

Currently, only one university (University of Florida) offers a Ph.D. in plant medicine (University of Florida 2002), but a number of universities continue to offer both undergraduate and graduate degrees in IPM (See Textbox 13.1). Facing continuing budget reductions in Extension service, the University of Florida's program and similar curricula could be expanded to embrace long-distance teaching or learning technologies. The ongoing development of a number of National IPM Centers in the United States undoubtedly will contribute toward providing the much-needed pool of IPM trainers (See next section on government policy). The real challenges for the future will be helping farmers sift through the glut of data available in the Information Age and providing them with what they need.

The Texas Pest Management Association's (TPMA) Internet site (TPMA 2002) is just one example of how complex the delivery of information can be. This statewide, multicommodity, nonprofit, producer organization relies on approximately 150 field monitors each season to collect important information for farmers making pest and crop management decisions. The organization's web site offers the downloadable software package Scout Master (formerly, Cotton Data Entry Analysis Tool), a database management program used primarily to record scouting but also crop management data. Many other useful links appear on the site, from weather data to government agency links, to news and market data. The challenge is to convince growers to use the information. Roughly 10% of IPM growers in various regions of the United States use computers for pest management decisions (Sorensen, Day, and Stewart 2000). This percentage is greater (35%) among TPMA members but still comparatively low. Helping growers become computer literate and take advantage of the information available on the Internet is one of the many challenges faced by IPM educators.

Information is the substructure on which IPM recommendations are developed (Nowak, Padgett, and Hoban 1996). Producers need to know about markets, weather, new technologies, inputs, input prices, farm programs, and community activities. As an intensive and information-based whole-system approach, IPM provides multiple options for pest control based on accurate data inputs. The information base from which producers can make the best possible decisions regarding pest control is generated from on-farm, site-specific observations. As more agricultural production technologies become available, and as more and more pests become resistant to conventional pesticides, producers must have enhanced access to information.

Integrated pest management systems must rapidly incorporate new technologies such as GMOs and precision agricultural tools into evolving pest management systems. The technology push has several implications. Even though the research community provides data regarding cutting-edge technologies, it may no longer have enough lead time to produce the necessary educational materials, especially for biotechnologies (USEPA/USDA 1999). As new IPM technologies become increasingly complex, they may be used inappropriately or be discarded. Many new technologies involve unfamiliar concepts and require technical skills that producers may not possess. Information regarding costs, investments, and effects of new technologies on net returns may be unavailable.

With IPM evolving toward more complex integrated systems, approaches to the education of IPM practitioners, delivery of information to producers, and evaluation of related effects will continue to change (See Chapter 13). The increasing rate of development and the introduction of new technologies will challenge the agricultural community's capacity to provide the necessary training and information to incorporate new tools into existing IPM programs. This situation is especially true in a time of dwindling financial support for Extension programming. The importance of the role of the private sector in IPM training and delivery systems likely will increase, and, it is hoped, complement public programs.

Government Policy and Regulations

Although significant progress has been made in IPM during the last three decades (Fernandez-Cornejo and Jans 1999; Kennedy and Sutton 2000; Madden 1992), much discussion and debate has focused on the level of related accomplishments (GAO 2001; Goodell and Zalom 2000; Merrigan 2000). Government polices including bans, regulations, resource allocations, and special programs have shaped and will continue to shape IPM programs, especially those related to the 1993 IPM Clinton Administration/USDA 1994 IPM Initiative to place 75% of crop acreage under IPM by 2000 and to implement the FQPA (Anderson and Milewski 2000; Jacobsen 1996). The 1993/1994 Clinton policy included the executive policy goal for governmental departments and agencies, which was to decrease pesticide use by 50% by implementing IPM. This objective has been extremely challenging for the Departments of Agriculture, Defense, Health and Human Services, Housing and Urban Development, Interior, and affiliated agencies such as the U.S. Forest Service, Bureau of Land Management, and Federal Wildlife Service. On properties under their management, these agencies use significant quantities of pesticides for noncropland weed control and for mosquito, ant, flea, fly, cockroach, silverfish, and termite control. Significant progress has been made by all departments and affiliated agencies in decreasing herbicide use for weed control and insecticide use for structural and urban insect pests. The specific goals of the Clinton IPM Initiative and the history of government policy relative to IPM are detailed in Chapter 1. Published assessments of IPM, especially those related to the 1994 IPM initiative, differ considerably in their conclusions. The discussion herein will address briefly the effects, assessments, and challenges of federal, international, and other governmental unit policies on IPM.

At the state level, several universities and state departments of education and/or health have created policy statements regarding implementation of IPM in schools and on other public properties. Koehler, Fasulo, and Scherer (2002) provide information about IPM policy implementation by local school boards. Many states (e.g., New York, Pennsylvania, California, and Texas) use IPM for pest control in parks and on other state property and have provided direct IPM implementation funding. Schools represent one of the most rapidly growing areas of IPM adoption. Implementation of IPM for urban pests likely will be the policy for most public properties at the federal, state, and local levels.

Although government IPM programs and policies have contributed to implementation at the grower level, many agencies and groups also have encouraged IPM adoption. These include the USDA, the EPA, other government agencies, agricultural Extension services, universities, consumer groups, private industries, private consultants, and environmental organizations (Fernandez-Cornejo and Jans 1999).

Various facets of federal and international regulation and IPM certification were discussed in Chapter 2. The USDA is responsible for issuing permits and notices related to regulated technologies (Bridges 2002). The EPA is involved in pesticide regulation and also promotes the use of IPM, both in agriculture and in urban environments, particularly in schools, on right-of-ways, in home gardens, and on golf courses. A special division within the EPA's Office of Pesticide Programs facilitates and accelerates the registration of biopesticides compatible with IPM systems. The EPA's Pesticide Applicator Training Program, conducted by the land-grant university system, acquaints pesticide applicators with IPM principles. Although national and international regulations and labels have the most significant effects on IPM programs, local governments also can play a key role. A few towns and cities have become so concerned about the possible negative aspects of pesticides that they have banned their use (Lutz 2000).

The pending international phase-out of methyl bromide undoubtedly offers one of the greatest IPM challenges ever encountered by growers and scientists (EPA 2000). But an equally difficult challenge is posed by the phasing in of the FQPA. This act redefines how pesticides are regulated and will restrict greatly the future availability of pesticides, organophosphate and carbamate insecticides, and several fungicides. Fruits and vegetables will be especially carefully evaluated by the FQPA inasmuch as they are prevalent in children's diets. Regrettably, because these crops are grown on relatively small acreages, they represent relatively small markets for new pesticide products. Because of the related risks that new pesticide developments pose to the crop protection industry, there has been little incentive to develop new chemical pest management tools for these "minor use" crops.

Integrated pest management implementation also has been recognized as a crucial paradigm in implementation of the FQPA. Integrated pest management considerations are mentioned seven times in the FQPA legislation as mitigating factors to be considered in potential decisions to cancel pesticides under the FQPA. In the reregistration process, the EPA must consider issues such as pest RM, and damage to established IPM programs. Thus, FQPA law encourages IPM implementation in word; in practice, however, the FQPA, by restricting whole classes of pesticides, sharply restricts farmers' flexibility in managing pests. In particular, the EPA's implementation of the FQPA by restricting pesticide registrations for specified uses, when combined with the act's proviso that all pesticide use be limited to what would be safe for the most-sensitive consumers, constitutes a significant barrier to more flexible, lower-cost IPM methods that might use risky pesticides in small amounts under special circumstances (Swinton and Batie 2001). Also, the FQPA might negatively affect the availability of quality food (Ragsdale 2000).

In response to fundamental crop protection questions posed by implementation of the FQPA, the USDA in concert with the EPA recently developed three new grant programs related to IPM (Coble, H. 2001. Personal communication). These are the Pest Management Alternatives Program (PMAP); the Crops at Risk from FQPA Implementation (CAR); and the FQPA Risk Avoidance and Mitigation Program (RAMP). The PMAP addresses short-term (1- to 2-yr) needs for development of replacement tactics for pesticides targeted for cancellation. The CAR addresses intermediate-term (2- to 3-yr) research and Extension work to provide the basis for transitions for crops or cropping systems affected negatively by FQPA implementation. The RAMP funds intermediate to long-term (3- to 5-yr) biointensive research and Extension projects to enhance stability and sustainability of pest management systems within crop regions, generally multistate. In FY 2000, these three programs supplemented the USDA Regional IPM Research and Extension grant programs and the National Research Initiative—Biology and Management of Pests and Beneficial Organisms, Biologically Based Pest Management, Biology of Plant Microbe Associations, and Biology of Weedy and Invasive Plants grant programs.

In addition to establishing the aforementioned funding programs and other ongoing programs such as the Regional IPM and National Research Initiative, the USDA in 2000 established a network of regional pest management centers that focus on major regional cropping patterns, pests, and environmental issues. Four of these IPM centers were initiated late in 2000, with eight others envisioned. Each of the four centers is located in a region with characteristic cropping systems, weather patterns, pest problems, and geographies: the western region, the north central region, the southern region, and the northeastern region of the United States. The purposes of these centers are to foster coordination of program activities and to help identify research and education priorities. The centers also facilitate more efficient and less repetitive pest management programming for regions characterized by specific production systems. With funding coming from the Pesticide Impact Assessment Programs (PIAP) and other sources, each center is funded at $1,000,000/yr for 3 yr. Although the four new centers must use funds to continue the PIAP data-gathering functions, the master plan will direct funds, including current formula IPM funds and some regional IPM dollars, through these centers to encourage IPM activities at the ecosystem or regional level. Currently, the regional centers place a high priority on fulfilling the requirements of the FQPA.

The Pest Management Information Decision Support System (PMIDSS), a web-based program managed by the NSF Pest Management Center (Center for Integrated Pest Management 2002) promises additional support for these centers and others. This support will include the provision of information from numerous databases that should facilitate the development of improved IPM strategies and tactics as well as address IPM-related aspects of the FQPA (Zurek and Henry 2002).

Whereas IPM funding for research continues to be limited, total IPM funding is considerable. For example, IPM funds by Cooperative State Research, Education, and Extension Service in FY 2002 are as follows: PMAP—$1.625 million; CAR—$1.5 million; RAMP—$4.9 million; Extension (formula and special)—$10.783 million; RIPM—$2.731 million; and Alternatives to Methyl Bromide—$2.5 million (Coble, H. 2002. Personal communication). Although this funding amounts to more than $20 million, it is inadequate to support a sorely needed systems approach to IPM.

Assessments of Integrated Pest Management

An intensive, survey-based study for the year 1996 by Fernandez-Cornejo and Jans (1999) documented striking progress in the adoption of IPM for field crops. For example, scouting for weeds in acres planted included 72% of cotton and 94% of fall potatoes. Overall, scouting for insects averaged 67%. (Note: Scouting may indicate more precise timing of pesticide application—more important would be information about alternative practices such as biocontrol.) Greater than 50% of all acreage planted to field crops was scouted for disease. Percentage of fields scouted for fruit pests ranged from 71% for peach to 98% for strawberry. An example of a key IPM practice in fruit production was alternating pesticides to minimize pest resistance. This practice was followed in 75% of apples, compared with 36% of grapes. Pest resistant varieties also were used at fairly high rates: peach—44%, tomato and strawberry—37%. (See Chapter 4 for more examples of IPM implementation, especially for field crops; see Chapter 12 for details regarding IPM adoption). But in 2000, Merrigan stated that only by "some absurd calculations" (p. 497) could one suggest that the 75% goal of IPM implementation had been reached. In a 2001 FAO report, one conclusion indicated that only 18% of farms had achieved a biointensive level of IPM. Because of the "collective failure to meet the 75% goal," Merrigan also suggested that regardless of the differing assessments, "we can agree that we are not where we had hoped to be when the IPM Initiative was announced ... " (p. 497). She also concluded that one of the greatest shortcomings in most current IPM programs is the limited use of a systems approach. Although much progress has been made within given pest types (and often-associated pest antagonists/pathogens/crops) (Kennedy and Sutton 2000; Madden 1992), integration across pest types for given crops and for overall production systems continues to be limited.

Government policies regarding specific groups of foods or products such as organic foods and GMOs continue to evolve. For organics, the Final Rule excludes the use of GMOs as well as most pesticides (See Chapter 4). The ongoing national and international debates on the future of GMOs undoubtedly will affect policies related to production, marketing, and use of these new products (See earlier section on Gene Technology Constraints).

Risk management has become a central topic in farm policy discussions (Jacobsen 2000). Because insurance companies require a production history, farmers generally cannot secure crop insurance. Crop insurance for best management practices, IPM, and IPM labeling could have important effects on pest management (Jacobsen 2000). On December 12, 2001, the USDA's Risk Management Agency endorsed an insurance policy that allows farmers to decrease their use of fertilizer without fear of losing money if yields drop. The new policy, designed by American Farmland Trust's Agricultural Conservation Innovation Center (ACIC), insures farmers who follow university recommendations for applying nitrogen and phosphorus on corn against any potential losses in yield. The new insurance will be available to farmers in Iowa, Minnesota, and Wisconsin in 2003. The ACIC also is developing IPM policies for corn rootworm and for potatoes.

Improvements Needed to Further the Goals of Integrated Pest Management

In an August 2001 report, the U.S. General Accounting Office (GAO) assessed the success of the USDA's 1994 IPM Initiative and offered recommendations that could facilitate the development of *IPM*. That report identified several IPM facets needed to promote pest management. One key issue involved the use of a definition of IPM that differentiated between practices tending to decrease chemical use (those that are biologically based) and those that do not. A federal entity with authority to lead the IPM initiative—including the coordination of IPM programs sponsored by the six agencies, the land-grant universities, and the EPA—is crucial to the development of IPM. A related recommendation involves the establishment of effective department-wide leadership, coordination, and management of federally funded IPM efforts. Additional needs and shortcomings in the current IPM programs are enumerated in the GAO report (2001).

In response to the GAO report, the USDA agreed to the following (Coble, H. 2001. Personal communication):

1. to work with all relevant and involved agencies and stakeholders to develop a comprehensive, authoritative, and focused road map for IPM;
2. to prioritize the results the department wants to

achieve by developing a strategic plan finalized by a national workshop; and

3. to work with all agencies to set measurable goals for the IPM initiative and to devise methods for measuring progress toward goals.

The most recent legislation to affect IPM adoption is the 2002 Farm Bill. The new IPM initiative will focus on decreasing both economic and environmental risk (Coble, H. 2002. Personal communication; Kelley 2002; USDA–RMA 2003). The 1996 Federal Agricultural Improvement and Reform (FAIR) Act has provided incentives to use IPM practices in certain regions under the Environmental Quality Incentives Program. This approach could be extended significantly under the 2002 legislation (U.S. Senate 2000), which could foster IPM adoption, possibly by including "green payments" to farmers for environmental stewardship. The new initiative, known as the "National Roadmap for IPM," was introduced at the Fourth National IPM Symposium in Indianapolis in April 2003 (Agriculture Texas 2002; Coble, H. 2002. Personal communication). If the criteria for receiving payments include adoption of new IPM practices, the act might meaningfully expand the scope and diversity of IPM practices on U.S. farms. As indicated earlier, however, decisions regarding the use of related funds will be made at the state level. The federal act on agricultural bioterrorism (Hutchinson 2001) and a related bill recently approved (CID 2002; Hawks 2002), which describe new programs and expand others to address the threat of agricultural bioterrorism, likely will impact IPM programs. Rapid and reliable pest diagnostic identification methods and effective countermeasures are needed urgently (Schaad, N. 2002. Personal communication).

In conclusion, IPM, even with very limited resources, has progressed phenomenally since the 1982 CAST report was published. And with new technical resources becoming available, the prospects for IPM and related cropping systems are excellent. Examples of emerging technologies and issues include genetic engineering; precision agriculture; a growing understanding of soil microbiology/ecology and of soil microbes to enhance plant growth while suppressing crop pests; and new, safer, and often "natural" pesticide chemistries. In addition, rapidly increasing computer capacity should facilitate the use of a systems approach for the harmonic deployment of improved IPM strategies and tactics in continually improved crop and animal production/management systems. Major challenges such as public perceptions of new technologies, limited financial resources, and an inadequate infrastructure for IPM will accompany development and deployment of these improved and, often, new systems. As the earth's carrying capacity for humankind is stretched further each year, and as thousands of invasive pests are encountered across the country (and potentially augmented in unforeseen ways through agricultural bioterrorism), research, agriculture, industry, government, and communities must work together to address the quality of food and fiber. In this regard, IPM provides a robust and effective framework for resolving current as well as unforeseen pest-related problems.

Appendix A. Abbreviations and Acronyms

a.	acre
Bt	*Bacillus thuringiensis*
CAR	Crops at Risk from the Food Quality Protection Act
DNA	deoxyribonucleic acid
EBMP	economic best management practices
EPA	U.S. Environmental Protection Agency
FQPA	Food Quality Protection Act
GAO	General Accounting Office
GMO	genetically modified organism
ICM	integrated crop management
IPM	integrated pest management
NAS	National Academy of Sciences
NRC	National Research Council
PIAP	Pesticide Impact Assessment Programs
PMAP	Pest Management Alternatives Program
RAMP	Risk Avoidance and Mitigation Program
RM	resistance management
TPMA	Texas Pest Management Association
USDA	U.S. Department of Agriculture
yr	year

Appendix B. Glossary

Cross-resistance. Whereby a pest becomes resistant to two or more pesticides without being exposed to all chemistries involved.

Pest-protected plants. Plants resistant to pests.

Pleiotropic. Related to unanticipated effects on plant physiology.

Literature Cited

Agriculture Texas. 2002. Fourth National IPM Symposium/Workshop, April 8–10, 2003, <http://www.texagnet.net/ag-texas/reports/12112002_9.html> (22 January 2003)

Anderson, J. L. and E. Milewski. 2000. Regulation of plant-pesticides: Current status. Pp. 154–162. In G. G. Kennedy and T. B. Sutton (eds.). *Emerging Technologies for Integrated Pest Management: Concepts, Research, and Implementation*. APS Press, American Phytopathological Society, St. Paul, Minnesota.

Benbrook, C., E. Groth, III, J. M. Halloran, M. K. Hansen, and S. Marguardt. 1996. *Pest Management at the Crossroads*. Consumers Union, Yonkers, New York. 272 pp.

Bridges, D. C. 2002. Implications of pest-resistant/herbicide tolerant plants for IPM. Pp. 141–162. In G. G. Kennedy and T. B. Sutton (eds.). *Emerging Technology for Integrated Pest Management: Concepts, Research, and Implementation*. APS Press, American Phytopathological Society, St. Paul, Minnesota.

Carroll, N. B. 2000. Development in transition. Pp. 339–354. In G. G. Kennedy and T. B. Sutton (eds.). *Emerging Technologies for Integrated Pest Management: Concepts, Research, and Implementation*. APS Press, American Phytopathological Society, St. Paul, Minnesota.

Cavigelli, M. A., S. R. Deming, L. K. Probyn, and D. R. Mutch (eds.). 2000. *Michigan Field Crop Pest Ecology and Management*. Extension Bulletin E–2704. Michigan State University, East Lansing.

Center for Infectious Disease Research and Policy (CID). 2002. Congress approves wide-ranging bioterrorism bill, <http://www1.umn.edu/cidrap/content/fs/biosecurity/news/btbill.html> (23 January 2003)

Center for Integrated Pest Management. 2002. Homepage, <http://cipm.ncsu.edu/index.html> (9 September 2002)

Council for Agricultural Science and Technology (CAST). 1982. *Integrated Pest Management*. Report Number 93. Council for Agricultural Science and Technology, Ames, Iowa.

Council for Agricultural Science and Technology (CAST). 2001. *Evaluation of the U.S. Regulatory Process for Crops Developed through Biotechnology*. Issue Paper Number 19. Council for Agricultural Science and Technology, Ames, Iowa.

Fernandez-Cornejo, J. and S. Jans. 1999. *Pest Management in U.S. Agriculture*. Agriculture Handbook Number 177. Economic Research Service, U.S. Department of Agriculture, Washington, D.C.

Ferro, D. N. 2000. Success and failure of *Bt* products: Colorado potato beetle. A case study. Pp. 177–189. In G. G. Kennedy and T. B. Sutton (eds.). *Emerging Technologies for Integrated Pest Management: Concepts, Research, and Implementation*. APS Press, American Phytopathology Society, St. Paul, Minnesota.

Fitt, G. P. and L. J. Wilson. 2000. Genetic engineering in IPM: Bt cotton. Pp. 108–125. In G. G. Kennedy and T. B. Sutton (eds.). *Emerging Technologies for Integrated Pest Management: Concepts, Research, and Implementation*. APS Press, American Phytopathological Society, St. Paul, Minnesota.

Geden, C. J. and J. A. Hogsette (eds.). 2001. Research and extension needs for integrated pest management for arthropods of veterinary importance, <http://www.cmave.usda.ufl.edu/lincoln.html> (18 October 2002)

Goodell, P. B. and F. G. Zalom. 2000. Delivering IPM: Progress and challenges. Pp. 483–496. In G. G. Kennedy and T. B. Sutton (eds.). *Emerging Technologies for Integrated Pest Management: Concepts, Research, and Implementation*. APS Press, American Phytopathological Society, St. Paul, Minnesota.

Hawks, B. 2002. Rules and Regulations—Agricultural Bioterrorism Protection Act of 2002: Listing of biological agents and toxins and requirements and procedures for notification of possession. *Fed Regist* 67-155. U.S. Department of Agriculture, Animal and Plant Health Inspection Service, Washington, D.C.

Hutchinson, K. B. 2001. The Agricultural Bioterrorism Countermeasures Act of 2001, October 17, <Hutchison.senate.gov/speec212.htm> (18 October 2002)

Jacobsen, B. 1996. USDA Integrated Pest Management Initiative. National IPM Network, University of Minnesota, St. Paul.

Jacobsen, B. 2000. Summary: International conference on emerging technologies for integrated pest management. Pp. 505–513. In G. G. Kennedy and T. B. Sutton (eds.). *Emerging Technologies for Integrated Pest Management: Concept, Research, and Implementation*. APS Press, American Phytopathological Society, St. Paul, Minnesota.

Kelley, S. 2002. United States Farm Bills. <http://www.nationalaglawcenter.org/farmbill.htm> (23 January 2003)

Kennedy, G. G. 2000. Perspectives on progress in IPM. Pp. 2–11. In G. G. Kennedy and T. B. Sutton (eds.). *Emerging Technologies for Integrated Pest Management: Concepts, Research, and Implementation*. APS Press, the American Phytopathological Society, St. Paul, Minnesota

Kennedy, G. G. and T. B. Sutton (eds.). 2000. *Emerging Technologies for Integrated Pest Management: Concepts, Research, and Implementation*. APS Press, American Phytopathological Society, St. Paul, Minnesota.

Koehler, P. G., T. R. Fasulo, and C. W. Scherer. 2002. School IPM, <http://schoolipm.ifas.ufl.edu> (9 September 2002)

Kogan, M. 1998. Integrated pest management: Historical perspectives and contemporary developments. *Annu Rev Entomol* 43:243–270.

Liebman, M., C. L. Mohler, and C. P. Staver. 2001. *Ecological Management of Agricultural Weeds*. Cambridge University Press, United Kingdom.

Lutz, K. 2000. California town bans pesticides on city property, March 16, <http://www.pmac.net/ banning_pesticides.html> (9 September 2002)

Madden, J. P. (ed.). 1992. Beyond pesticides: Biological approaches to pest management in California. Division of Agriculture and Natural Resources, University of California, Oakland.

Magdoff, F. and H. van Es. 2000. *Building Soils for Better Crops*. (2nd ed.). Sustainable Agriculture Network, Burlington, Vermont.

Merrigan, K. A. 2000. Politics, policy, and IPM. Pp. 497–504. In G. G. Kennedy and T. B. Sutton (eds.). *Emerging Technologies for Integrated Pest Management: Concepts, Research, and Implementation*. APS Press, American Phytopathological Society, St. Paul, Minnesota.

National Center for Food and Agricultural Policy (NCFAP). 2002. Plant biotechnology: Current and potential impact for improving pest management in U.S. agriculture. An analysis of 40 case studies, <http://www.ncfap.org/40 Case Studies.htm> (December 15, 2002)

National Research Council (NRC). 1996. *Ecologically Based Pest Management: New Solutions for a New Century*. National Academy Press, Washington, D.C. 144 pp.

National Research Council (NRC). 2000. *Genetically Modified Pest-Protected Plants: Science and Regulation*. Committee on Genetically Modified Pest-Protected Plants, Board on Agricultural and Natural Resources, National Academy Press,

Washington, D.C. 261 pp.

Nowak, P., S. Padgett, and T. J. Hoban. 1996. Practical considerations in assessing barriers to IPM adoption. Pp. 93–114. In S. Lynch, C. Greene, and C. Kramer-LeBlanc (eds.). *Proceedings of the Third International IPM Symposium / Workshop. Broadening Support for 21st Century IPM.* Economic Research Service, U.S. Department of Agriculture, Washington, D.C.

Rabb, R. L. and F. E. Guthrie (eds.). 1970. *Concepts of Pest Management.* North Carolina State University, Raleigh.

Ragsdale, N. N. 2000. The impact of the Food Quality Protection Act on the future of plant disease management. *Annu Rev Phytopathol* 38:577–596.

Sorensen, A., E. Day, and P. Stewart. 2000. Factors affecting the adoption of new technologies. Pp. 12–31. In G. G. Kennedy and T. B. Sutton (eds.). *Emerging Technologies for Integrated Pest Management: Concepts, Research, and Implementation.* APS Press, American Phytopathological Society, St. Paul, Minnesota.

Swinton, S. M. and S. S. Batie. 2001. FQPA: Pouring out (in?) the risk cup. *Choices* 16:14–17.

Texas Plant Management Association (TPMA). 2002. Homepage, <http://www.tpma.org> (9 September 2002)

University of Florida. 2002. Plant medicine, July 11, <http://www.dpm.ifas.ufl.edu> (9 September 2002)

University of Guelph. 2002. Homepage, <http://www.uoguelph.ca> (9 September 2002)

U.S. Department of Agriculture–Risk Management Agency (USDA–RMA). 2003. RMA online. 2002 Farm Bill, <http://www.rma.usda.gov/news/2002farmbill> (23 January 2003)

U.S. Environmental Protection Agency (EPA) and U.S. Department of Agriculture (USDA). 1999. *Conference on* Bt *corn*, Chicago, Illinois, June 18.

U.S. Environmental Protection Agency (EPA). 2000. Methyl bromide phaseout changes in the U.S. U.S. Environmental Protection Agency. Washington, D.C.

U.S. General Accounting Office (GAO). 2001. *Agricultural Pesticides: Management Improvements Needed to Further Promote Integrated Pest Management.* Report GAO–01–815. U.S. General Accounting Office, Washington, D.C.

U.S. Senate. 2000. Conservation Security Act of 2000. S.3223/S.3260/HR5511.IH.

Zurek, E. and D. Henry. 2002. Welcome to the USDA regional pest management centers information system, <http://pmcenters.org> (9 September 2002)

Related Web Sites

Appropriate Technology Transfer for Rural Areas. 2002. Homepage, <www.attra.org> (14 October 2002)

Integrated Pest Management Practitioners Association. 2002. Homepage, September 14, <http://www.efn.org/~ipmpa/ipmpa.html> (14 October 2002)

Invasivespecies.gov. 2002. Homepage, September 30, <www.invasivespeciesgov> (14 October 2002)

University of California Integrated Pest Management Program. 2002. Homepage, <www.ipm.ucdavis.edu> (14 October 2002)

Index

G

H

I

M

N

O

P

W

X

Y

Z